Fault Analysis and its Impact on Grid-connected Photovoltaic Systems Performance

Fault Analysis and its Impact on Grid-connected Photovoltaic Systems Performance

Edited by

Ahteshamul Haque
Senior Member IEEE
Department of Electrical Engineering
Jamia Millia Islamia (A Central University)
India

Saad Mekhilef
IEEE Fellow
School of Science, Computing and Engineering Technologies
Swinburne University of Technology
Melbourne, Australia

IEEE PRESS

WILEY

Published by John Wiley & Sons, Inc., Hoboken, New Jersey.
Published simultaneously in Canada.

For general information on our other products and services or for technical support, please contact our Customer Care Department within the United States at (800) 762-2974, outside the United States at (317) 572-3993 or fax (317) 572-4002.

Wiley also publishes its books in a variety of electronic formats. Some content that appears in print may not be available in electronic formats. For more information about Wiley products, visit our web site at www.wiley.com.

Library of Congress Cataloging-in-Publication Data

Names: Haque, Ahteshamul, editor. | Mekhilef, Saad, editor.
Title: Fault analysis and its impact on grid-connected photovoltaic systems
   performance / edited by Ahteshamul Haque, Saad Mekhilef.
Description: Hoboken, New Jersey : Wiley, [2023] | Includes bibliographical
   references and index.
Identifiers: LCCN 2022040364 (print) | LCCN 2022040365 (ebook) | ISBN
   9781119873754 (hardback) | ISBN 9781119873761 (adobe pdf) | ISBN
   9781119873778 (epub)
Subjects: LCSH: System failures (Engineering) | Photovoltaic power systems.
Classification: LCC TA169.5 .F38 2023 (print) | LCC TA169.5 (ebook) | DDC
   620/.00452–dc23/eng/20221006
LC record available at https://lccn.loc.gov/2022040364
LC ebook record available at https://lccn.loc.gov/2022040365

Cover Design: Wiley
Cover Image: © peterschreiber.media/Shutterstock

Set in 9.5/12.5pt STIXTwoText by Straive, Chennai, India

I would like to dedicate this book to the Last Prophet of humanity, Prophet Mohammad, who introduced our creator Almighty ALLAH (SWT) to us.

Ahteshamul Haque

I would like to dedicate this book to my wife who supported me during my academic journey.

Saad Mekhilef

Contents

Preface

Automatic fault detection with machine learning techniques has been extensively used to assist the decision-making process during abnormal conditions. However, these approaches were majorly constrained in the fields of medical and image processing-based applications. This is because of the complexities due to the unavailability and uncertainty of the input data of real-world and mainly industrial applications. In light of the above observation, this book proposes a failure mode effect classification approach for distributed generation systems and their components operating in a grid-connected environment, which are widely established around the world. This book specifically adheres to the faults in grid-connected photovoltaic systems and measures to classify them for achieving proactive monitoring of the system. Generally, failure mechanisms are a critical aspect of determining the reliability of the power system. The failure mechanisms deal with the physical, chemical, and electrical processes through which the failure occurs in the system. Based on the type of failure process, these failure mechanisms can be modelled when appropriate material and environmental information are available. Moreover, with the advancements in machine learning approaches, the failure data along with the modelled mechanisms can be used to identify the operating state of the power system. This helps monitor the operation of the system, perform a risk analysis, estimate the reliability of the product, and reduce the probability that a customer is exposed to a potential failure and/or process problem.

We would like to thank our research team and authors who have contributed to this book. Researchers from Academia who are working in the field of photovoltaic systems and power electronics converters, and distributed generation companies, which are commonly facing issues in operation and maintenance. This book will focus on helping them for developing efficient proactive monitoring approaches for the day-to-day operation and maintenance of grid-connected photovoltaic systems and their components.

The power module-manufacturing units, photovoltaic module development companies, power electronics converter assembly units, and various testing and certification centers can refer to this book for understanding the issues regarding the integration of power electronics converters with distributed generation systems in grid-connected environment. The Research and Development units in various organizations can use the book for developing and implementing advanced controllers with photovoltaic systems.

Ahteshamul Haque
Advance Power Electronics Research Lab
Department of Electrical Engineering
Jamia Millia Islamia (A Central University)
New Delhi

Saad Mekhilef
School of Science, Computing and
Engineering Technologies
Swinburne University of Technology
Hawthorn, VIC
Australia

List of Contributors

Ibrahim Alhamrouni
British Malaysian Institute
Universiti Kuala Lumpur
Selangor
Malaysia

Pooya Davari
AAU Energy
Aalborg University

Ahteshamul Haque
Advance Power Electronics
Research Lab
Department of Electrical Engineering
Jamia Millia Islamia (A Central
University)
New Delhi
India

Barry P. Hayes
School of Engineering and
Architecture
University College Cork
Cork
Ireland

Irfan Khan
Clean and Resilient Energy Systems
(CARES) Lab
Texas A&M University
Galveston, TX
USA

Mohammed Ali Khan
Department of Electrical Power
Engineering
Faculty of Electrical Engineering and
Communication
Brno University of Technology
Brno
Czech Republic

V S Bharath Kurukuru
Advance Power Electronics
Research Lab
Department of Electrical Engineering
Jamia Millia Islamia (A Central
University)
New Delhi
India

Tarmo Korõtko
Department of Electrical Power
Engineering and Mechatronics
Tallinn University of Technology
Smart City Center of Excellence
(Finest Twins)
Tallinn
Estonia

Azra Malik
Advance Power Electronics
Research Lab
Department of Electrical Engineering
Jamia Millia Islamia (A Central
University)
New Delhi
India

Saad Mekhilef
School of Science, Computing and
Engineering Technologies
Swinburne University of Technology
Hawthorn, VIC
Australia

Huai Wang
AAU Energy
Aalborg University

Zahraoui Younes
Department of Electrical Power
Engineering and Mechatronics
Tallinn University of Technology
Smart City Center of Excellence
(Finest Twins)
Tallinn
Estonia

Zhaoyang Zhao
Institute of Smart City and Intelligent
Transportation
Southwest Jiaotong University

About the Editors

Dr. Ahteshamul Haque is a Senior Member of IEEE. He is working as Associate Professor, Department of Electrical Engineering, Jamia Millia Islamia (A Central University), New Delhi, India. His area of research is power electronics and its applications, control of power electronics converters, intelligent techniques in power electronics. He has authored and co-authored around 150 publications in journals and proceedings. He has authored 1 book and 17 book chapters. He was working in R&D labs of world-reputed multinational companies. His inventions are patented, published, and awarded in the United States, Europe, and India. He has established state-of-the-art Advance Power Electronics Research Lab. More than 20 PhD and masters have graduated under his supervision. He serves in the editorial team of IEEE Journal of Emerging and Selected Topics in Power Electronics as an associate guest editor. Dr. Haque has reviewed around 160 research papers of reputed journals.

Dr. Haque has been awarded with Outstanding Engineer Award for the year 2019 by IEEE Power and Energy Society. He is involved in industry consultancy of power electronics converters, solar PV plant design, grid integration, and MPPT design. He has research collaboration with Aalborg University – Denmark, Swinburne University – Australia, and West Florida University – USA.

Prof. Dr. Saad Mekhilef is an IEEE and IET Fellow. He is a distinguished professor at the School of Science, Computing and Engineering Technologies, Swinburne University of Technology, Australia, and an honorary professor at the Department of Electrical Engineering, University of Malaya. He authored and co-authored more than 500 publications in academic journals and proceedings and 5 books with more than 34 000 citations, and more than 70 PhD students graduated under his supervision. He serves as an editorial board member for many top journals such as IEEE Transaction on Power Electronics, IEEE Open Journal of Industrial Electronics, IET Renewable Power Generation, Journal of Power Electronics, and International Journal of Circuit Theory and Applications.

Prof. Mekhilef has been listed by Thomson Reuters (Clarivate Analytics) as one of the highly cited (world's top 1%) engineering researchers in the world in 2018, 2019, 2020, and 2021. He is actively involved in industrial consultancy for major corporations in the power electronics and renewable energy projects. His research interests include power conversion techniques, control of power converters, maximum power point tracking (MPPT), renewable energy, and energy efficiency.

1

Overview and Impact of Faults in Grid-Connected Photovoltaic Systems

Mohammed Ali Khan

Department of Electrical Power Engineering, Faculty of Electrical Engineering and Communication, Brno University of Technology, Brno, Czech Republic

1.1 Introduction

Climate change has made renewable energy more important in recent years. The rapid increase in the share of renewable energy has made it possible to decentralize power generation. It has helped consumers further reduce their energy costs and help utilities meet their ever-increasing energy needs. To further support the rise, various countries have developed national strategies and initiatives to promote the introduction of renewable energy [1]. To promote sustainable energy projects, the United Nations has agreed on specific Sustainable Development Goals [2]. Even the European Union has promised to reduce greenhouse gas emissions by about 80–95% by 2050 [3]. In 2017, the total installed capacity of solar energy projects exceeded the net installed capacity of coal, gas, and nuclear power combined [4].

Given the projected growth in distributed generation (DG) production, effective coordination, monitoring, and maintenance tools are required to adapt to existing grid infrastructure. This improves the performance of grid-connected DG systems to ensure stable power generation and optimal energy harvesting. Abnormal behavior on a particular DG can cause an entire power system failure, which can lead to a major blackout. Failure detection and monitoring of photovoltaic (PV) systems with grid connections have been intensively studied [5–7]. If the grid fails, the PV system connected to the grid needs remote island prevention detection. If the DG cannot recover the network from a failure condition via ride-through, Remote Island Protection protects the DG from the network, providing the necessary security for both the utility and the DG, and avoiding a complete network failure. The main goal of preventive island protection is to keep the power grid

Fault Analysis and its Impact on Grid-connected Photovoltaic Systems Performance, First Edition.
Edited by Ahteshamul Haque and Saad Mekhilef.

running and to prevent accidental islands in the area for security reasons. In case of inconsiderate isolation, general management tends to power the local PCC site [8]. This isolates the mesh area locally and the network is unaware of the isolation that has occurred. As a result, the life of the utility employee working online can be in danger as they can be electrocuted undetected by ordinary management. In the event of an accidental single trip, even improper grounding can cause significant temporary overvoltage due to sudden loss of the load [9]. Most studies in the literature use PV panel voltage (VI) monitoring for troubleshooting, while techniques such as panel thermal imaging and inverter output are monitored to identify faults. However, the time until failure was discovered was one of the major drawbacks of such methods. Therefore, various intelligent error detection methods have been introduced to achieve fast response times [10–13]. Many fuzzy algorithms have been used to identify defects [14, 15]. However, reliance on a specific set of rules can lead to misclassification. In addition, the increasing number of data loggers and data acquisition devices in various parts of DG units actually provides information on system operation [16]. This data essentially processes information about the health of system components and provides a new approach to evaluating the performance of power systems with great economic potential for operation and maintenance. This data essentially processes information about the health of system components and provides a new approach to evaluating the performance of power systems with great economic potential for operation and maintenance. These aspects have prompted research to monitor conditions and increase the output of solar systems [17–19].

If a malfunction is identified, the malfunction must be managed, and an appropriate remedial mechanism must be provided. This can be done by developing a fault ride-through (FRT) mechanism [20]. In some studies, FRT mechanisms have been discussed using controller switching or other reactive power injection strategies [21–25]. A grid-connected PV plant is considered to be in an FRT state if certain criteria in the grid code are not met. At the point of common coupling (PCC), monitoring of various parameters such as operating frequency, operating voltage, power factor and reactive power is performed. If the malfunctioning is severe, maintenance must be scheduled more than the FRT limit. And when the fault is corrected by FRT, the fault can be notified through scheduled inspection, so that deep testing can be performed during the inspection period. During maintenance, diesel generators can operate in autonomous mode [26].

It is essential to understand the system in operation and fault associated with the system as the system may witness a small fault out of the wide spectrum of possibilities, and localization of the faults is necessary for faster response and smooth operation [27, 28]. In this chapter a brief about the fault in a grid-connected PV system is discussed along with it impact on the system and the method to identify such faults.

1.2   Grid-Connected PV System

The grid-connected PV system operates in coordination with the operation taking place on the DC and the AC sides of the inverter. The inverter acts as an isolation between the two sides and aims to maintain a constant AC output to the DC input received. The solar panels convert the irradiation into the electrical output. But the output of panel is varying based on the variation taking place in the irradiance and the temperature. To regulate the output of the solar panel a DC-DC converter is connected to the system which regulates the power generation and extracts the maximum outcome possible from a solar panel by adopting the maximum power point tracking (MPPT) algorithm [29]. The regulated DC is supplied to the inverter and the inverter is controlled by monitoring the voltage and current at PCC and DC link. The inverter presents an AC output with some harmonics. The filter is used after the inverter to reduce the harmonic and make the system.

The control structure of the inverter plays a very important role in controlling the operation and maintaining a stable operation [30]. To achieve symmetrical power transfer from two interconnected power supplies, the intermediate circuit voltage is regulated by the inverter's capacitor feedback internal loop control. The proportional integrator (PI) controller acts as an active current exchange in the network, but by improving the controller's transient response, you can get a reference to the active power in the test bench. In addition, the feedforward controller coordinates reactive current injection into the network. Steady-state frame current control ($\alpha\beta$) is used in this study because of its fixed current range, low reliance on network impedance, and easy harmonic compensation for low-frequency components. This also reduces the effect of harmonics on the mains voltage on the current regulator. Resonant integration reduces the effect of line current harmonics present on the current return. Phase-locked loops (PLLs) are used to filter harmonics from line voltages and extract positive sequences for synchronization.

1.2.1   Inverter Control

For operating the inverter, it is necessary to vary the switching scheme of the power electronics switches involved depending on the load change and unexpected interferences. Inverter control can be divided into two configurations as explained later [31, 32]:

1.2.1.1   Grid-Connected Inverter Control

From Figure 1.1, it can be realized that V_{in} is the voltage at the DC link and v_g represents the grid voltage. The voltage monitored at PCC is represented by v_{pcc}. The impedance of the grid is denoted by Z_g. The summary of the grid-connected inverter controller is shown in Figure 1.2. On multiplying the sampling current

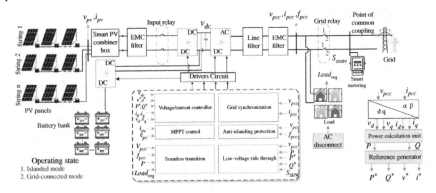

Figure 1.1 Overview of grid-connected PV system.

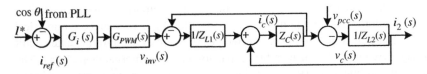

Figure 1.2 Block diagram of grid-connected inverter control.

amplitude (I^*) along with grid phase angle obtained by phase lock loop (PLL), a reference current (i_{ref}) is obtained. The current controller is denoted by $G_i(s)$. Considering the digital control delay [33], $G_{PWM}(s)$ denotes equivalent gain for the inverter. The mathematical expression can be represented as

$$G_{PWM}(s) = K_{PWM}\, G_d(s) = \frac{V_{in}}{V_{tri}}e^{-1.5sT_s} \tag{1.1}$$

where a gain for the modulated wave (v_m) is represented by K_{PWM}. Inverter bridge voltage is represented by V_{inv} which is equal to the ratio between DC link voltage V_{in} and carrier triangular wave V_{tri}. The delay transfer function for digital control is represented by $G_d(s)$. The sampling of the control system is denoted by T_s. For simplification of analysis $G_d(s)$ is subjected to second order pade approximation [34] as presented as follows:

$$G_{d(s)} = e^{-s1.5T} \approx \frac{1 - 0.75sT_s + 0.083(1.5sT_s)^2}{1 + 0.75sT_s + 0.083(1.5sT_s)^2} \tag{1.2}$$

The expression of the current controller [35] can be represented as:

$$G_{i(s)} = K_p + \frac{K_i}{s} \tag{1.3}$$

where K_p and K_i are proportional and integral gains.

1.2.1.2 Standalone Inverter Control

For controlling a stand-alone inverter, a dual-based control loop is implemented [36–38]. Current is regulated using the inner loop, whereas the voltage across the filter capacitor is controlled by the outer loop. The designing of the loops is explained as follows:

- *Designing of the inner current loop*

For designing the inner current loop, the capacitor current (i_c) is considered as feedback value. By using the i_c as feedback, the system can achieve better performance in case of load variation. The load current is assumed to be zero [39]. The control diagram for the inner loop is illustrated in Figure 1.3.

For digital implementation, one switching period delay (T_s) needs to be considered as until the next switching cycle, the modulation signal will not get updated [40]. The pure switching cycle delay is represented by a delay block (G_d).

$$G_d = e^{-sT_s} = \frac{12 - 6sT_s + (sT_s)^2}{12 + 6sT_s + (sT_s)^2} \tag{1.4}$$

As per the small signal analysis in [41], the standalone controller plant can be represented by G_p in Eq. (1.5)

$$G_p = \frac{1}{sL_1} \tag{1.5}$$

As per the control diagram represented in Figure 1.3, the following equation represents an uncompensated [42] inner loop

$$G_{ip} = K_{pwm}G_dG_p \tag{1.6}$$

where, G_{ip} represents the uncompensated controller plant and K_{pwm} is the gain value for Pulse width modulation. On applying PI controller for compensation [42] of the inner loop (G_{ip-c}), the expression can be deduced as

$$G_{ip-c} = \frac{k_p s + k_i}{s} G_{ip} \tag{1.7}$$

where, K_p and K_i are proportional and integral gains for the PI controller.

- *Designing of the outer current loop*

The voltage across the filter capacitor is regulated by the outer control loop. The representation of the outer control loop is present in Figure 1.4.

$$v_o = G_v + G_{ip-c} + \frac{1}{C} \tag{1.8}$$

Figure 1.3 Block diagram of the inner current loop for standalone control.

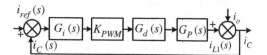

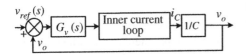

Figure 1.4 Block diagram of the outer current loop for standalone control.

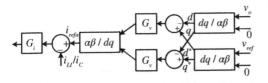

Figure 1.5 A DQ-based control diagram for inverter control.

where, G_v is the compensated outer voltage loop, and C is the filter capacitor.

The presence of a PI controller at the outer loop helps in achieving a satisfactory value of phase margin and also helps in attaining system stability for the uncompensated plants. In case of a dual loop-compensated system, the crossover frequency must be approximately ten times larger than the grid frequency and lesser when related to the inner loop crossover frequency. The voltage error is tracked and can be eliminated by implementing direct quadrature (DQ) control with a synchronous rotating frame [43]. For DQ rotor frame control, the AC quantity present in the $\alpha\beta$ stationary frame acts as a DC quantity as the rotating frequency for the DQ frame is like the fundamental angular frequency for AC quantities [44]. Both current and voltage loops are performed in a synchronous rotor frame for the DQ control method. The benefit of this method is that the controller can be implemented without the generation of β component. Figure 1.5 depicts the DQ-based controller representation along with state space representation in Eqs. (1.9) and (1.10).

$$\begin{bmatrix} I_d \\ I_q \end{bmatrix} = \begin{bmatrix} \cos(\omega t) & \sin(\omega t) \\ -\sin(\omega t) & \cos(\omega t) \end{bmatrix} \begin{bmatrix} V_\alpha \\ V_\beta \end{bmatrix} \tag{1.9}$$

$$\begin{bmatrix} I_\alpha \\ I_\beta \end{bmatrix} = \begin{bmatrix} \cos(\omega t) & -\sin(\omega t) \\ \sin(\omega t) & \cos(\omega t) \end{bmatrix} \begin{bmatrix} V_d \\ V_q \end{bmatrix} \tag{1.10}$$

where, I_d, I_q, V_d, and V_q are the direct and quadrature current and voltage components respectively, and V_α, V_β, I_α, and I_β are the $\alpha\beta$ components of the single-phase voltage and current, respectively.

1.3 Overview of Module Faults

An abnormal operation which takes place during the normal operation of the PV module can cause a system failure. Manufacturing defects are believed to be the main cause of instability in the performance of some modules. Some cases of manufacturing defects are single-crystal and polycrystalline solar cells, which can be observed in the form of striped rings or medium crystalline defects. Mismatches in mixing ratios cause performance degradation due to light, called a manufacturer's failure to PV failure. Due to modular technology failures, amorphous silicon modules are susceptible to light degradation, resulting in a 10–30% reduction in performance during the early stages of installation [45]. Although this deterioration can be retrieved to some degree by thermal annealing [46], it is only applicable in hot summer, and the characteristics of the module deteriorate due to seasonal fluctuations. The module fault can be categorized into an external issue and a manufacturing fault.

1.3.1 External Issue

In addition to manufacturing defects, there are many others that are usually difficult to classify as manufacturing defects or other issues. Some module failures are due to external causes such as shipping errors, clamping, cabling errors, connector errors, and lightning strikes [47]. Damage to the glass cover and laminate of some modules has been observed to be due to impact and vibration during transport. This defect is incompatible with manufacturing defects and is one of the main external causes of module defects. Most transport failures cannot be seen visually or by observing the nominal power. Of the installation issues correlated with modules, clamps are the most common mistake, especially for frameless PV modules, which leads to glass breakage as illustrated in Figure 1.6.

Sharp-edged clamp designs, narrow clamps, improper placement, and overtightening of the module clamp screws can stress the PV module and lead to breakage. The effects of broken glass can cause corrosion due to the ingress of moisture through cracks, which creates electrical safety issues and degrades performance during long-term operation. The resulting cracks also lead to hotspots that lead to module overheating [49]. Cables and connectors connect to the PV system and provide electrical connectivity to other components of the PV system, including solar modules and inverters. Connectors are crucial elements and perform a vital role in securing the safety and reliability of power generation and transmission. In addition to various connectors, low-voltage DC connectors have been widely

Figure 1.6 Poor clamp PV module design.
Source: Köntges et al. [48]/IEA.

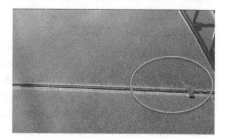

discussed for use in electric vehicles and solar systems [50]. A connector can show defects because of the different metals associated with them, which causes corrosion when exposed to the atmospheric humidity in combination with gases [51].

Any limitations due to cables or connectors used in solar systems are not considered manufacturing defects when all impacts and consequences are considered. Connector errors are usually caused by improper cable selection or incorrect connections between PV modules and related components. These failures can result in a complete loss of power to the string and cause arcing and fire. The effect of lightning strikes on the DC side is observed as a bypass diode fault in the module [52]. This effect is usually an external cause that leads to subsequent security breaches. Most of these external causes lead to failure of the module due to the open circuit of the shunt diode or the direct effect of lightning.

Most expected failure modes and degradation mechanisms are associated with junction box failures [53], glass breakage [54], connection failures [55], and delamination [56]. Improperly designed junction boxes allow moisture to penetrate and cause corrosion of the junction box terminals. This causes a wiring failure and an internal electrical arc. There is also a potential for soldering failure, i.e. solder joint fatigue and silver (Ag) leaching, in junction boxes. When the soldering point encounters the Ag electrode of the solar cell, it dissolves in the soldering electrode (tin-lead (Pb – Sn)) and is observed as an Ag_3Sn compound. This Ag-leaching effect causes the thermal expansion to crack the soldered interface and break the connection. The delamination effect occurs due to adhesion contamination or environmental factors, i.e. moisture, humidity, and corrosion in PV module panels. Lamination failure results in optical reflections, which further causes subsequent loss of power generated from the modules. As reported in [56], more than 90% of PV modules can be delaminated. The requirement of adhesion, which is to be met by the manufactures can result in the delamination of the panel [57].

PV back panels deliver reliable operation at high voltages and safeguard electronic components from the severe field condition [58]. Backsheets are made of a variety of materials such as polymers, glass and metal leaf. Backseats are usually made of a very stable laminated structure and UV-resistant polymer. The material was selected based on the required mechanical strength, cost, and electrical insulation. There have been some serious problems with this structure as glass can break through improper installation or mechanical stress. Despite the concerns, glass/backsheet structures can deliver 2–3% more power than standard backsheet modules. Polymer laminates are the most used backsheet-building materials. They have multiple layers, which have the effect of separating the interfaces due to high physical and chemical stresses. The only benefit I have seen with polymer laminate breakage is that it does not create immediate safety concerns when delamination occurs. Peeling the backsheet near the junction box is usually a serious problem as it results in an unrestricted junction box. These interruptions lead to failure of the shunt diode connection, which is further complicated by the formation of an unresolved arc across the total system voltage.

1.3.2 Failure Due to Manufacturing Issue

1.3.2.1 Silicon Wafer-based PV Fault

PV modules based on crystalline silicon wafers occupy a dominant position in the world of PV modules due to their widespread use [59]. The market share of these modules is 95% (as of 2017) [60], which is the most common type of solar cell. Despite their extensive application and use, these modules are susceptible to failures such as potential and light degradation and snail marks. Discoloration of ethylene-vinyl acetate (EVA) encapsulant was originally observed at the location of the Carrizo Plain in California in the early 1990s and has emerged as a major problem [49, 61]. The developed single crystalline silicon PV modules which use polyvinyl butyral encapsulant and Tedlar/aluminum/Tedlar backsheet structure [62] need a robust electrical insulation layer in the middle of the cell and the film, which causes many protection concerns. Apart from the insulation, the metal leaf also functions as a high-voltage capacitor, and the disturbance of the electrical insulation on the foil surface charges the foil with the system voltage. The factor affecting the EVA polymer encapsulation degradation was discussed in [63]. Analysis showed that the field was degraded from yellow to dark brown EVA to understand chemical and physical damage. EVA is manufactured with additives such as crosslinking agents, antioxidants, UV absorbers, hindered amine light stabilizers and adhesives. Discoloration results were observed in various samples, and difficulties due to the diffusion of oxygen and acetic acid due to additive reactions were investigated. The origin of the chromophore creates a transparent EVA ring at the edge of the plate-based cell. In some EVA discoloration scenarios, one cell of the module is observed to be darker than the other. This usually means that the temperature sensitivity of a particular cell is higher than that of an adjacent cell either due to low photocurrent or because the cell is located above a junction box [64]. In severe cases, EVA discoloration is consistent with EVA embrittlement and concomitant dissolved oxygen corrosion [65]. Although most module failures cause complete failure, discoloration and delamination do not lead to failure, but degrade functionality with a very slow rate of degradation of ~0.5 %/year [66]. There is an overall loss of ~10% in severe discoloration, which means that EVA discoloration is absurd for a complete failure of the silicon module.

Observations in [67] revealed different types of cracks: multidirectional cracks, diagonal cracks, and cracks parallel and perpendicular to the tire. When a cell cracks, the device cannot be completely detached from the cell, but resistance occurs between the cell device and the number of cycles present in the deformed module. It is stated that cracking criticality and cracking of solar modules result in low power output stability during artificial aging. Experimental analysis was performed on 667 cracked cells in 27 PV modules and the results showed that 50% of the cracks were oriented parallel to the busbar cracks considered critical. Differences in solar module manufacturing processes cause cracks in the cells, especially in the stretching process [68]. Problems with the transportation and installation of PV modules have also been identified as major causes of cell cracking. For several months, [69] investigated PV modules in which snail tracks were formed due to external influences of faulty modules. Detailed microscope images of the discoloration helped in the identification of the problem. The snail trajectory was observed to be mainly located near the edge of the cell or microcracks. Not all microcracks need to change to a snail trajectory, but whenever a trajectory is observed in a module or cell, a microcrack is found in the same location. The optical impression of the footprint is due to the location of the mesh fingers being discolored brown and imprinted on the EVA foil. The early discoloration process due to snail marks is not well documented in the literature. The visual impression of the snail track varies from module to module and destroys cell fragments and cell edges. What the module traces is called the electromechanical degradation process, but it has never been considered a direct cause of power loss.

Potentially induced degradation (PID) [70] is caused by polarization effects. The level of degradation depends on the polarity and the level of potential difference between the photocell and the earth. The PID is also responsible for the durability issue of the module. Typically, PID occurs when a high voltage causes sodium ions to spread out of the glass through encapsulation and accumulate on the cell surface. This leads to surface recombination, lower filling factor, and increased local fractionation [71]. There are two types of PIDs: irreversible and reversible. Irreversible PID is caused by an electrochemical reaction that leads to electro-corrosion of transparent conductive oxides. Reversible PID, also known as surface polarization, creates a positive charge in the PV cell, resulting in a leakage current. The leakage current value is determined by the PV array grounding configuration. This lowers the power-generation capacity of the solar cell. The general situation in the development of PID is observed at various levels such as environmental factors, counting factors, systemic factors, and the cellular level. The electrochemical corrosion of the string connection of the cell due to encapsulation leads to deterioration of the PV module [72, 73]. This phenomenon increases the series resistance of the PV module and lowers the parallel resistance [74]. PV cells typically have front and back contacts connected by bus strips to power external circuits. Defects in the strings cause thermal expansion (contraction and mechanical stress), resulting in loss of output power [75]. The electrochemical corrosion [72, 73] of the string connection of the cell due to encapsulation leads to deterioration of the PV module. This phenomenon increases the series resistance of the PV module and lowers the parallel resistance

[74]. PV cells typically have front and back contacts connected by bus strips to power external circuits. Defects in the strings cause thermal expansion (contraction and mechanical stress), resulting in loss of output power [75].

1.3.2.2 Thin Film Module Fault

Thin film PV modules have been developed to save material (and therefore cost). These modules have lower conversion efficiency than traditional crystal modules. However, this is offset by low cost (production is less material and usually less technically required) and improved properties at low irradiation. From a material and manufacturing process perspective, the modules are divided into CuInSe2 (CIS), Cu (In, Ga) Se2 (CIGS), CuGaSe2 (CGS) modules, CdTe modules, amorphous and micromorphos silicon modules, etc. Other thin-film cells such as multi-junction cells, nanostructured cells, and organic cells. The design process for all the above modules is different as they deal with multiple connections and configurations. However, regardless of the design procedure, the causes and consequences of malfunctioning of a given module have the potential to deteriorate the module. Some of the potential causes and effects observed for thin film modules are spliced connectors, micro arcs, hotspot shunts [76], windshield breakage, and back contact degradation [77]. Initial degradation of 4–7% can be expected within the first 1–3 years depending on climate and system interconnection factors. The various defects that can be viewed without the aid of the tool are shown in Table 1.1.

1.4 Overview of Converter Faults

Grid-connected PV systems are considered highly reliable. However, like any other complex electrical system, it is vulnerable to failure. These PV systems have a modular structure. Their output can vary from a few watts to megawatts. Therefore, it can have different types of topologies and configurations [79], further complicating the assessment of system failure modes. Researchers have observed that most of the grid-connected system failure are related to inverters. The operation of the inverter is different from that of a rotating machine such as an induction machine. Their fault currents decay rapidly because they have no inductive characteristics, which are largely dependent on the time constant of the circuit [80]. However, it can be controlled so that the error reaction time can be changed under program control. The inverter can use a voltage or control algorithm depending on the failure that occurs. During the transition period, using the voltage adjustment algorithm will increase the error contribution. With current control algorithms, the rate of increment or decrement is much lower. Inverters are mainly composed of switching devices such as insulated gate bipolar transistors (IGBTs) and metal

Table 1.1 Visual defects related to PV modules [69, 78].

Fault type	Power loss	Safety issue	Visual fault
Short circuit of wires in module and diode	Power loss (<3%)	May cause fire hazard	
Fragments of cell laminated	Power loss (<3%)	Can cause electrical shock, fire hazard and physical risk	
Cell cracks (<10% of cell area)	Power loss, degradation with overtime saturation	No impact on safety of individual	
Delamination	Power loss, degradation with step time	Safety concern due to electrical shock	
Marks of burning on backsheet	Power loss, degradation with step time	Can cause electrical shock, fire hazard, and physical risk	
Discoloration of front panel due to overheating of metallic interconnections	Power loss, degradation with step time	Can cause electrical shock, fire hazard, and physical risk	
Delamination of multicrystalline Si module	Power loss, degradation with step time	Physical risk due to failure	
Delamination of thin film module	Power loss, degradation with step time	Physical risk due to failure	
Glass breakage in thin film modules	Power loss, degradation with step time	Physical risk due to failure	

Table 1.1 (Continued)

Fault type	Power loss	Safety issue	Visual fault
EVA browning	Power loss degradation over linear time	No risk initially but with growth in the defect it may result into fire hazard	
Snail trails	Power loss degradation over linear time	May cause fire hazard	
Delamination of back sheet	Power loss degradation over linear time	May cause fire hazard	

oxide semiconductor field effect transistors (MOSFETs). Studies show that nearly 34% of power electronics system failures can be traced back to switching devices and solder joint failure [81, 82]. DC link capacitors are also another most susceptible component in systems [83]. There are two types of errors in these switching devices. The first type is a failure that occurs suddenly, such as a sudden overvoltage or overcurrent or a sudden temperature rise. The second type is failure due to gradual wear over a long period of time.

There are DC/AC power electronic converters, i.e. inverters for the flexible and efficient conversion of power from DC to AC. Malfunctions can occur which may cause trouble in system operation. These tasks include maintaining optimal power quality in case of a failure, maintaining reliability, eliminating energy losses that occur, and detecting and determining failures. Therefore, to improve reliability, it is necessary to have a tool to predict the occurrence of failures in advance. Error detection may be delayed, or errors may not be detected, and both events can have devastating outcomes. The system must consist of a fault-tolerant mechanism to prevent it from shutting down in the case of an unforeseen failure [84].

A failure may occur on the DC side or AC side. Failure on the DC side is caused by a failure in the PV panel, capacitor, or boost DC/DC converter. The AC side is affected by the failure of the inverter and the failure of the filter element. Since PV operates in intermittent environmental conditions, aging can cause material

Table 1.2 IGBT failure and causes.

Failure	Failure categories	Reason for failure
Catastrophic failure	Breakdown caused by high voltage	Spikes in high voltage
	Failure related to Latch-up	During the turn off high dv/dt
	Electrostatic Discharge (ESD)	The gate at high voltage application
	Secondary Breakdown	Stress due to high current
	Overstress (Mechanical)	Metallization at the surface
Wear-out failure	Liftoff occurring at bond-wire	Al and Silicon CTEs mismatch
	Crack in the bond-wire heel	Aging
	Failure in Solder joint	Copper and Silicon CTEs mismatch
	Degradation of Gate-oxide	High Electric Field and Temperature

deformation and seriously affect PV performance. Man-made errors such as wiring errors, careless handling, and manufacturing defects are other possible causes of poor PV performance and can lead to failure events as discussed in Table 1.2. Arc failures, local hotspots, and temperature sensitivity of cell materials are a variety of other possible failure events [85]. However, the most dangerous obstacle to solar power is the ground fault. PV systems should use the concept of careful ground protection. They can depend on several factors, such as the size of the system, the type of installation, and the location of the system. Regardless of the ground protection method used, ground faults may occur for obvious reasons. These reasons can be caused by the damage to cable insulation, aging, corrosion, unintentional short circuits, and so on. These error events may lead to the system shutting down. Therefore, an appropriate fault tolerance mechanism is required along with a ground protection strategy. Bypass diode failures, circuit breaks, line-to-line failures, double ground faults, etc., are other failure events that can occur in PV systems [86]. Of these, line-to-line errors are extremely dangerous. An unintended low impedance path between two points on a PV module can lead to line-to-line failures. This will change the current route and, in the worst case, could cause a fire [87]. Fuse and overcurrent protection schemes are commonly used to prevent line failure events. Arc failures can also impair the efficient operation of PV systems. Arc failures can be divided into series arc failure events and parallel arc failure events. Issues such as unwinded screws, poor cable connections, thermal stress increment, etc., can cause such events [88]. PV is deemed as a promising energy source which is used not just in the energy sector, but also has its application in the field of transportation, lighting, construction, and housing sectors. It is free of pollutants, environmentally friendly and available in abundance. However, an

DC/AC conversion is required as the output of the panel is in DC. PV can be connected to grids and other AC loads using an inverter as an interface. The inverter serves to provide a clean sinusoidal output at the required voltage and frequency. An efficient grid connection is required for an adequate grid synchronization. In some cases, the system may require a boost DC/DC converter to improve the performance of the PV array. Both power electronics converters, i.e. DC/DC converters and DC/AC converters (inverters) are key components of semiconductor-based power-switching devices. With advances in semiconductor technology, IGBTs have become the best switching devices for power electronics applications [89]. However, because of the intermediate circuit capacitors, it is one of the most susceptible components in grid-connected PV systems. Most power electronics converter failures are caused by IGBTs and intermediate circuit capacitors. Therefore, it is of utmost importance to study the electrical and thermal properties of IGBTs. In recent years, IGBTs have made tremendous developments in terms of size, cost, thermal efficiency, and reliability. Innovative assembly and packaging technology have increased the power density of the IGBT. It is expected that the performance of IGBTs will continue to improve in the future. Today, IGBTs have not only made structural and operational advances but also have increasingly better cooling capacity, making them suitable for power electronics converters. IGBTs are widely used in many important applications such as switched-mode power supply (SMPS), uninterruptible power supply (UPS), and inverters. However, while the characteristics of IGBTs continue to improve, there are still concerns about reliability [90]. IGBTs are prone to failure due to electrical, mechanical, and thermal overloads. There are two types of IGBT failures: wearout failure and catastrophic failure. Wearout defects may occur due to bond wire lift and tear, solder joint failure, avalanche breakdown, etc. These wearout defects are caused by the long operation of the device and power cycle [91]. Catastrophic failure of an IGBT is caused by an overload event such as a sudden rise in voltage, current, or temperature.

1.4.1 Wearout Failure

Investigation of wear defects in IGBTs is very important to ensure the reliability of power electronics converters. IGBT materials experience these defects due to thermomechanical fatigue stress [92]. These errors can be further categorized as follows:

Bond-wire lift-off – Bond wire lift-off errors are caused by mechanical events. Inconsistencies in the coefficients of thermal expansion (CTE) of aluminum (Al) and silicon (Si) can cause the bond wire to tear. The temperature change and the difference between the CTEs cause wire bond cracks, and the growth of wire bond cracks causes bond failures. The difference in elongation between the two materials creates stress, which is temperature-dependent. This leads to the formation of cracks and is continued by stress-strain energy loss. [93]. This hysteresis energy loss is expressed by temperature fluctuations. If the temperature fluctuations in the barrier layer are large enough, the cracks will move from the sides of the Al wire to the center.

Bond-wire heel crack – Bond-wire heel cracks rarely occur in advanced IGBTs. After the long-term operation of the IGBT, heel cracks in the bond wire are observed. This is believed to be the main cause of the failure mechanism of power electronics converters [94]. This can be confirmed by measuring the saturated collector-emitter voltage ($V_{CE_{sat}}$). The criterion for detecting this fault is a $V_{CE_{sat}}$ increase of about 5%.

Solder-joint failure – Defects in solder joints are very common and are one of the leading causes of wear failure in IGBTs. The CTE of copper is higher than silicon. The difference in CTE causes shear stress between the silicon chip and the copper substrate, which further causes cracks and voids in the joints. These cavities lead to a reduction in the effective heat dissipation area, which results in heat concentration on the surface [95]. This further accelerates the formation of cavities and increases thermal resistance, resulting in very localized heating. This, along with the power cycling process, causes solder fatigue. Solder joint failure adversely affects the collector-emitter on-voltage ($V_{CE_{on}}$), reducing its value over the linear operating range of the IGBT.

Gate-oxide degradation – The gate oxide of IGBTs can degrade over time because of the high temperatures and strong electric fields. This is caused because of the die electric breakdown taking place over time [96]. The hot electrons damage the oxide film of the gate due to impact ionization. Damage to the gate oxide increases the gate leakage current. It also negatively affects the threshold voltage $V_{CE_{th}}$ between collector and emitter.

1.4.2 Catastrophic Failure

In addition to wear failures in IGBTs, there are catastrophic failures caused by excessive electrical, mechanical, and thermal stresses. It negatively affects the operation of the IGBT and can sometimes permanently damage the device. These failures can be further divided into the following categories:

High voltage breakdown – Because of the high voltage conditions, the heating effect is large and may further affect the operation of the IGBT. High voltage peaks can occur due to the high rate of decrease in collector current. Repeated peaks can destroy the IGBT when switched off [96]. This can also lead to increased leakage current and high local temperature. High collector-emitter voltage (V_{CE}) or gate-emitter voltage (V_{GE}) values when switched off can cause short circuits. One must then provide a safe operating area (SOA) suitable for the specific power unit to meet the desired radiator performance.

Latch-up failure – Latch-up occurs when the gate loses control of the collector current. High dv/dt in the switched-off state can trigger the IGBT's parasitic transistors and cause further latch-up. High collector currents can also cause latch-up by turning on the parasitic transistor [97]. Loss of gate control due to latch-up damages the IGBT. However, the latch-up resistance of new devices has been improved.

Electrostatic discharge (ESD) – ESD comprises similar characteristics to the high voltage breakdown failure. This causes partial penetration of the gate oxide, which can cause the device to function properly over a certain period of operation and eventually fail the device after a certain period [97]. Applying too high a voltage to the gate will short the gate and further affect the operation of the device.

Secondary breakdown – High voltage can lead to localized thermal breakdown of the device when turning it on and off. Secondary failure is a kind of thermal failure. As the current increases, the space charge density of the collector-base junction also increases and the breakdown voltage decreases, which in turn increases the current density [97]. This cycle continues until the area of high current density decreases to the minimum area of stable current filaments. As a result, the temperature of the filament rises sharply due to self-heating, resulting in a sharp drop in voltage across the IGBT.

Mechanical overstress – Mechanical overvoltage events can lead to atomic migration, which can affect device reliability in the long run. Surface metallization is observed in the IGBT device, which may be due to the reconstruction of the Al contact pad [97]. In addition, the thermal resistance is unevenly distributed, which can increase the temperature of the device. Failure to take appropriate remedial measures may result in localized burnout of the device.

1.4.3 IGBT Thermal Modeling

Continued further development of switching frequencies and power densities of IGBT components makes electric heating and thermomechanical analysis inevitable. These multidimensional studies are needed to achieve optimized device operation. One type of failure can cause another type of failure on the device. This may depend on the various electrical characteristics of the device. Therefore, an optimal overall thermal or electric or thermomechanical model, i.e., mechanical overload events, high junction temperature events, bond wire failures, high dielectric events, heat sink problems, and all possible causes of failure, need to develop. The most used model is the thermal modeling, which makes it easy to integrate device properties via the RC network model. They are easy to simulate electrically and work well. Modeling should be done so that the model is accurate, and the reliability parameters can also be analyzed. The thermal model can be further categorized as:

Analytic models – Mathematical thermal models were initially developed for reliability analysis. However, these models have been continuously developed over the years and are now very efficient. Based on the two-port theory, they have developed a computationally efficient tool for calculating effective transition temperature rises and heat flow distributions. The boundary element method (BEM) was used to simulate a high-performance IGBT with power cycling [98]. It was able to considerably decrease the difficulty of the device shape and solve the heat loss calculation. The 3D finite difference method (FDM) has also been proposed to solve the heat conduction equations of the IGBT thermal model. A thermal interface material (TIM) mechanism was also introduced to optimize the geometry-based operating conditions of the model and analyze the heat sink heat dissipation.

Numeric models – Based on numerical simulation, a finite element method (FEM) was proposed to attain the temperature distribution for the device with precision. It was based not only on the detailed geometry of the device but also on the material properties. Over the years, they have made tremendous progress and are more advanced than they used to be. The coupling approach can be used to calculate the power loss and device temperature that occur on the device. Along with the sensitivity of the model parameters, the thermal characteristics can be modeled correctly. It also helps identify the characteristics of the heat sink and provide an efficient cooling mechanism for the device.

Network models – These are often used for dynamic and steady-state thermal analysis of power devices [99]. This can be expressed using basic thermal resistance (R_{th}) and capacitance (C_{th}). Initially, A charge control model was used that could be analyzed in $1D$ and could be implemented with an IGBT circuit simulator. The calculation method was developed for the thermal diffusion equation and temperature distribution. This method was easy to connect to an electrical network and was even more useful for high-speed dynamic calculations as well as estimation of electric heating characteristics. This method can accurately measure the temperature rise of the IGBT during the switching cycle of the device. The two networks that fall into this model are the Foster and Cauer RC networks [100]. Although these networks have some structural differences, they reflect the same behavior in terms of accuracy. These models consume very little time and are extremely accurate and can handle the thermal, electrical, and physical properties of power devices.

1.4.4 Measurement for Cooling

Choosing the right cooling strategy can improve the reliability of switching devices such as IGBTs. One of the most important design considerations for IGBTs is the heatsink consideration. This is because it impacts thermal management and thus affects the reliability of the IGBT. Improved cooling measures can prevent many failures of IGBT components. The operation of the heat sink is not affected by the geometry. Rather, its function is mainly influenced by the material used for cooling. The air-cooling method is mainly used because it is inexpensive and ready to use. However, the cooling efficiency is inferior to that of the liquid cooling method. The literature proposes direct liquid-based cooling solutions such as heatsink cooling and spray cooling. These cooling methods are far superior to air cooling methods because they can reduce thermal resistance by 30% [101].

Traditional secondary cooling was centered around a heat sink with a power module mounted using thermal paste, resulting in high thermal resistance. However, the thermal paste is not required when direct liquid cooling is used, which results in reduced thermal resistance and improves cooling efficiency. Microchannel cooling solutions have been proposed in the literature. They require minimal coolant, are compact in size and exhibit unique cooling behavior. Two-phase forced convection cooling solutions with better heat transfer coefficients compared to other methods have also been proposed. These solutions have excellent properties for maintaining a uniform surface temperature on the device, making them suitable when reliability is required. Jet collision and spray cooling processes are primarily used in high-performance applications. They have amazing cooling properties and can easily dissipate large amounts of heat, making them one of the best solutions for larger surfaces. One of the latest innovative cooling methods is a solid and liquid hybrid cooling solution for isothermal switchgear [102].

1.4.5 Failure of DC-Link Capacitor

Following semiconductor-based switching devices, capacitors are one of the least reliable components of electrical systems. Electrolytic capacitors are inexpensive and are commonly used in power electronics converter applications. However, there are some major issues such as narrow filter bandwidth, sensitivity to temperature fluctuations, and short service life. Metallized polypropylene film (MPPF) capacitors (metallized polypropylene film) are better suited for power electronics applications [103]. It has low filter bandwidth, low equivalent series resistance (ESR), and self-healing characteristics. In addition to these advantages, it is largely unaffected by fluctuations in ambient temperature. Nevertheless, these two capacitors are prone to failure.

Capacitors can fail in a variety of ways. Manufacturing process issues can lead to premature failure. Random failure events are extremely rare in capacitors. In most cases, capacitor failures are wear-out failures. Long-term temperature fluctuations can lead to capacitor failure due to partial discharge and aging. In addition to common environmental degradation events, electrical and mechanical overload events can also cause failures. Gradually breaking the dielectric reduces leakage current and capacitance, which can eventually lead to full failure. Capacitors can also be generated by improper operation, i.e. Persistent overvoltage or overcurrent events will fail. In the event of an open failure, the failure does not spread further, but a short-circuit failure event is considered to have a further negative effect on the state of the capacitor. Therefore, it is important to take some measures in the early stages. Otherwise, short-circuit failures will spread further and can lead to the failure of adjacent devices. For capacitors mounted on printed circuit boards (PCBs), external materials can cause line corrosion and lead to failure. Repeated overload events can have a very negative effect on the capacitor, as they increase the ESR value and can increase heat loss.

Electrolytic capacitors are amongst the most vulnerable components of grid-connected PV systems. An equivalent circuit consists of a capacitor and an inductor connected in series along with an equivalent resistor. These capacitors are vulnerable to system malfunction and can be impacted by ambient temperature changes. Primary wear failure can be caused by electrolyte evaporation between the cathode and anode leads of a capacitor. This evaporation effectively reduces the electrolyte area, reducing capacitance and increasing ESR. This causes the temperature to rise and the electrolyte to evaporate further, repeating a similar process. There are standards that suggest that for an electrolytic capacitor to fail entirely, the rise in ESR must be double the value of ESR, and the value of the capacitor must decrease by 20%. To identify and diagnose the failure mechanism of capacitors, various failure detection methods have been proposed in the literature. These methods are based on identifying ESR and corresponding changes in capacitance and temperature changes.

1.4.6 Failure of Power Diode

Power silicon diodes are considered an essential component for a power electronic converter. However, it can fail due to excessive voltage or current. The main failure of power diodes occurs due to voltage surges. Even if a surge of voltage or current does not have sufficient energy to raise the temperature of the device, damage can occur [104]. This type of fault can be detected by analyzing the current density. It will be higher at some point. Another serious diode failure is a secondary breakdown. The characteristic of the power diode VI is that it is composed of positive resistance in the low current region and negative resistance in the high current region. The current densities in the two regions are balanced. If an imbalance of current density is observed for any reason, a higher current density will appear in a particularly concentrated area, leading to secondary failure. This secondary failure ultimately leads to diode failure. Therefore, the failure mechanism of the power diode must be considered for the correct operation of a networked PV system.

1.4.7 Power Semiconductors' Failure Mechanisms

Failure Mode Impact Analysis (FMEA) is a method to develop a comprehensive list of failure modes for a system and analyze the impact of failure modes on a larger system. Failure mode effect and criticality analysis (FMECA) extends FMEA by introducing a severity metric that ranks failure modes based on their severity, occurrence, and detectability. Criticality analysis allows engineers to focus on these critical failure modes defined by risk priority numbers to reduce the impact on end users or system manufacturers. In literature [105] publishes FMMEA for Silicon Power Devices. See Table 1.3. However, this FMMEA is limited to discrete IGBT components where the junction temperature fluctuates widely, and power cycles occur when the average temperature is high [105]. FMMEA was developed with the purpose of identifying error precursor parameters for predictive applications and has fully fulfilled this function. FMMEA like this is limited in several ways. First, as explained in the Failure Mechanism Severity Analysis, FMMEA requires knowledge of the application. So, it is not a true FMMEA and does not attempt to explain the importance of the failure mechanism. Second, Patil, etc. Not all failure mechanisms are associated with Silicon Power devices under reasonably expected operating conditions. Once the lifecycle profile is identified, FMMEA should create a list of failure modes and mechanisms. This can be derived from the stresses present in the life cycle profile. Power semiconductor devices have many prospective causes of thermal, mechanical, and electrical stress, but the mechanisms that can be eliminated depend on the physical properties of the power semiconductor device. The dimetalization of power semiconductors is usually Al. If the board is a DBA, the Al is further metallized on the ceramic board. Thermal cycles of components due to Joule heat, switching losses, changes in ambient temperature, and inconsistencies in the CTE between Al and silicon and ceramic substrates create thermomechanical stresses in the housing. These stresses cause the metallization of the Al and can be large enough to bend or hit the Al. This mechanism is called the reconstruction of Al [106–108].

Besides, the moisture inside the package can cause corrosion of Al wires and contact pads [109, 110]. In the presence of dampness, Al reacts to form $Al(OH)_3$, which passivates the surface and Al. This passivation coating can dissolve in the existence of impurities such as halogens. When the component is turned on and off, the sealing layer can peel off the base plate, creating a path for damp and impurities to enter the component. Like thermomechanical fatigue, corrosion can enhance the internal resistance of power semiconductors. Further, the fatigue in the bond wire is also a thermomechanical failure mechanism. Al bond wire, unlike gold or copper bond wire commonly used in low-power applications with smaller wires connected to ball bonds due to the large diameter wires used to handle high current densities in power devices. A further spot of thermodynamic stress due to

Table 1.3 Possible failure modes, causes, and mechanisms.

Failure modes	Causes	Mechanisms
Increased leakage current, short circuit, and gate control loss (oxide)	Overvoltage, excessive temperature, and high electric field	Time-based dielectric breakdown (V_{th}, g_{th})
Device burnout, loss of gate control (silicon die)	Ionizing radiation, high electric field, overvoltage	Latch-up ($V_{CE(ON)}$)
Leakage currents (oxide, oxide/substrate interface) are quite high.	High current densities, overvoltage	Extremely hot electrons (V_{th}, g_m)
The circuit is open (bond wire)	High current densities, high temperature	Bond wire is crumbling, therefore, it is time to get rid of it. $V_{CE(ON)}$
The circuit is open (die attach)	High current densities, high temperature	Delamination of die attachment, voiding ($V_{CE(ON)}$)

CTE mismatch is substrate attachment that attaches the substrate to the substrate. Since the die is vertically conductive, the die attachment needs to be conductive as it is a section of the power semiconductor component's electrical path. Comparable to metallization of Al and wire bonding, the die attach is in close contact with the die and is subject to thermodynamic fatigue and delamination potential [111, 112]. Delamination occurs when the matrix itself and the matrix adherent material separate. However, fatigue can appear and transmit through the punching tool. In addition to this, the failures due to substrate cracking [113], melting of bond-wire [114], and voiding of die attach [115–117] are also widely identified failure modes.

Furthermore, as the semiconductors and metals have imperfect material properties, the latching is observed as a short circuit between the conducting leads [118]. When the current exceeds the so-called latching current of the device, the thyristor is activated. However, if the latching event is not identified, a thermal fault can occur and burn out the device. Latching is an overvoltage failure mechanism. Further, the avalanche Breakdown mechanisms [119, 120], and the dielectric breakdown mechanisms [121, 122] can often occur during switching when the drain or collector voltage surpasses the breakdown voltage of the power supply. Besides, the silicone gel containing the metallization and connecting wires of the power module is used to enhance the breaking strength of these conductors. However, because of the high voltage and geometry of the conductor, the electric field inside the housing can still be amplified and initiate a partial discharge inside the silicone gel [123, 124]. Over time, these discharges can create carbonized conductive pathways within the gel, which can increase leakage. A local partial discharge can form bubbles in the gel due to local heating. Partial discharge inside the gel is observed as an increase in leakage, which develops into a short circuit and becomes a wear mechanism.

1.5 Detection Strategies for PV System

Fault detection of the PV system required precision to make the detection process easy and more accurate. If the fault is detected accurately with proper identification of the fault location, then the section under question can be tested further and adequate actions can be taken to improve the system operation. For the PV panel It is vital to undertake not just to diagnostics of obvious faults, but also protective diagnostics, which aid in the early detection of potential problems and hence the avoidance of economic losses, such as potentially dangerous microcracks. There are a variety of ways for evaluating parameters while a PV device is in operation, some of which are supported by international standards. The diagnostics of a panel can be performed by following techniques:

• Thermography

When evaluating parameters obtained through electric measuring, it is required to isolate the cells and the modules from the rest of the device and examine them individually, often in a laboratory. With less accessible installations, this procedure can be quite costly and time-consuming in the case of PV modules. If only for this reason, any approach that allows for flaw detection without disassembly must be used. Thermography is one such technology that is commonly employed in the diagnosis of PV system defects [125]. The method is based on the detection of heat irradiation with the use of an appropriate detector. The module's cells are connected in a serial manner. When all the cells are identical in an ideal state, the same current flows through them while the voltage is constant. The voltage is zero in case of the short circuit current flow through the cell. The variation in current and the voltage of the cell based on the single fault is expressed in [126].

Since the total voltage must be kept at zero in a short circuit, the damaged cell is polarized in the opposite direction of the other cells and the voltage contained therein is the forward total voltage of the other cells. In actual operation, the cell operates at the highest level of performance, but the heating mechanism of the damaged area is the same [127]. This effect can be caused by cell defects or shadowing. So-called bypass diodes are used to reduce this effect. These diodes open when the reverse voltage drop exceeds the threshold voltage VB. Bypass diodes are not used in the actual modules of individual cells, but there are antiparallel

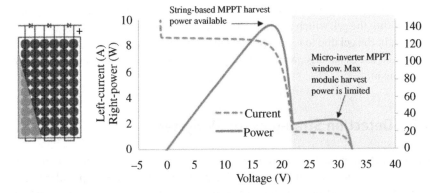

Figure 1.7 Shading impact on PV curve.

connections to groups of cells connected in series (typically 20–24 cells) [128]. You can also see how the bypass diode works from the IV curve where the "stairs" occur. This can make MPP detection more difficult due to the multiple limits on the curve as illustrated in Figure 1.7 [129].

Shading of submodule strings with shaded illumination of approximately 150 W/m² and unshaded illumination of approximately 1000 W/m². The temperature distribution can be recorded using the appropriate device. The thermal field reading was done using a thermometer (contact or noncontact). Infrared cameras are now widely used due to falling prices. Detection by the thermal imager allows for the detection of a large number of errors [130]. However, it is disadvantageous that the thermogram does not provide a quantitative evaluation of the module under investigation [131].

- Electroluminescence

Electroluminescence (EL) is another valuable diagnostic technique. This method is used for cell and module defects such as cracks (usually abbreviated as ELCD EL crack detection test), technical defects, and other nonuniformity assessments. As a result, it may also be used to assess the modules visually. It operates on the premise of detecting EL radiation generated by recombined charge carriers during the radiative recombination process. The device should be installed in a location with appropriately low illumination (darkroom is best) and the sensor should be a special sensor that allows for detection in the near infrared range. Commonly used sensors are cooled CCD, CMOS, or InGaAs sensors. The third advantage is that it is much more sensitive in the available wavelength range. This allows for shorter exposure times. A good EL image takes about five minutes for a CCD camera and only a few milliseconds for an InGaAs camera [132]. The PV module is connected to the power supply and the current must not exceed the ISC value. The intensity of the emitted radiation depends on this current, and low current intensity causes various defects. Areas affected by some damage or areas with high defect density appear as dark areas on the EL image. Such places do not contribute to power generation. In that case, the radiant intensity is on the scale of PV module function [133].

- Photoluminescence

This procedure is used to diagnose PV cells. In contrast to EL, the radiation recombination excitation is excited by a strong optical pulse. So, the contact system does not require a sample. So, this method and control method can be used during the manufacturing process. The disadvantage of this method is that it requires a special sensor (such as an EL housing) and much more complex equipment for radiation excitation [134, 135].

- Microplasma Luminescence

This technique provides knowledge about short circuits in the structure. In contrast to EL measurements, the module is switched in the opposite direction, but the reverse polarization must not exceed the cell breakdown voltage [136]. Microplasma appears due to the reverse polarization of the area affected by the defect. Microplasma appears like noise or luminescence. Luminescence causes spots of light in the resulting image. This means that these images are actually the opposite of EL images [137].

- Static Characteristics Measurement – Current-Voltage Characteristics

The current-voltage characteristic (IV curve) measurement method, also known as IV analysis or voltammetry, is the most widely used method for diagnosing PV cells and modules. This allows you to determine the basic parameters of the PV component. Measurements can be made directly in the laboratory or at the installation site, but mainly accurate measurements under standard test conditions specified by international standards such as irradiation $G = 1000 \text{ W/m}^2$ and cell temperature 25 ° C. This can be achieved by using a continuous solar simulator or a lightning solar simulator. Measurements are usually performed in a lightning solar simulator (lightning tester). This is much more complex and cheaper than a continuous solar simulator, and most technologies have sufficient pulse width. For irradiation from PV components, the entire IV curve is measured by electronic load control. The continuous solar simulator is only needed for special cases such as concentrator modules and solar collectors. The solar analyzer is used for external measurements. This also allows you to usually connect an external sensor to measure ambient conditions [138]. The most problematic part is testing thin films for seasonal glow effects and high capacity.

- Effect of Season Annealing

When measuring thin-film modules, it was observed relatively early on that performance and efficiency were strongly dependent on operational history. This means that the MPP of the running module is 100 W (under STC). After storage in the dark and new measurements, only 90 W will be measured. It was shown that temperature and irradiation caused a metastable state of the PV module, which needed to be removed or calculated during the measurement. The annealing effect of the season consists of two phenomena: the light immersion (LS) effect and the temperature annealing (TA) effect. The former is particularly well known for amorphous silicon-based PV modules that produce the so-called Staebler–Wronski effect [139]. In the process, the optically excited charge carriers break and release the weak Si-Si bonds, creating recombination centers and shortening the service life of the charge carriers. This decline occurs during the first hundred hours of operation and can reach a 30% reduction. Hydrogen is used for suppression and tandem cells also show a low LS effect. Other thin film technologies behave similarly. In many experiments, CdTe PV modules often significantly improve device performance in the range of a few percent to 10% within the first few hours of exposure to light. This power reduction can also be achieved by applying a forward-biased current in the dark. The positive effects of light immersion can also be observed in the copper indium gallium selenide (CIGS) module. In contrast to the CdTe module, there is not yet a sufficient theory for the CIGS module to explain this effect. Annealing always has a positive effect on the performance of thin film modules. Higher temperatures can also restore the Staebler–Wronski effect.

- Capacitance Effects

Large-capacity modules, such as thin film modules, can easily be erroneously evaluated when measured with a flash tester. This is due to capacity loading or unloading. When measuring the module from a short-circuit condition, the measured power value may be lower than the actual value. That is, the capacity is charged. During reverse measurement (from idle to short circuit), this capacitance is discharged, which can lead to a substantial increase in measured power. If there is no difference between the forward and reverse regimes, the pulse width can be considered long enough [140]. The effects of pulse width can be eliminated by either long-enough pulse width, multiflash measurements, or PASAN's so-called "Dragon Back Pulse" tester, which has developed a special measurement method for a single high-performance module. Controlled 10 ms pulse [46].

Most literature on IV curve measurements, such as [141–144], shows that capacitance is the main problem with distortion of the output characteristics of solar cells. The problem with this behavior is that it does not work when modeled like a typical capacitor. The more solar cells you have, the less capacity you will get as a result, but the opposite is true. These effects can be simulated relatively easily using SPICE (the diode parameters can be edited to achieve PV cell operation). In contrast to the traditional diode model, the SPICE diode model covers these effects by "borrowing" the parameter transition time (TT) from the transistor theory. It's very useful because it explains a lot in the context of the "strange" behavior of capacitance in a structure. During irradiation, the capacitance is ideally infinite, so the electrodes of the virtual capacitors are infinitely close to each other. In that case, the entire region is filled with nonequilibrium carriers. These charge carriers require some time to re-form the depletion region apart to restore their previous equilibrium when the module is re-immersed in darkness. This time it can only be characterized by the TT constraint. So, the resulting value of the series connection of the capacitors in the module can be thought of as a battery connection and a

Table 1.4 Fault identification in PV modules using thermography.

Investigation	Description	Cause of failure	Electrical properties	Safety concern
	Some cells of a module are warmer than other cells due to random distribution of individual cells- (Patch Work Pattern)	Cells connected incorrectly	Short circuit of complete module	Short circuits may cause fire
	Warmer module than the other.	Module open circuited	Normal operation	Safety is not the main concern
	Cell temperatures are varying throughout the array	Delaminated and defect in cell, shadowing effect	Power loss may occur, but not for every instant of this failure.	Fire may be caused by extreme conditions.
	A certain part of the cell in a module has a different temperature than the cell	This may occur because of broken cell, or failure in interconnection	Drastic reduction in power and form factor	Failure may lead to fire.
	The extremely hot Sub-string part of the module.	Cell string are short circuit	Reduction in power with high short circuit current.	Fire may be caused by extreme conditions.

Source: Tsanakas et al. [130]/with permission of Elsevier.

Table 1.5 Cell failure detection using I–V characteristics.

	Failure	Broken cell innterconnect ribbons	Cracked cells	Short-circuited cells
	Power loss	Reduce power loss in stages over time	Power loss degradation in steps over time	Degradation of power in steps over time
Characteristic	Safety issue	Failure might result in a fire, electric shock, or physical harm.	There is no impact on safety.	There is no impact on safety.
I_{SC} P_{max}		✓	✓ ✓	✓
V_{OC}				✓
R_{OC}		✓		✓
R_{SC}				

Note: NF, no failure; F,failure condition.

kind of transition charge. It can achieve the same effect by modifying the capacitance value, but the problem is that you need to change this capacitance each time you change the pulse width to get the correct value. This is due solely to the fact that in this case, the capacitance cannot be represented by the simple capacitor here. Tables 1.4, 1.5, and 1.6 show the various errors that can be identified from the deviations in the IV properties at both the cell and module levels.

1.6 Summary

This chapter provides an overview of the different types of faults that may occur in grid-connected PV systems. Fault related to the PV panel and power electronics converter was discussed in detail. A common fault in the inverters and the

Table 1.6 Module failure detection using I–V characteristics.

Failure	Bypass diode (short-circuit)	Delamination		Induced degradation		Solder corrosion
		Homogenous	Heterogenous	Potential	Light	
Power loss	Power loss degradation in steps over time	Power loss degradation that saturates over time	Degradation of power loss which saturates over time	Degradation of power loss linearly over time	Degradation of power Loss in steps over time	Degradation of power loss linearly over time
Safety issue	Failure might result in a fire, electric shock, or physical harm.	Failure might result in a fire, electric shock, or bodily harm.	Failure might result in a fire, electric shock, or bodily harm.	There is no impact on safety.	There is no impact on safety.	There is no impact on safety.
Characteristic						
P_{max}	✓	✓	✓	✓	✓	✓
I_{sc}		✓	✓		✓	
V_{oc}	✓			✓	✓	
R_{oc}					✓	✓
R_{sc}			✓			

most vulnerable component, i.e. power electronics switch was also discussed in the chapter. There has been a lot of advancement in fault identification and localization as well such as the impact of the fault on the system has been analyzed in the chapter.

References

1 Kurbatova, T. and Perederii,T. (2020). Global trends in renewable energy development. *2020 IEEE KhPI Week on Advance Technology. KhPI Week 2020 – Conference Proceedings,* Kharkiv, Ukraine (13 November 2020), pp. 260–263. doi: https://doi.org/10.1109/KhPIWeek51551.2020.9250098.

2 REN21 Renewable Now (2016). Renewable Energy Network Policy for the 21st Century (REN21). *Renewables 2016 Global Status Report (2016).*

3 Duque-escobar, P.G. (2011). Communication from The Commission to The European Parliament, The Council, The European Economic and Social Committee and The Committee of The Regions. Copenhagen: European Environmental Policy.

4 AlOtaibi, Z.S., Khonkar, H.I., AlAmoudi, A.O., and Alqahtani, S.H. (2020). Current status and future perspectives for localizing the solar photovoltaic industry in the Kingdom of Saudi Arabia. *Energy Transitions* 4 (1): 1–9. https://doi.org/10.1007/s41825-019-00020-y.

5 Chine, W., Mellit, A., Pavan, A.M., and Kalogirou, S.A. (2014). Fault detection method for grid-connected photovoltaic plants. *Renew. Energy* 66: 99–110. https://doi.org/10.1016/j.renene.2013.11.073.

6 Spataru, S., Sera, D., Kerekes, T., and Teodorescu, R. (2015). Diagnostic method for photovoltaic systems based on light I-V measurements. *Sol. Energy* 119: 29–44. https://doi.org/10.1016/j.solener.2015.06.020.

7 Gautam, S. and Brahma, S.M. (2013). Detection of high impedance fault in power distribution systems using mathematical morphology. *IEEE Trans. Power Syst.* 28 (2): 1226–1234. https://doi.org/10.1109/TPWRS.2012.2215630.

8 Haeberlin, H. (2001). Evolution of inverters for grid connected PV-systems from 1989 to 2000. *17th European Photovoltaic Solar Energy Conference,* Munich, Germany (22–26 October 2001), p. 5.

9 Baghaee, H.R., Mlakic, D., Nikolovski, S., and Dragicevic, T.D. (2019). Support vector machine-based Islanding and grid fault detection in active distribution networks. *IEEE J. Emerg. Sel. Top. Power Electron.* 1-1. https://doi.org/10.1109/JESTPE.2019.2916621.

10 Ducange, P., Fazzolari, M., Lazzerini, B., and Marcelloni, F. (2011). An intelligent system for detecting faults in photovoltaic fields. *Int. Conf. Intell. Syst. Des. Appl. ISDA* 1341–1346. https://doi.org/10.1109/ISDA.2011.6121846.

11 Angelov, P.P. and Filev, D.P. (2004). An approach to online identification of Takagi-Sugeno fuzzy models. *IEEE Trans. Syst. Man, Cybern. Part B Cybern.* 34 (1): 484–498. https://doi.org/10.1109/TSMCB.2003.817053.

12 Hong, Y.-Y., Gu, J.-L., and Hsu, F.-Y. (2018). Design and realization of controller for static switch in microgrid using wavelet-based TSK reasoning. *IEEE Trans. Ind. Informatics* 14 (11): 4864–4872. https://doi.org/10.1109/TII.2018 .2804896.

13 Ahmad, S., Hasan, N., Bharath Kurukuru, V. S., Ali Khan, M., and Haque, A. (2018). Fault classification for single phase photovoltaic systems using machine learning techniques. *2018 8th IEEE India International Conference on Power Electronics (IICPE)* (December 2018), pp. 1–6. doi: 10.1109/IICPE.2018.8709463.

14 Dhimish, M., Holmes, V., Mehrdadi, B., and Dales, M. (2018). Comparing Mamdani Sugeno fuzzy logic and RBF ANN network for PV fault detection. *Renew. Energy* 117: 257–274. https://doi.org/10.1016/j.renene.2017.10.066.

15 Chine, W., Mellit, A., Lughi, V. et al. (2016). A novel fault diagnosis technique for photovoltaic systems based on artificial neural networks. *Renew. Energy* 90: 501–512. https://doi.org/10.1016/j.renene.2016.01.036.

16 Yang, T. (2019). ICT technologies standards and protocols for active distribution network. In: *Smart Power Distribution Systems*, 205–230. Elsevier.

17 Khan, M.A., Bharath Kurukuru, V.S., Haque, A., and Mekhilef, S. (2021). Islanding classification mechanism for grid-connected photovoltaic systems. *IEEE J. Emerg. Sel. Top. Power Electron.* 9 (2): 1966–1975.

18 Khan, M.A., Haque, A., Kurukuru, V.S.B., and Saad, M. (2022). Islanding detection techniques for grid-connected photovoltaic systems – a review. *Renew. Sustain. Energy Rev.* 154: 111854. https://doi.org/10.1016/j.rser.2021 .111854.

19 Kurukuru, V.S.B., Haque, A., Khan, M.A. et al. (2021). A review on artificial intelligence applications for grid-connected solar photovoltaic systems. *Energies* 14 (15): 4690. https://doi.org/10.3390/en14154690.

20 Khan, M., Haque, A., and Kurukuru, V.S.B. (2021). Dynamic voltage support for low-voltage ride-through operation in single-phase grid-connected photovoltaic systems. *IEEE Trans. Power Electron.* 36 (10): 12102–12111. https://doi .org/10.1109/TPEL.2021.3073589.

21 Kirtley, J.L., El Moursi, M.S., and Xiao, W. (2013). Fault ride through capability for grid interfacing large scale PV power plants. *IET Gener. Transm. Distrib.* 7 (9): 1027–1036. https://doi.org/10.1049/iet-gtd.2013.0154.

22 Khan, M. A., Haque, A., and Bharath Kurukuru, V. S. (2019). Enhancement of Fault ride through strategy for single-phase grid-connected photovoltaic systems. *2019 IEEE Industry Applications Society Annual Meeting* (September 2019), pp. 1–6. doi: 10.1109/IAS.2019.8911895.

23 Mirhosseini, M., Pou, J., and Agelidis, V.G. (2015). Single- and two-stage inverter-based grid-connected photovoltaic power plants with ride-through capability under grid faults. *IEEE Trans. Sustain. Energy* 6 (3): 1150–1159. https://doi.org/10.1109/TSTE.2014.2347044.

24 Easley, M., Jain, S., Shadmand, M.B., and Abu-Rub, H.A. (2020). Autonomous model predictive controlled smart inverter with proactive grid fault ride-through capability. *IEEE Trans. Energy Convers.* 1–1. https://doi.org/10.1109/TEC.2020.2998501.

25 Khan, M.A., Haque, A., Kurukuru, V.S.B., and Saad, M. (2020). Advanced control strategy with voltage sag classification for single-phase grid-connected photovoltaic system. *IEEE J. Emerg. Sel. Top. Ind. Electron.* 1–1. https://doi.org/10.1109/JESTIE.2020.3041704.

26 Fatama, A.Z., Khan, M.A., Kurukuru, V.S.B. et al. (2020). Coordinated reactive power strategy using static synchronous compensator for photovoltaic inverters. *Int. Trans. Electr. Energy Syst.* 30 (6): https://doi.org/10.1002/2050-7038.12393.

27 Kurukuru, V.S.B., Haque, A., Padmanaban, S., and Khan, M.A. (2021). Rule-based inferential system for microgrid energy management system. *IEEE Syst. J.* 1–10. https://doi.org/10.1109/JSYST.2021.3094403.

28 Kurukuru, V.S.B., Haque, A., Khan, M.A., and Blaabjerg, F. (2021). Resource management with kernel-based approaches for grid-connected solar photovoltaic systems. *Heliyon* 7 (12): e08609. https://doi.org/10.1016/j.heliyon.2021.e08609.

29 Haque, A. (2014). Maximum power point tracking (MPPT) scheme for solar photovoltaic system. *Energy Technol. Policy* 1 (1): 115–122. https://doi.org/10.1080/23317000.2014.979379.

30 Khan, M.A. and Sangwongwanich, A. (2021). Control strategy for grid-connected solar inverter for IEC standards. In: *Reliability of Power Electronics Converters for Solar Photovoltaic Applications*, 141–188. Institution of Engineering and Technology.

31 Khan, M.A., Haque, A., and Kurukuru, V.S.B. (2021). Intelligent transition control approach for different operating modes of photovoltaic inverter. *IEEE Trans. Ind. Appl.* 1–1. https://doi.org/10.1109/tia.2021.3135250.

32 Khan, M.A., Haque, A., Blaabjerg, F. et al. (2021). Intelligent transition control between grid-connected and standalone modes of three-phase grid-integrated distributed generation systems. *Energies* 14 (13): 3979. https://doi.org/10.3390/en14133979.

33 Li, X., Fang, J., Tang, Y., and Wu, X. (2018). Robust design of LCL filters for single-current-loop-controlled grid-connected power converters with unit PCC voltage feedforward. *IEEE J. Emerg. Sel. Top. Power Electron.* 6 (1): 54–72. https://doi.org/10.1109/JESTPE.2017.2766672.

34 Wayne Bequette, B. (2003). Dynamic behavior. In: *Process Control: Modeling, Design, and Simulation*, 114–118. Upper Saddle River, NJ: Prentice Hall PTR.

35 Singh, M., Khadkikar, V., Chandra, A., and Varma, R.K. (2011). Grid interconnection of renewable energy sources at the distribution level with power-quality improvement features. *IEEE Trans. Power Deliv.* 26 (1): 307–315. https://doi.org/10.1109/TPWRD.2010.2081384.

36 Khan, M.A., Haque, A., Kurukuru, V.S.B. et al. (2021). Standalone operation of distributed generation systems with improved harmonic elimination scheme. *IEEE J. Emerg. Sel. Top. Power Electron.* 9 (6): 6924–6934. https://doi.org/10.1109/JESTPE.2021.3084737.

37 Khan, M.A., Haque, A., and Kurukuru, V.S.B. (2020). Intelligent control of a novel transformerless inverter topology for photovoltaic applications. *Electr. Eng.* 102 (2): 627–641. https://doi.org/10.1007/s00202-019-00899-2.

38 Khan, M.A., Haque, A., and Kurukuru, V.S.B. (2020). Performance assessment of stand-alone transformerless inverters. *Int. Trans. Electr. Energy Syst.* 30 (1): https://doi.org/10.1002/2050-7038.12156.

39 Dong, D. (2009). Modeling and control design of a bidirectional PWM converter for single-phase energy systems. *Response* 1–114.

40 Li, X., Fang, J., Tang, Y. et al. (2018). Capacitor-voltage feedforward with full delay compensation to improve weak grids adaptability of LCL-filtered grid-connected converters for distributed generation systems. *IEEE Trans. Power Electron.* 33 (1): 749–764. https://doi.org/10.1109/TPEL.2017.2665483.

41 Chen, C.-L., Lai, J.-S., Wang, Y.-B., Park, S.-Y., and Miwa, H. (2008). Design and control for LCL-based inverters with both grid-tie and standalone parallel operations. *2008 IEEE Industry Applications Society Annual Meeting* (October 2008), pp. 1–7. doi: https://doi.org/10.1109/08IAS.2008.326.

42 Sha, D., Deng, K., Guo, Z., and Liao, X. (2012). Control strategy for input-series–output-parallel high-frequency AC link inverters. *IEEE Trans. Ind. Electron.* 59 (11): 4101–4111. https://doi.org/10.1109/TIE.2011.2174538.

43 Zhang, Y., Umuhoza, J., Liu, H., Farrell, C., and Mantooth, H. A. (2015). Optimizing efficiency and performance for single-phase photovoltaic inverter with dual-half bridge converter. *2015 IEEE Applied Power Electronics Conference and Exposition (APEC)* (March 2015), pp. 1507–1511. doi: https://doi.org/10.1109/APEC.2015.7104547.

44 Zhang, Y., Umuhoza, J., Liu, H., et al. (2015). Realizing an integrated system for residential energy harvesting and management. *2015 IEEE Applied Power Electronics Conference and Exposition (APEC)* (March 2015), pp. 3240–3244. doi: https://doi.org/10.1109/APEC.2015.7104816.

45 Shah, A.V., Schade, H., Vanecek, M. et al. (2004). Thin-film silicon solar cell technology. *Prog. Photovoltaics Res. Appl.* 12 (23): 113–142. https://doi.org/10.1002/pip.533.

46 Fanni, L., Virtuani, A., and Chianese, D. (2011). A detailed analysis of gains and losses of a fully-integrated flat roof amorphous silicon photovoltaic plant. *Sol. Energy* 85 (9): 2360–2373. https://doi.org/10.1016/j.solener.2011.06.029.

47 Köntges, M., Kajari-Schröder, S., Kunze, I., and Jahn, U. (2011). "Crack statistic of crystalline silicon photovoltaic modules." *26th EUPVSEC*, Hamburg, Germany (September), pp. 3290–3294.

48 Köntges, M., Kurtz, S., Packard, C. et al. (2014). *Review of Failures of Photovoltaic Modules*. Paris, France: IEA International Energy Agency.

49 Wohlgemuth, J., Cunningham, D.W., and Nguyen, A. (2010). Failure modes of crystalline Si modules. *PV Modul. Reliab. Work.* Technical Report (NREL/TP-5200-60171): 347–380.

50 Leader, W., Tsoutsos, T., Gkouskos, Z., and Tournaki, S. (2011). Catalogue of common failures and improper practices on PV installations and maintenance. *Eur. Photovolt. Ind. Assoc.* June: 1–18.

51 Ferrara, C. and Philipp, D. (2012). Why do PV modules fail? *Energy Procedia* 15 (2011): 379–387. https://doi.org/10.1016/j.egypro.2012.02.046.

52 N. H. Zaini M. Z. A. Ab-Kadir; M. Izadi; N. I. Ahmad; M. A. M Radzi; N. Azis; W. Z. Wan Hasan (2016). On the effect of lightning on a solar photovoltaic system. *2016 33rd International Conference on. Lightning Protection, ICLP 2016*, Estoril, Portugal (25–30 September 2016). doi: https://doi.org/10.1109/ICLP.2016.7791421.

53 Kalejs, J. (2014). Junction box wiring and connector durability issues in photovoltaic modules. *SPIE Sol. Energy + Technol.* 9179: 91790S. https://doi.org/10.1117/12.2063488.

54 Singh, J.P., Guo, S., Peters, I.M. et al. (2015). Comparison of glass/glass and glass/backsheet PV modules using bifacial silicon solar cells. *IEEE J. Photovoltaics* 5 (3): 783–791. https://doi.org/10.1109/JPHOTOV.2015.2405756.

55 Itoh, U., Yoshida, M., Tokuhisa, H. et al. (2014). Solder joint failure modes in the conventional crystalline si module. *Energy Procedia* 55 (3): 464–468. https://doi.org/10.1016/j.egypro.2014.08.010.

56 Kleiss, G., Kirchner, J., and Reichart, K. (2015). Quality and reliability – sometimes the customer wants more. *Proceedings of the NREL PV Module Reliability Workshop*, Denver, CO, USA, p. 760.

57 Zhu, J., Wu, D., Montiel-Chicharro, D. et al. (2017). Realistic adhesion test for photovoltaic modules qualification. *IEEE J. Photovoltaics* 1–6. https://doi.org/10.1109/JPHOTOV.2017.2775149.

58 Jorgensen, G.J., Terwilliger, K.M., DelCueto, J.A. et al. (2006). Moisture transport, adhesion, and corrosion protection of PV module packaging materials. *Sol. Energy Mater. Sol. Cells* 90 (16): 2739–2775. https://doi.org/10.1016/j.solmat.2006.04.003.

59 Guo, S. (2015). Analysis and Optimization of Crystalline Silicon Wafter Based PV Modules. ScholarBank@NUS Repository. NUS Repository.

60 Ise and PSE (2022). Photovoltaics Report, Prepared by Fraunhofer Institute for Solar Energy Systems, ISE with support of PSE Projects GmbH (24 February 2022).

61 Rosenthal, A.L. and Lane, C.G. (1991). Field test results for the 6 MW Carrizo solar photovoltaic power plant. *Sol. Cells* 30 (1–4): 563–571. https://doi.org/10 .1016/0379-6787(91)90088-7.

62 Chianese, D., Realini, A., Cereghetti, N. et al. (2003). Analysis of weathered c-Si PV modules. *3rd World Conference onPhotovoltaic Energy Conversion, 2003. Proceedings*, vol. 3, pp. 2922–2926.

63 Pern, F. J., Branch, C., and Renewable, N. (1994). Factors that affect the eva encapsulant upon accelerated exposure. *Proceedings of 1994 IEEE 1st World Conference on Photovoltaic Energy Conversion – WCPEC (A Joint Conference of PVSC, PVSEC and PSEC)*, Waikoloa, HI, USA, pp. 897–900.

64 Kaplani, E. (2012). Detection of degradation effects in field-aged c-Si solar cells through IR thermography and digital image processing. *Int. J. Photoenergy* 2012: https://doi.org/10.1155/2012/396792.

65 Weber, N. L. U., Eiden, R., Strubel, C., et al. (2012). Acetic acid production, migration and corrosion effects in Ethylene-Vinyl-Acetate- (EVA-) based PV modules. *27th European Photovoltaic Solar Energy Conference and Exhibition*, Frankfurt, Germany (24–28 September 2012), pp. 2992–2995. doi: 10.4229/27thEUPVSEC2012-4CO.9.4.

66 Jordan, D.C. and Kurtz, S.R. (2013). Photovoltaic degradation rates – an analytical review. *Prog. Photovoltaics Res. Appl.* 21 (1): 12–29. https://doi.org/10 .1002/pip.1182.

67 Dhimish, M., Holmes, V., Mehrdadi, B., and Dales, M. (2017). The impact of cracks on photovoltaic power performance. *J. Sci. Adv. Mater. Devices* 2 (2): 199–209. https://doi.org/10.1016/j.jsamd.2017.05.005.

68 Gabor, A.M., Ralli, M., Montminy, S. et al. (2006). Soldering induced damage to thin Si solar cells and detection of cracked cells in modules. *21st European Photovoltaic Solar Energy Conference*, Dresden (4–8 September 2006).

69 Meyer, S., Richter, S., Timmel, S. et al. (2013). Snail trails: root cause analysis and test procedures. *Energy Procedia* 38: 498–505. https://doi.org/10.1016/j .egypro.2013.07.309.

70 Pingel, S., Frank, O., Winkler, M. et al. (2010). Potential induced degradation of solar cells and panels. *35th IEEE PVSC*, Honolulu, HI, USA (20–25 June 2010), pp. 2817–2822. doi: https://doi.org/10.1109/PVSC.2010.5616823.

71 Luo, W., Khoo, Y.S., Hacke, P. et al. (2017). Potential-induced degradation in photovoltaic modules: a critical review. *The Royal Society of Chemistry* 10 (1): 43–68.

72 Cristaldi, L., Faifer, M., Lazzaroni, M. et al. (2015). Diagnostic architecture: a procedure based on the analysis of the failure causes applied to photovoltaic plants. *Meas. J. Int. Meas. Confed.* 67: 99–107. https://doi.org/10.1016/j .measurement.2015.02.023.

73 Cristaldi, L., Faifer, M., Lazzaroni, M., et al. (2014). Failure modes analysis and diagnostic architecture for photovoltaic plants. *13th IMEKO TC10 Workshop on Technical Diagnostics Advanced Measurement Tools in Technical Diagnostics for Systems' Reliability and Safety*, Warsaw, Poland, pp. 206–211.

74 Saly, V., Ruzinsky, M., Packa, J., and Redi, P. (2002). Examination of solar cells and encapsulations of small experimental photovoltaic modules vladimir. *no.* 1: 137–141.

75 Alers, G. (2010). Photovoltaic Module Reliability: Enduring a storm. *2010 IEEE International Integrated Reliability Workshop Final Report*, South Lake Tahoe, CA, USA (17–21 October 2010).

76 Buerhop, C. and Bachmann, J. (2007). Infrared analysis of thin-film photovoltaic modules. *J. Phys. Conf. Ser.* 214: https://doi.org/10.1088/1742-6596/ 214/1/012089.

77 Jenkins, C., Pudov, A., Gloeckler, M. et al. (2003). CdTe back contact: response to copper addition and out-diffusion, *NCPV and Solar Program Review Meeting Proceedings*, Denver, Colorado (CDROM) (24–26 March 2003).

78 Veldman, D., Bennett, I.J., Brockholz, B., and De Jong, P.C. (2011). Non-destructive testing of crystalline silicon photovoltaic back-contact modules. *Conf. Rec. IEEE Photovolt. Spec. Conf.* 8 (April): 003237–003240. https:// doi.org/10.1109/PVSC.2011.6186628.

79 Wang, H., Zhou, D., and Blaabjerg, F. (2013). A reliability-oriented design method for power electronic converters. *2013 Twenty-Eighth Annual IEEE Applied Power Electronics Conference and Exposition (APEC)* (March 2013), pp. 2921–2928. doi: 10.1109/APEC.2013.6520713.

80 Bahman, A.S., Iannuzzo, F., and Blaabjerg, F. (2016). Mission-profile-based stress analysis of bond-wires in SiC power modules. *Microelectron. Reliab.* 64: 419–424. https://doi.org/10.1016/j.microrel.2016.07.102.

81 Kurukuru, V. S. B., Haque, A., Khan, M. A., and Tripathy, A. K. (2019). Reliability analysis of silicon carbide power modules in voltage source converters. *2019 International Conference on Power Electronics, Control and Automation (ICPECA)* (November 2019), pp. 1–6. doi: https://doi.org/10.1109/ ICPECA47973.2019.8975617.

82 Kurukuru, V.S.B., Haque, A., Tripathi, A.K., and Khan, M.A. (2021). Condition monitoring of IGBT modules using online TSEPs and data-driven approach. *Int. Trans. Electr. Energy Syst.* June: 1–24.

83 Lee, K.-W., Kim, M., Yoon, J. et al. (2008). Condition monitoring of DC-link electrolytic capacitors in adjustable-speed drives. *IEEE Trans. Ind. Appl.* 44 (5): 1606–1613. https://doi.org/10.1109/TIA.2008.2002220.

84 Ristow, A., Begovic, M., Pregelj, A., and Rohatgi, A. (2008). Development of a methodology for improving photovoltaic inverter reliability. *IEEE Trans. Ind. Electron.* 55 (7): 2581–2592. https://doi.org/10.1109/TIE.2008.924017.

85 Haque, A., Bharath, K.V.S., Khan, M.A. et al. (2019). Fault diagnosis of photovoltaic modules. *Energy Sci. Eng.* 7 (3): 622–644. https://doi.org/10.1002/ese3.255.

86 Kurukuru, V.S.B., Blaabjerg, F., Khan, M.A., and Haque, A. (2020). A novel fault classification approach for photovoltaic systems. *Energies* 13 (2): 308. https://doi.org/10.3390/en13020308.

87 Mahmud, N., Zahedi, A., and Mahmud, A. (2017). A cooperative operation of novel PV inverter control scheme and storage energy management system based on ANFIS for voltage regulation of grid-tied PV system. *IEEE Trans. Ind. Informatics* 13 (5): 2657–2668. https://doi.org/10.1109/TII.2017.2651111.

88 Zhao, Y., Lehman, B., De Palma, J. F., Mosesian, J., and Lyons, R. (2011). Challenges to overcurrent protection devices under line-line faults in solar photovoltaic arrays. *IEEE Energy Conversation Congress and Expo. Energy Conversation Innovation for a Clean Energy Future. ECCE 2011, Proceeding*, Phoenix, AZ, USA (17–22 September 2011), pp. 20–27. doi: 10.1109/ECCE.2011.6063744.

89 Vernica, I., Wang, H., and Blaabjerg, F. (2018). Uncertainties in the lifetime prediction of IGBTs for a motor drive application. *2018 IEEE International Power Electronics and Application Conference and Exposition (PEAC)*, Shenzhen, China (November 2018), pp. 1–6. doi: https://doi.org/10.1109/PEAC.2018.8590417.

90 Wu, R., Blaabjerg, F., Wang, H., Liserre, M., and Iannuzzo, F. (2013). Catastrophic failure and fault-tolerant design of IGBT power electronic converters – an overview. *IECON 2013 – 39th Annual Conference of the IEEE Industrial Electronics Society* (November 2013), pp. 507–513. doi: https://doi.org/10.1109/IECON.2013.6699187.

91 Gao, Z., Cecati, C., and Ding, S.X. (2015). A survey of fault diagnosis and fault-tolerant techniques-part 1&2: Fault diagnosis with knowledge-based and hybrid/active approaches. *IEEE Trans. Ind. Electron.* 62 (6): 3768–3774. https://doi.org/10.1109/TIE.2015.2419013.

92 Kurukuru, V.S.B., Khan, M.A., and Malik, A. (2021). Failure mode classification for grid-connected photovoltaic converters. In: *Reliability of Power Electronics Converters for Solar Photovoltaic Applications*, 205–249. Institution of Engineering and Technology.

93 Yang, S., Xiang, D., Bryant, A. et al. (2010). Condition monitoring for device reliability in power electronic converters: a review. *IEEE Trans. Power Electron.* 25 (11): 2734–2752. https://doi.org/10.1109/TPEL.2010.2049377.

94 Devi Vidhya, S. and Balaji, M. (2019). Failure-mode analysis of modular multilevel capacitor-clamped converter for electric vehicle application. *IET Power Electron.* 12 (13): 3411–3421. https://doi.org/10.1049/iet-pel.2018.6101.

95 Choi, U.-M. and Blaabjerg, F. (2018). Separation of wear-out failure modes of IGBT modules in grid-connected inverter systems. *IEEE Trans. Power Electron.* 33 (7): 6217–6223. https://doi.org/10.1109/TPEL.2017.2750328.

96 Diaz Reigosa, P., Wang, H., Yang, Y., and Blaabjerg, F. (2015). Prediction of bond wire fatigue of IGBTs in a PV inverter under a long-term pperation. *IEEE Trans. Power Electron.* 1–1. https://doi.org/10.1109/TPEL.2015.2509643.

97 Gorecki, K., Gorecki, P., and Zarebski, J. (2019). Measurements of parameters of the thermal model of the IGBT module. *IEEE Trans. Instrum. Meas.* 68 (12): 4864–4875. https://doi.org/10.1109/TIM.2019.2900144.

98 Khatir, Z. and Lefebvre, S. (2001). Thermal analysis of power cycling effects on high power IGBT modules by the boundary element method. *Seventeenth Annual IEEE Semiconductor Thermal Measurement and Management Symposium (Cat. No.01CH37189)*, San Jose, CA, USA (22–22 March 2001), pp. 27–34. doi: 10.1109/STHERM.2001.915141.

99 Anurag, A., Yang, Y., and Blaabjerg, F. (2015). Thermal performance and reliability analysis of single-phase PV inverters with reactive power injection outside feed-in operating hours. *IEEE J. Emerg. Sel. Top. Power Electron.* 3 (4): 870–880. https://doi.org/10.1109/JESTPE.2015.2428432.

100 Khan, M.A., Mishra, S., and Haque, A. (2018). A present and future state-of-the-art development for energy-efficient buildings using PV systems. *Intell. Build. Int.* 1–20. https://doi.org/10.1080/17508975.2018.1437709.

101 Stevanovic, L. D., Beaupre, R. A., Gowda, A. V., Pautsch, A. G., and Solovitz, S. A. (2010). Integral micro-channel liquid cooling for power electronics. *2010 Twenty-Fifth Annual IEEE Applied Power Electronics Conference and Exposition (APEC)* (February 2010), pp. 1591–1597. doi: 10.1109/APEC.2010.5433444.

102 Wang, P., McCluskey, P., and Bar-Cohen, A. (2013). Hybrid solid- and liquid-cooling solution for isothermalization of insulated gate bipolar transistor power electronic devices. *IEEE Trans. Components, Packag. Manuf. Technol.* 3 (4): 601–611. https://doi.org/10.1109/TCPMT.2012.2227056.

103 Mohamed, S., Jeyanthy, P., Devaraj, D. et al. (2019). DC-link voltage control of a grid-connected solar photovoltaic system for fault ride-through capability enhancement. *Appl. Sci.* 9 (5): 952. https://doi.org/10.3390/app9050952.

104 Zhu, B., Tan, C., Farshadnia, M., and Fletcher, J.E. (2019). Postfault zero-sequence current injection for open-circuit diode/switch failure in open-end winding PMSM machines. *IEEE Trans. Ind. Electron.* 66 (7): 5124–5132. https://doi.org/10.1109/TIE.2018.2868007.

105 Patil, N., Das, D., Yin, C., et al. (2009). A fusion approach to IGBT power module prognostics. *EuroSimE 2009 – 10th International Conference on Thermal, Mechanical and Multi-Physics Simulation and Experiments in Microelectronics and Microsystems* (April 2009), pp. 1–5. doi: https://doi.org/10.1109/ESIME.2009.4938491.

106 Smet, V., Forest, F., Huselstein, J.-J. et al. (2011). Ageing and failure modes of IGBT modules in high-temperature power cycling. *IEEE Trans. Ind. Electron.* 58 (10): 4931–4941. https://doi.org/10.1109/TIE.2011.2114313.

107 Nguyen, T.A., Lefebvre, S., Joubert, P.-Y. et al. (2015). Estimating current distributions in power semiconductor dies under aging conditions: bond wire liftoff and aluminum reconstruction. *IEEE Trans. Components, Packag. Manuf. Technol.* 5 (4): 483–495. https://doi.org/10.1109/TCPMT.2015.2406576.

108 Martineau, D., Mazeaud, T., Legros, M. et al. (2009). Characterization of ageing failures on power MOSFET devices by electron and ion microscopies. *Microelectron. Reliab.* 49 (9–11): 1330–1333. https://doi.org/10.1016/j.microrel.2009.07.011.

109 Ciappa, M. (2002). Selected failure mechanisms of modern power modules. *Microelectron. Reliab.* 42 (4–5): 653–667. https://doi.org/10.1016/S0026-2714(02)00042-2.

110 Huang, J., Hu, Z., Gao, C., and Cui, C. (2014). Analysis of water vapor control and passive layer process effecting on transistor performance and aluminum corrosion. *Proceedings 2014 Prognostics and Systems Health Management Conference PHM 2014*, no. 61201028, Zhangjiajie, China (24–27 August 2014), pp. 26–30. doi: 10.1109/PHM.2014.6988126.

111 Navarro, L.A., Perpiñà, X., Vellvehi, M. et al. (2012). Thermal cycling analysis of high temperature die-attach materials. *Microelectron. Reliab.* 52 (9–10): 2314–2320. https://doi.org/10.1016/j.microrel.2012.07.022.

112 Lai, W., Chen, M., Ran, L. et al. (2016). Low stress cycle effect in IGBT power module die-attach lifetime modeling. *IEEE Trans. Power Electron.* 31 (9): 6575–6585. https://doi.org/10.1109/TPEL.2015.2501540.

113 McCluskey, P. (2012). Reliability of power electronics under thermal loading. *2012 7th International Conference on Integrated Power Electronics Systems, CIPS 2012*, Nuremberg, Germany (6–8 March 2012), vol. 9.

114 Teasdale, K., Engineer, P., and Corporation, I.R. (2009). Application Note AN-1140 Continuous dc Current Ratings of International Rectifier's Large Semiconductor Packages Table of Contents, 1–17. Infineon Technologies AG.

115 Fleischer, A.S., Chang, L., and Johnson, B.C. (2006). The effect of die attach voiding on the thermal resistance of chip level packages. *Microelectron. Reliab.* 46 (5–6): 794–804. https://doi.org/10.1016/j.microrel.2005.01.019.

116 Katsis, D.C. and van Wyk, J.D. (2003). Void-induced thermal impedance in power semiconductor modules: Some transient temperature effects. *IEEE Trans. Ind. Appl.* 39 (5): 1239–1246. https://doi.org/10.1109/TIA.2003.816527.

117 Zhu, N. (1999). Thermal impact of solder voids in the electronic packaging of power devices. *Fifteenth Annual IEEE Semiconductor Thermal Measurement and Management Symposium (Cat. No.99CH36306)*, San Diego, CA, USA (9–11 March 1999), pp. 22–29. doi: 10.1109/STHERM.1999.762424.

118 Benbahouche, L., Merabet, A., and Zegadi, A. (2012). A comprehensive analysis of failure mechanisms: latch up and second breakdown in IGBT(IXYS) and improvement. In: *2012 19th International Conference on Microwaves, Radar & Wireless Communications* (May 2012), 190–192. https://doi.org/10.1109/MIKON.2012.6233539.

119 Jahdi, S., Alatise, O., Bonyadi, R. et al. (2015). An analysis of the switching performance and robustness of power MOSFETs body diodes: a technology evaluation. *IEEE Trans. Power Electron.* 30 (5): 2383–2394. https://doi.org/10.1109/TPEL.2014.2338792.

120 Spirito, P., Maresca, L., Riccio, M. et al. (2015). Effect of the collector design on the IGBT avalanche ruggedness: a comparative analysis between punch-through and field-stop devices. *IEEE Trans. Electron Devices* 62 (8): 2535–2541. https://doi.org/10.1109/TED.2015.2442334.

121 Duvvury, C., Rodriguez, J., Jones, C., and Smayling, M. (1994). Device integration for ESD robustness of high voltage power MOSFETs. *Proceedings of 1994 IEEE International Electron Devices Meeting*, San Francisco, CA, USA (11–14 December 1994), pp. 407–410. doi: 10.1109/IEDM.1994.383381.

122 Anolick, E.S. and Nelson, G.R. (1980). Low-field time-dependent dielectric integrity. *IEEE Trans. Reliab.* R-29 (3): 217–221. https://doi.org/10.1109/TR.1980.5220804.

123 Sato, M., Kumada, A., Hidaka, K. et al. (2016). Surface discharges in silicone gel on AlN substrate. *IEEE Trans. Dielectr. Electr. Insul.* 23 (1): 494–500. https://doi.org/10.1109/TDEI.2015.005412.

124 Fabian, J.-H., Hartmann, S., and Hamidi, A. (2005). Analysis of insulation failure modes in high power IGBT modules. *Fourtieth IAS Annual Meeting. Conference Record of the 2005 Industry Applications Conference,* Hong Kong, China (2–6 October 2005), vol. 2, pp. 799–805. doi: https://doi.org/10.1109/IAS.2005.1518425.

125 Kumar, S., Jena, P., Sinha, A., and Gupta, R. (2017). Application of infrared thermography for non- destructive inspection of solar photovoltaic module. *Journal of Non-Destructive Test* 25–32.

126 Daliento, S., Chouder, A., Guerriero, P. et al. (2017). Monitoring, diagnosis, and power forecasting for photovoltaic fields: a review. *Int. J. Photoenergy* 2017: 1–13. https://doi.org/10.1155/2017/1356851.

127 van Dyk, E., Meyer, E., Vorster, F., and Leitch, A.W. (2002). Long-term monitoring of photovoltaic devices. *Renew. Energy* 25 (2): 183–197. https://doi.org/ 10.1016/S0960-1481(01)00064-7.

128 Spagnuolo, G., Petrone, G., Lehman, B. et al. (2015). Control of photovoltaic arrays: dynamical reconfiguration for fighting mismatched conditions and meeting load requests. *IEEE Ind. Electron. Mag.* 9 (1): 62–76. https://doi.org/ 10.1109/MIE.2014.2360721.

129 Stapleton, G. and Neill, S. (2012). Grid-connected Solar Electric Systems, 17. London: Earthscan.

130 Tsanakas, J.A., Ha, L., and Buerhop, C. (2016). Faults and infrared thermographic diagnosis in operating c-Si photovoltaic modules: A review of research and future challenges. *Renew. Sustain. Energy Rev.* 62: 695–709. https://doi.org/10.1016/j.rser.2016.04.079.

131 N. P. Avdelidis, D. P. Almond, A. Dobbinson, and B. C. Hawtin, *Pulsed Thermography: Philosophy, Qualitative Quantitative Analysis on Aircraft Materials & Applications.* 2006.

132 Haunschild, J., Glatthaar, M., Kasemann, M. et al. (2009). Fast series resistance imaging for silicon solar cells using electroluminescence. *Phys. Status Solidi – Rapid Res. Lett.* 3 (7–8): 227–229. https://doi.org/10.1002/pssr .200903175.

133 Hofierka, J. and Kaňuk, J. (2009). Assessment of photovoltaic potential in urban areas using open-source solar radiation tools. *Renew. Energy* 34 (10): 2206–2214. https://doi.org/10.1016/j.renene.2009.02.021.

134 Lenchyshyn, L.C., Thewalt, M.L.W., Sturm, J.C. et al. (1992). High quantum efficiency photoluminescence from localized excitons in Si1-xGex. *Appl. Phys. Lett.* 60 (25): 3174–3176. https://doi.org/10.1063/1.106733.

135 Currie, M.J., Mapel, J.K., Heidel, T.D. et al. (2007). Concentrators for Photovoltaics. *Science (80-.).* 1 (2005): 2007–2009.

136 Vanek, J., Koktavy, P., Kubickova, K. et al. (2007). Usage of microplasma signal noise for solar cells diagnostic. *Proceedings, Noise and Fluctuations in Circuits, Devices, and Materials* 6600: 660017. https://doi.org/10.1117/12 .724585.

137 Bauer, J., Wagner, J.-M., Lotnyk, A. et al. (2009). Hot spots in multicrystalline silicon solar cells: Avalanche breakdown due to etch pits. *Phys. Status Solidi - Rapid Res. Lett.* 3 (2–3): 40–42. https://doi.org/10.1002/pssr.200802250.

138 Cerna, L., Benda, V., and MacHacek, Z. (2012). A note on irradiance dependence of photovoltaic cell and module parameters. *2012 28th International Conference on Microelectronics – Proceedings, MIEL 2012*, no. Miel, pp. 273–276. doi: 10.1109/MIEL.2012.6222852.

139 Staebler, D.L. and Wronski, C.R. (1977). Reversible conductivity changes in discharge-produced amorphous Si. *Appl. Phys. Lett.* 31 (4): 292–294. https://doi.org/10.1063/1.89674.

140 Taylor, N. (2010). Guidelines for PV power measurement in industry. *European Commission Joint Research Centre Institute for Energy* April: 1–80.

141 Merhej, P., Dallago, E., and Finarelli, D. (2010). Effect of capacitance on the output characteristics of solar cells. *Ph. D. Research in Microelectronics and Electronics (PRIME), 2010 Conference*, Berlin, Germany (18–21 July 2010), pp. 1–4.

142 Gao, Q., Zhang, Y., Yu, Y. et al. (2018). Effects of I–V measurement parameters on the hysteresis effect and optimization in high-capacitance PV module testing. *IEEE Journal of Photovoltaics* 8 (3): 710–718. https://doi.org/10.1109/JPHOTOV.2018.2810852.

143 Monokroussos, C., Gottschalg, R., Tiwari, A.N. et al. (2007). The effects of solar cell capacitance on calibration accuracy when using a flash simulator. *2006 IEEE 4th World Conference on Photovoltaic Energy Conference*, Waikoloa, HI, USA (7–12 May 2006) 2: 2231–2234. https://doi.org/10.1109/WCPEC.2006.279953.

144 Edler, A., Schlemmer, M., Ranzmeyer, J., and Harney, R. (2012). Understanding and overcoming the influence of capacitance effects on the measurement of high efficiency silicon solar cells. *Energy Procedia* 27: 267–272. https://doi.org/10.1016/j.egypro.2012.07.062.

2

Aging Detection for Capacitors in Power Electronic Converters

Zhaoyang Zhao[1], Pooya Davari[2], and Huai Wang[2]

[1] *Institute of Smart City and Intelligent Transportation, Southwest Jiaotong University, Chengdu, China*
[2] *AAU Energy, Aalborg University, Aalborg, Denmark*

2.1 Introduction

Capacitors are widely used in dc-links of grid-connected photovoltaic (PV) converters to balance power, suppress harmonics, and switch frequency ripples. However, capacitors are sensitive to thermal and electrical stresses and have the main disadvantage of finite lifespan and high degradation failure rate, making them considered one of the most vulnerable parts in power electronic systems. Usually, its equivalent series resistance (ESR) increases and the capacitance (C) decreases as the degradation of a capacitor happens. Because of the increase of ESR, the ripple current flowing past will introduce higher thermal stress, which will increase the ripple voltage and accelerate the degradation process of the capacitors. At the same time, the dc-link voltage will further increase due to the decrease of capacitance, which results in oscillations in the PV-operating point around its maximum power point (MPP). Generally, the degradation of capacitors will reduce the average output power and the reliability risk of the system will increase. With regard to this, monitoring the degradation state of capacitors and scheduling maintenance before severe degradation or breakdown occurs have great significance for ensuring the reliable operation of PV systems and optimal earnings.

2.1.1 Capacitors for PV Applications

The typical structure of PV systems with dc-link capacitors is shown in Figure 2.1, which can be configured as single-stage or double-stage [1]. Usually, the PV-side capacitor bank C_{pv} is used to suppress the switching ripples and power pulsation

Fault Analysis and its Impact on Grid-connected Photovoltaic Systems Performance, First Edition.
Edited by Ahteshamul Haque and Saad Mekhilef.

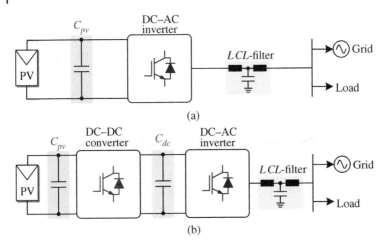

(a)

(b)

Figure 2.1 Typical configurations of PV systems with dc-link capacitors. (a) Single-stage configuration. (b) Double-stage configuration.

Table 2.1 Performance comparison of capacitors (++++ superior, + inferior).

Type	Frequency characteristics	Temperature characteristics	High voltage	High capacity	Long life	Cost per capacity
Al-Cap (electrolytic)	+	+	+++	++++	+	++++
Al-Cap (solid)	++++	+++	++	++++	++	+++
MPPF-Cap	++++	++++	++++	++	++++	+
MLC-Cap	++	+++	++++	+	++++	+

between PV and DC–DC converters or DC–AC inverters [2, 3]. The capacitor bank C_{dc} in the double-stage system is used to realize the power decouples of two converters and serves as an energy storage element [4, 5].

Generally, three types of capacitors are used in power electronic converters, which are aluminum electrolytic capacitors (Al-Caps), metallized polypropylene (PP) film capacitors (MPPF-Caps), and multilayer ceramic capacitors (MLC-Caps) [6]. According to the difference in dielectric materials, the Al-Caps can be divided into electrolytic capacitors and solid capacitors. Table 2.1 summarizes the performance of different types of capacitors [7]. It is found that the capacity of Al-Caps and MPPF-Caps is usually larger than that of MLC-Cap, which results in Al-Caps and MPPF-Caps being widely used in dc-links of PV converters.

Furthermore, taking the Al-Caps (electrolytic) and MPPF-Caps manufactured by TDK as examples, Figure 2.2 shows the capacitance and voltage ranges of

Figure 2.2 Typical capacitance and voltage ranges of Al-Caps (electrolytic) and MPPF-Caps manufactured by TDK.

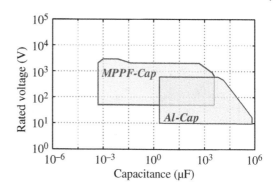

capacitors [8]. It can be seen that the rated voltage of Al-Caps is usually smaller than that of MPPF-Caps. However, its capacity is relatively large. Therefore, considering capacitors' voltage and current limitation, Al-Cap bank, MPPF-Cap bank, or hybrid banks are usually placed at dc-link based on series-parallel configuration.

2.1.2 Basic Characteristics of Capacitors

Figure 2.3a shows a simplified model of the Al-Caps and MPPF-Caps, where C is the capacitance, R_{ESR} and L_{ESL} represent the equivalent series resistance (ESR) and the equivalent series inductance (ESL), and R_p denotes the insulation resistance. According to the model, the dissipation factor (DF) is defined as $\tan \delta = \omega \cdot R_{ESR} \cdot C$. The impedance Z_C of capacitors is expressed as

$$Z_C = \sqrt{R_{ESR}^2 + \left(2\pi f \cdot L_{ESL} - \frac{1}{2\pi f \cdot C}\right)^2}. \tag{2.1}$$

Using Eq. (2.1), Figure 2.3b shows the impedance characteristics of capacitors [6]. It is found that the electrical parameters (e.g. Z_C and R_{ESR}) of capacitors are easily influenced by the frequency. Generally, the impedance is dominated by the capacitance C in Region I (i.e. low-frequency band, $f \leq f_1$), R_{ESR} in Region II (i.e. mid-frequency band, $f_1 \leq f \leq f_2$), and L_{ESL} in Region III (i.e. high-frequency band, $f \geq f_2$). Taking an Al-Cap and an MPPF-Cap as examples, Figure 2.3c gives the typical values of f_1 and f_2 at 25 °C [9, 10], where the types of capacitor are SLPX (Al-Cap, 470 μF/450 V) and B32778-JX (MPPF-Cap, 480 μF/450 V), respectively.

Besides the frequency, the parameters of capacitors are easily influenced by ambient temperature. Taking the Al-Cap (Type: SLPX 470 μF/450 V) as an example, Figure 2.4a and b show R_{ESR} and C variations versus temperature, respectively [8], which are given from a manufacturer datasheet. It is found that C increases as the ambient temperature T_a increases and R_{ESR} decreases as T_a

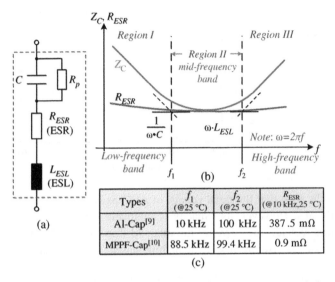

Figure 2.3 Equivalent circuit model and impedance characteristics of capacitors. (a) Equivalent circuit model. (b) Impedance characteristics. (c) Typical values of f_1 and f_2, where the types of Al-Cap and MPPF-Cap are SLPX [9] (470 µF/450 V), and B32778-JX [10] (480 µF/450 V), respectively.

Types	f_1 (@25 °C)	f_2 (@25 °C)	R_{ESR} (@10 kHz,25 °C)
Al-Cap[9]	10 kHz	100 kHz	387.5 mΩ
MPPF-Cap[10]	88.5 kHz	99.4 kHz	0.9 mΩ

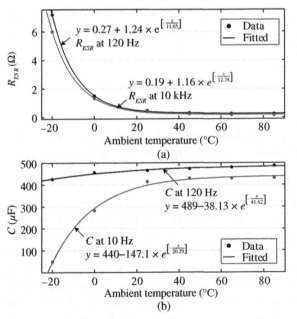

Figure 2.4 R_{ESR} and C variations versus temperature of a new capacitor (Type: SLPX 470 µF/450 V, datasheet [5]). (a) R_{ESR} versus temperature. (b) C versus temperature. Source: Zhao [11]/with permission of IEEE.

increases. The relationship between R_{ESR}, C of a new capacitor and ambient temperature is given as [11]

$$\begin{cases} R_{ESR}(T_a) = \alpha_{Al} + \beta_{Al}e^{-T_a/\gamma_{Al}} \\ C(T_a) = \chi_{Al} + \lambda_{Al}e^{-T_a/\nu_{Al}} \end{cases}, T_{min} < T_a < T_{max} \tag{2.2}$$

where α_{Al}, β_{Al}, γ_{Al}, χ_{Al}, λ_{Al}, and ν_{Al} are characteristic coefficients of capacitors, which are determined experimentally. T_{min} and T_{max} represent the minimum and maximum operating temperatures of capacitors, respectively, defined by the manufacturers. Assuming the operating frequencies of capacitors are 120 Hz and 10 kHz, Figure 2.4 also shows the detailed parameters of α_{Al}, β_{Al}, γ_{Al}, χ_{Al}, λ_{Al}, and ν_{Al}, which are obtained through data fitting.

Generally, the ESR of MPPF-Caps is relatively small. The capacitance is the most important electrical parameter, as shown in Figure 2.3c. The temperature also influences the capacitance C of MPPF-Caps. For MPPF-Caps, the effects of temperature on C are dependent on the capacitor materials. For PP capacitors, C decreases as the ambient temperature increases. For polyethylene terephthalate (PET) and polyethylene naphthalate (PEN) capacitors, C increases as the ambient temperature increases. Generally, the relationship between ambient temperature T_a and capacitance is [12]

$$C(T_a) = \alpha_{MPPF}C_{20\,^\circ C}(T_a - T_{test}) + C_{test} \tag{2.3}$$

where C_{test} and T_{test} are the capacitance and temperature under the testing condition, $C_{20\,^\circ C}$ is the reference capacitance at 20 °C. α_{MPPF} is the temperature coefficient, α_{MPPF} of PP, PET, and PEN capacitors are -250×10^{-6}/K, 600×10^{-6}/K, and 200×10^{-6}/K, respectively. The capacitance of MPPF-Caps is also influenced by the humidity H_a. Usually, the capacitance C increases as the ambient humidity increases [12], i.e.

$$C(H_a) = \frac{\beta_{MPPF}C_{test}(H_a - H_{test}) + C_{test}}{2 + \beta_{MPPF}H_{test} - \beta_{MPPF}H_a} \tag{2.4}$$

where C_{test} and H_{test} are the capacitance and humidity under the testing condition, respectively. β_{MPPF} is the humidity coefficient, β_{MPPF} of PP, PET, and PEN capacitors are 40–100×10^{-6}%RH, 500–700×10^{-6}%RH, and 700–900×10^{-6}%RH, respectively.

2.2 Laws of Aging for Capacitors

2.2.1 Aging Mechanisms and Indicators of Capacitors

Generally, the failure of capacitors can be divided into two categories. One is the catastrophic failure, including short-circuit and open-circuit failures caused by single-event overstress such as extreme stress and manufacturing defects. Another

Table 2.2 Typical failure modes, failure mechanisms, and indicators of Al-Caps and MPPF-Caps.

Types	Critical stressors	Failure mechanisms	Electrical indicators	Nonelectrical indicators
Al-Caps	V_C, i_C, T_a	• Reduction in anode and cathode foil capacitance • Deterioration of oxide film • Electrolyte evaporation	• R_{ESR}, DF increase • C decreases • Z_C, R_p, i_{LC} change	• Internal pressure increases • Internal temperature increases • Weight decreases • Structure change
MPPF-Caps	V_C, i_C, T_a, humidity	• Deterioration of quality of electrode metallization • Dielectric loss due to moisture absorption by film • Self-healing • Electrochemical corrosion	• R_{ESR}, DF increase • C decreases • i_{LC} increases • Z_C, R_p, i_{LC} change	• Internal pressure increases • Internal temperature increases • Structure change

C – capacitance, R_{ESR} – equivalent series resistance, i_{LC} - leakage current, Z_C – impedance, R_p – insulation resistance, V_C – capacitor ripple voltage, i_C – capacitor ripple current, T_a – ambient temperature

is the wear-out failure due to the long-time degradation of capacitors. Usually, aging detection is used to predict the wear-out failure of capacitors.

The typical wear-out failure mechanisms and indicators of Al-Caps and MPPF-Caps have been presented in [6, 13–15], as summarized in Table 2.2, where V_C, i_C, and T_a represent the capacitor ripple voltage, capacitor ripple current, and ambient temperature, respectively. It can be seen that a series of physical and chemical changes occur in the inside of capacitors with the degradation of capacitors, which will cause electrical parameters (e.g. R_{ESR}, C, Z_C, DF, and R_p), and nonelectrical parameters (e.g. weight, structure, internal temperature, and internal pressure) to be changed. Based on this, the health status of capacitors can be estimated.

2.2.2 Aging Laws of Capacitors Based on Electrical Parameters

Taking a NIPPON snap-in Al-Cap (400 V/470 μF) as an example, Figure 2.5 shows the accelerated aged test results, where the test temperature is 85 °C. The ripple

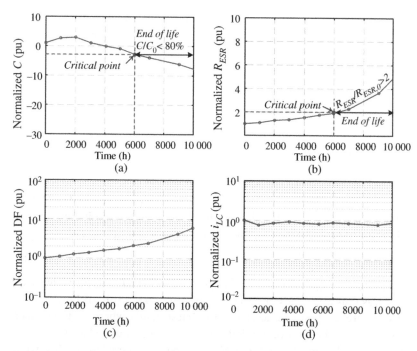

Figure 2.5 Degradation characteristic of a NIPPON snap-in Al-Cap (400 V/470 μF). (a) Normalized C. (b) Normalized R_{ESR}. (c) Normalized DF. (d) Normalized i_{LC}.

current and frequency are 2.51 A and 120 Hz, respectively. It can be seen that there exists a significant variation of C, R_{ESR}, and DF with the degradation of Al-Cap. However, the variation of i_{LC} is not obvious. Usually, C and R_{ESR} are chosen as the health status indicators of Al-Caps. Referring to Figure 2.5a and b, it is found that the rates of change of C and R_{ESR} increase after the critical point. Usually, one Al-Cap is considered to be failed when its C has reduced to 80% and/or ESR increased to two times the initial values. Therefore, the typical end-of-life criteria of Al-Caps are defined as $C/C_0 < 80\%$ and/or $R_{ESR}/R_{ESR0} > 2$, where C_0 and R_{ESR0} represent the initial values of C and R_{ESR}, respectively.

In order to estimate the remaining useful lifetime (RUL), the degradation model of Al-Caps can be built based on the variation curves of C and R_{ESR}. A simple degradation model of Al-Caps is [16]

$$\begin{cases} C(t) = A_{Al} \cdot \left(1 - e^{B_{Al} \cdot t}\right) \\ R_{ESR}(t) = D_{Al} \cdot e^{E_{Al} \cdot t} \end{cases} \tag{2.5}$$

where t represents the operation time, A_{Al} and D_{Al} are the parameters that depend on C_0 and R_{ESR0}, respectively. B_{Al} and E_{Al} describe temperature-dependent degradation rates, which can be determined experimentally. Taking the aging test results

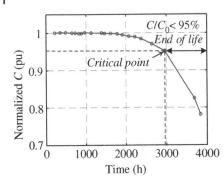

Figure 2.6 Degradation characteristic of a MPPF-Cap (1100 V/40 μF).

shown in Figure 2.5 as an example, A_{Al}, B_{Al}, D_{Al}, and E_{Al} are 0.343, 0.00033, 0.73, and 0.00018, respectively. Notice that the degradation model for different types of capacitors is different. Another simple linear model of Al-Caps is [17]

$$C(t) = A_{Al} \cdot (1 + B_{Al} \cdot t) \tag{2.6}$$

Considering that the ESR of MPPF-Caps is relatively small, C is the widely used health status indicator of MPPF-Caps. Taking an 1100 V/40 μF MPPF-Cap as an example, Figure 2.6 shows the accelerated aged test results, where the test temperature is 85 °C and the relative humidity is 55%. It is found that C decreases with the aging of MPPF-Caps. Usually, one MPPF-Cap is considered to be failed when its C has reduced to 95%, i.e. the end-of-life criterion is defined as $C/C_0 < 95\%$.

For MPPF-Caps, a simple degradation model is [18]

$$C(t) = A_{MPPF} - B_{MPPF} \cdot e^{D_{MPPF} \cdot t} \tag{2.7}$$

where A_{MPPF} is the parameter that depends on C_0, B_{MPPF}, and D_{MPPF} describe degradation rates that depend on temperature and humidity. In this example, A_{MPPF}, B_{MPPF}, and D_{MPPF} are 1.00102, 0.00073, and 0.0018, respectively.

2.2.3 Aging Laws of Capacitors Based on Nonelectrical Parameters

As mentioned earlier, R_{ESR} and C are the preferred electrical indicators of capacitors. The typical end-of-life criteria for Al-Caps and MPPF-Caps have been given in Figures 2.5 and 2.6. Unlike the electrical parameters, the nonelectrical parameters of capacitors depend on capacity, materials, rated voltages, and currents. Hence, there are no uniform end-of-life criteria for capacitors when choosing the nonelectrical parameters as the degradation indicators. Generally, two categories of methods are applied to define the end-of-life criteria of capacitors. One category is to construct the relationship between electrical and nonelectrical parameters, e.g. weight, temperature, volume, etc. Another is to identify the failure status of capacitors using the structure change of capacitors.

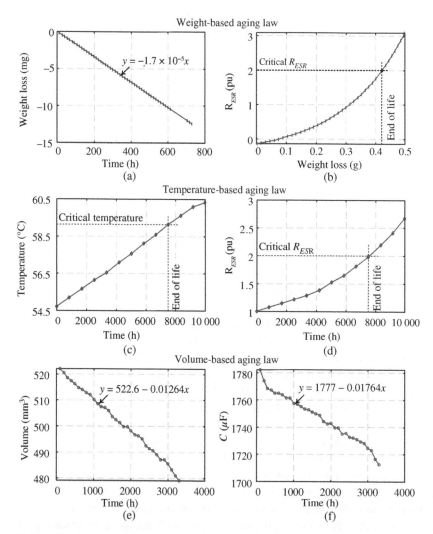

Figure 2.7 Aging laws of Al-Caps based on weight, temperature, and volume.
(a) Average weight loss of a type of Al-Cap (Nichicon PW series, 450 V/68 μF). (b) Weight loss versus R_{ESR} changes of Nichicon PW series Al-Cap. Source: Adapted from [19].
(c) Simulated capacitor temperature of a type of Al-Cap. (d) Simulated R_{ESR} changes of a type of Al-Cap Source: Adapted from [17]. (e) Average volume change of a type of Al-Cap (2200 μF). (f) Capacitance change of a type of Al-Cap (2200 μF). Source: Adapted from [20].

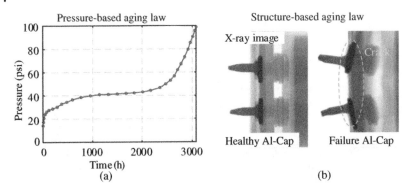

Figure 2.8 Aging laws of Al-Caps based on pressure and structure. (a) Internal pressure change of a type of Al-Cap. Source: Adapted from [21]. (b) X-ray images of a healthy Al-Cap and a failure Al-Cap. Source: Shrivastava et al. [22]/with permission of IEEE.

Capacitor weight, internal temperature, and internal pressure are suitable non-electrical indicators for Al-Caps. Taking weight loss as an example, Figure 2.7a shows the average weight loss of an Al-Cap (Nichicon PW series, 450 V/68 μF), and Figure 2.7b gives the relationship between weight loss and R_{ESR} changes [19]. It is found that the weight of capacitors decreases with the degradation of capacitors, the critical weight loss can be defined as a 200% increase in R_{ESR}.

Figure 2.7c and d show the simulated internal temperature and capacitance of an Al-Cap [17]. It is found that the internal temperature increases with the degradation of Al-Caps. Similar to that in Figure 2.7a and b, the end-of-life criteria can be derived based on the relationship between temperature and electrical parameters. Moreover, Figure 2.7e and f show the volume and capacitance change of an Al-Cap [20]. Generally, the volume of Al-Caps decreases with the evaporation of electrolytes. Therefore, the health status of Al-Caps can be evaluated using the relationship between volume and R_{ESR} or C.

Furthermore, Figure 2.8a shows the internal pressure change of an Al-Cap. It can be seen that the internal pressure increases with the aging of Al-Caps [21]. Therefore, the health status of Al-Caps can be estimated based on the relationship between pressure and critical electrical parameters, e.g. R_{ESR} and C. In addition, Figure 2.8b shows the X-ray images of a healthy Al-Cap and a failure Al-Cap [22]. It is found that there exists a structural difference between the health Al-Cap and the failure one. However, it is difficult to construct the relationship between the electrical parameters and structure change due to the uncertainty of structure change.

For MPPF-Caps, the nonelectrical parameters can also indicate the degradation level of capacitors. Takin a 0.47-μF MPPF-Cap as an example, Figure 2.9a and b show the weight and capacitance change curves [23]. It is found that the weight of the MPPF-Cap increases with its degradation. Therefore, the degradation model

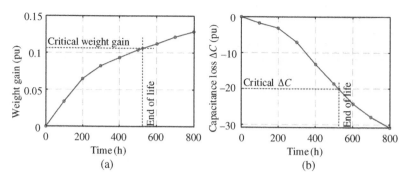

Figure 2.9 Aging laws of MPPF-Caps based on weight. (a) Average weight gain of a type of MPPF-Cap. (b) Capacitance change of a type of MPPF-Cap. Source: Li et al. [23]/with permission of IEEE.

and end-of-life criteria can be obtained by constructing the relationship between weight and electrical parameters.

2.2.4 Aging Detection Procedure

As discussed earlier, some electrical parameters (e.g. RESR and C), and nonelectrical parameters (e.g. weight, structure, internal temperature, and internal pressure) will change with the degradation of capacitors. Based on this, Figure 2.10 presents the condition monitoring (CM) procedure for capacitors. It mainly includes two steps, i.e. degradation indicators acquisition (Step I) and capacitor health status assessment (Step II). For Step I, Table 2.3 lists some commonly used parameter acquisition methods. Generally, some industrial instruments, including weighting meters, X-ray image meters, optical inspection meters, acoustic detection meters, thermography meters, etc., are widely used to obtain nonelectrical parameters. Moreover, internal temperature sensors and internal pressure sensors are employed to obtain the internal temperature and pressure of capacitors.

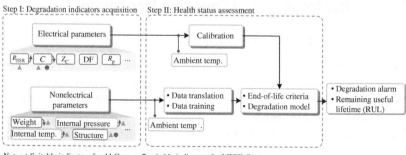

Figure 2.10 Aging detection procedure for capacitors.

Table 2.3 Typical degradation indicators and acquisition methods.

	Health status indicators	Indicator acquisition methods
Electrical parameters	• R_{ESR}, C, Z_C, etc.	• LCR meter, impedance analyzer, etc. • Parameter estimation (including physics-based methods and data-driven-based methods)
Nonelectrical parameters	• Weight	• Weighting meter
	• Internal temperature	• Internal temperature sensor • Thermography meter
	• Volume	• Ruler, etc.
	• Internal pressure	• Internal pressure sensor
	• Internal structure	• X-ray image meter • Optical inspection meter • Thermography meter • Acoustic detection meter

Referring to Table 2.3, although the electrical parameters can be measured using an LCR meter, impedance analyzer, etc., parameter estimation-based methods are widely used to estimate dc-link capacitors' electrical parameters because the capacitors do not need to be dismantled from converters in most cases. According to the dependence on the physical model of capacitors, two main categories of principles are generally used to estimate the electrical parameters of dc-link capacitors. One is the physical model-based method, and another is the data-driven-based method, which are discussed in detail in Sections 2.3 and 2.4, respectively.

2.3 Physical Model-based Condition Monitoring

2.3.1 Parameter Estimation Principles

According to the type of needed electrical signals, the physical model-based CM methods can be divided into two types, i.e. periodic small-signal ripples-based principle (Principle I) and nonperiodic large-signal charging/discharging profile-based principle (Principle II).

Figure 2.11a shows the equivalent circuit of dc-link capacitors, where v_{CAP} and i_C denote capacitors' voltage and current, respectively. Based on Ohm's law, the

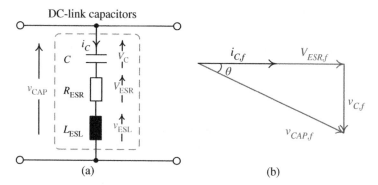

Figure 2.11 Equivalent circuit and phasor diagram of voltage. (a) Equivalent circuit of dc-links. (b) Phasor diagram of voltage.

capacitor voltage ripple Δv_{CAP} and capacitor current ripple Δi_C have the following relationship

$$\Delta v_{CAP}(t) = \frac{1}{C}\int_0^t i_C(t)dt + R_{ESR} \cdot \Delta i_C(t) + L_{ESL} \cdot \frac{di_C(t)}{dt} \tag{2.8}$$

Considering that L_{ESL} is very small (on the order of 10–100 nH) at the working frequency of converters (i.e. low- and mid-frequency bands in Figure 2.3), which can be ignored in dc-link capacitors, Eq. (2.8) can be written as

$$\Delta v_{CAP}(t) = \frac{1}{C}\int_0^t i_C(t)dt + R_{ESR} \cdot \Delta i_C(t). \tag{2.9}$$

Assuming the voltages across C, R_{ESR} are v_C, v_{ESR}, respectively. Figure 2.11b shows the phasor diagram of voltage, where $v_{CAP,f}$, $v_{C,f}$, $v_{ESR,f}$, and $i_{C,f}$ represent v_{CAP}, v_C, v_{ESR}, and i_C at arbitrary frequency f. θ denotes the phasor angle. Furthermore, Δv_{CAP} at the low-frequency band and the mid-frequency band can be simplified as

$$\Delta v_{CAP\_LF}(t) \approx \frac{1}{C}\int_0^t i_{C\_LF}(t)dt \tag{2.10}$$

$$\Delta v_{CAP\_MF}(t) \approx R_{ESR} \cdot \Delta i_{C\_MF}(t) \tag{2.11}$$

where Δv_{CAP_LF}, Δi_{C_LF} denote the ripples at the low-frequency band, and Δv_{CAP_MF}, Δi_{C_MF} indicate them at the mid-frequency band. From Eqs. (2.10) and (2.11), C and R_{ESR} can be estimated as

$$C \approx \frac{1}{\Delta v_{CAP\_LF}(t)}\int_0^t i_{C\_LF}(t)dt \tag{2.12}$$

$$R_{ESR} \approx \frac{\Delta v_{CAP\_MF}(t)}{\Delta i_{C\_MF}(t)} \tag{2.13}$$

From Eqs. (2.12) and (2.13), it is easily found that C is inversely proportional to the voltage ripple Δv_{CAP} and R_{ESR} is proportional to Δv_{CAP} for a given i_C. Hence, Eqs. (2.12) and (2.13) can be simplified as Eqs. (2.14) and (2.15) when i_C is given.

$$C \propto 1/\Delta v_{CAP\_LF} \tag{2.14}$$

$$R_{ESR} \propto \Delta v_{CAP\_MF}. \tag{2.15}$$

Referring to Eq. (2.11), the main power loss of capacitors is caused by R_{ESR}. Considering different harmonic frequencies, the total power loss P_{loss} of capacitors is

$$P_{loss} = \sum_{k=1}^{n} R_{ESR,k} I_{C\_RMS,k}^2 \tag{2.16}$$

where $R_{ESR,k}$ and $I_{C\_RMS,k}$ represent the ESR and the root mean square (RMS) current for each harmonic k. Although (2.16) can calculate ESR, it is challenging to determine $R_{ESR,k}$ for each harmonic k. Moreover, the end-of-life criteria (e.g. $R_{ESR}/R_{ESR0} > 2$) provided by manufacturers are usually based on a frequency region (i.e. mid-frequency band) without considering a specified frequency. As discussed in [24], R_{ESR} calculated by the average power loss $\overline{P}_{loss}$ and the total RMS current I_{C_RMS} can approximately represent the actual ESR of capacitors in power electronic converters. Therefore, R_{ESR} is approximately calculated as

$$R_{ESR} \approx \frac{\overline{P}_{loss}}{I_{C\_RMS}^2} = \frac{\int_0^t [\Delta v_{CAP}(t) \cdot i_C(t)] dt}{\int_0^t i_C^2(t) dt}. \tag{2.17}$$

Principle II is based on the large-signal discharging/charging profile as shown in Figure 2.12. Figure 2.12a shows the equivalent circuit of a capacitor discharging and charging schemes, where R_{eq} and R'_{eq} represent the equivalent resistances of discharging circuits and charging circuits, respectively. Figure 2.12b gives the discharging and charging profiles, where V_{dc} denotes the reference value of capacitor voltage. During the discharging period, the capacitor voltage is expressed as

$$v_{CAP}(t) = V_{dc} e^{-\frac{t}{(R_{eq}+R_{ESR})C}}. \tag{2.18}$$

During the charging period, the capacitor voltage is

$$v_{CAP}(t) = V_{dc} - V_{dc} \cdot e^{-\frac{t}{(R'_{eq}+R_{ESR})C}}. \tag{2.19}$$

By solving Eqs. (2.18) and (2.19), R_{ESR} and C can be estimated.

Based on (2.10)–(2.19), the CM methods for dc-link capacitors can be derived in a step-by-step manner. There exist 16 derived methods, as summarized in Table 2.4, where methods 1A–1E, 1E–1H, 1I–1K, 2A–2B, and 2C–2D are based on Eqs. (2.12), (2.13), (2.17), (2.18), and (2.19), respectively. The examples of these methods are discussed in Section 2.3.2.

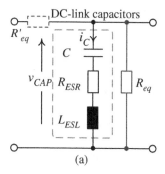 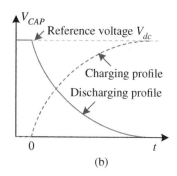

(a) (b)

Figure 2.12 Equivalent circuits and voltage profiles of RC charge and discharge schemes. (a) Equivalent circuits. (b) Voltage profiles.

2.3.2 Examples of Derived CM Methods

Referring to Figure 2.12a, the methods 1A and 1E use the voltage ripple to approximately estimate the capacitor parameters without dependency on the capacitor current [25, 26]. The methods 1B–1D, 1F–1H, and 1I–1K utilize sampled or estimated capacitor current to calculate R_{ESR} or C, illustrated using the examples in Figure 2.12.

Figure 2.13a shows an implementation example of methods 1B and 1F [27]. Here, the capacitor current is directly sampled by a current sensor. To accurately obtain R_{ESR} or C, a band-pass filter and a low-pass filter are required, respectively. To avoid using capacitor current sensors, methods 1C and 1G utilize the circuit operation model to indirectly obtain i_C. Referring to Figure 2.13b, taking a buck converter as an example, it is easily found that $i_C = i_L - i_o$, where i_L and i_o are inductor current and load current, respectively [28].

The implementation of schemes 1D and 1H can be divided into two categories. One is to directly inject current into the tested capacitor, as shown in Figure 2.13d [29]. Here, a signal generation circuit including a signal generator and a power amplifier injects current signals at given frequencies to the tested capacitor. The injected signal can be sinusoidal waves, square waves, triangle waves, etc. Another is to inject a perturbation signal into the power electronic systems. Taking a PV grid-connected inverter as an example, Figure 2.13e gives the implementation scheme [30]. Here, a current at hth harmonic frequency is injected into the grid, which causes $(h-1)$th and $(h+1)$th voltage and current ripples to appear on the dc-link capacitors. Then, R_{ESR} and C can be estimated using the voltage and current ripples at $(h-1)$th or $(h+1)$th.

According to (2.17), R_{ESR} can be calculated using power losses of capacitors. Figure 2.13c shows method 1I [24]. It is similar to method 1B; however, a band-pass filter is not required. The implementation of methods 1J and 1K are similar to 1G and 1H, respectively.

Table 2.4 Derived CM Methods Based on Principle I and Principle II.

Principles		Derived CM methods
Equation (2.12)	1A	Measure Δv_{CAP_LF}, and predict C based on $C \propto 1/\Delta v_{CAP\_LF}$
	1B	Measure Δv_{CAP_LF}, and obtain i_{C_LF} using a current sensor
	1C	Measure Δv_{CAP_LF}, estimate i_{C_LF} based on system operation model
	1D	Inject LF signal to capacitor/system, obtain the corresponding Δv_{CAP_LF}, i_{C_LF}
Equation (2.13)	1E	Measure Δv_{CAP_MF}, and predict R_{ESR} based on $R_{ESR} \propto 1/\Delta v_{CAP\_MF}$
	1F	Measure Δv_{CAP_MF}, and obtain i_{C_MF} using a current sensor
	1G	Measure Δv_{CAP_MF}, estimate i_{C_MF} based on system operation model
	1H	Inject MF signal to capacitor/system, obtain the corresponding Δv_{CAP_MF}, i_{C_MF}
Equation (2.17)	1I	Measure v_{CAP}, and obtain i_C using a current sensor
	1J	Measure v_{CAP}, estimate i_C indirectly based on the system operation model
	1K	Inject signal to capacitor/system, obtain corresponding Δv_{CAP}, i_C
Equation (2.18)	2A	Measure capacitor discharging profile during the shutdown process
	2B	Measure capacitor discharging profile during the system transient
Equation (2.19)	2C	Measure capacitor charging profile during the start-up process
	2D	Measure capacitor charging profile during the system transient

Notice that the methods 1A and 1E, 1B and 1F, 1C and 1G, and 1D and 1H can be grouped respectively when using (2.9). Here, C and R_{ESR} are estimated under the same frequency, such as low-frequency or mid-frequency.

Methods 2A and 2C obtain the discharging or charging profiles during the converters' shutdown process or start-up process. Referring to Figure 2.12, during the shutdown process of the converter, the dc-link capacitors discharge through the equivalent resistor R_{eq}. The capacitor parameters can be estimated based on the discharging profile and (2.18). During the start-up process of converter,

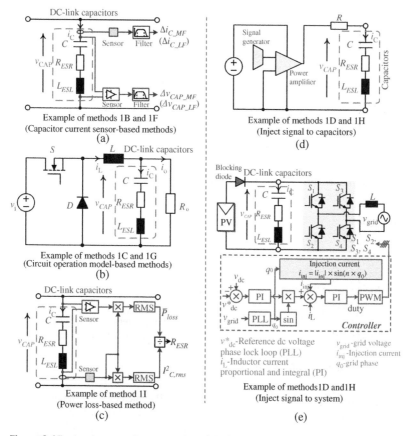

Figure 2.13 Implementation examples of derived methods based on Principle I. (a) Example of methods 1B and 1F. (b) Example of methods 1C and 1G based on a buck converter. (c) Example of method 1I. (d) Example of methods 1D and 1H, extra signals are injected into capacitors. (e) Example of methods 1D and 1H based on a PV inverter; here, extra signals are injected into systems.

the capacitors charge through R'_{eq}. The capacitor parameters can be estimated based on the charging profile and (2.19). The capacitors will be discharged or charged during transients, similar to the shutdown and start-up processes. Taking a full-bridge dc-dc converter as an example, Figure 2.14a shows the voltage profile during a loading transient [31]. The degradation status can be estimated based on the relationship between discharging profile and capacitor parameters. Furthermore, Figure 2.14b shows the unloading transient waveform of a buck converter [32]. Based on the relationship between charging profile and capacitor parameters, R_{ESR} and C can be estimated.

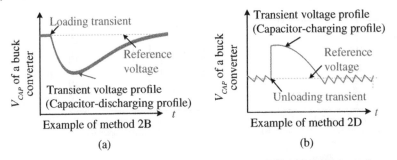

Figure 2.14 Implementation examples of derived methods based on Principle II. (a) Example of method 2B. (b) Example of method 2D.

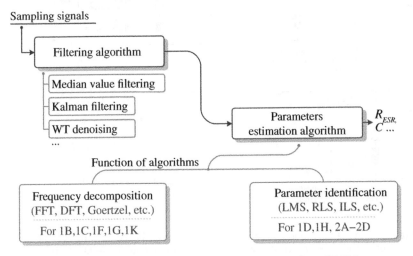

Note: FFT-fast Fourier transform, DFT-discrete Fourier transform, LMS-least mean squares, RLS-recursive least squares, ILS-iterative least squares, WT-wavelet transform.

Figure 2.15 Data-processing procedure for condition monitoring.

Data processing is a key link of the CM for dc-link capacitors. Figure 2.15 shows the general procedure of data processing for CM. First, the sampling signals are filtered using filtering algorithms including median value filtering, Kalman filtering, wavelet transform denoising, etc. Then parameter estimation algorithms are employed to obtain capacitor parameters.

For methods 1B, 1C, 1F, 1G, and 1K, the key issue is to obtain the low-frequency or mid-frequency components of capacitor voltage and capacitor current. Usually, frequency decomposition algorithms, including fast Fourier transform (FFT)

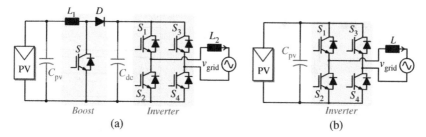

Figure 2.16 Typical structures of PV systems. (a) Single-stage configuration. (b) Double-stage configuration.

algorithm, discrete Fourier transform (DFT) algorithm, Goertzel algorithm, etc., are used to extract the low-frequency or mid-frequency components of v_{CAP} and i_C. For methods 1D, 1H, and 2A–2D, adaptive filter algorithms including least mean squares (LMS) algorithm, recursive least squares (RLS) algorithm, iterative least squares (ILS) algorithm, etc., are widely used to identify R_{ESR} and C in Eqs. (2.9)–(2.11), (2.18), and (2.19). Moreover, some optimization algorithms, such as particle swarm optimization (PSO) algorithm and genetic algorithm (GA) can be used to identify capacitor parameters.

2.3.3 Case Studies

Figure 2.16 shows two typical structures of PV converters, where C_{pv} and C_{dc} represent the PV-side and dc-link capacitors, respectively. Some impressive research has recently been done on the CM of capacitors in PV systems, as discussed in the following.

2.3.3.1 CM for C_{pv} and C_{dc} in Double-Stage Converter Systems
Generally, methods 1A and 1E can predict R_{ESR} and C without additional capacitor current sensors. However, it is impossible to obtain the detailed values of R_{ESR} and C. Considering this issue, reference [33] presents a capacitor current directly measured method (i.e. 1B and 1F) to realize the CM of C_{pv} in double-stage converter systems. Figure 2.17a shows the implementation scheme presented in [33], where two tunneling magnetoresistive (TMR) sensors TMR_1 and TMR_2 are used to measure the capacitor current $i_{C,pv}$ and inductor current i_L, respectively. Then, R_{ESR} and C at f can be estimated based on the phasor diagram shown in Figure 2.11b, i.e.

$$\begin{cases} R_{ESR,f} = (\Delta v_{CAP,f} Cos(\theta))/\Delta i_{C,f} \\ C_f = \Delta i_{C,f}/(2\pi f \Delta v_{CAP,f} Sin(\theta)) \end{cases}. \tag{2.20}$$

It is easily found that the CM of C_{pv} is based on methods 1B and 1F. The current $i_{C,dc}$ of C_{dc} is derived from the operation model of boost converters, i.e. methods

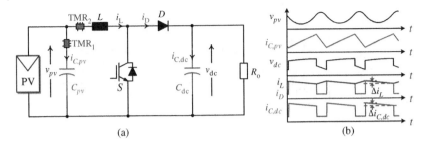

(a)   (b)

Figure 2.17 Capacitor current directly measured scheme (1B and 1F). (a) Implementation scheme. (b) Steady-state waveforms.

1C and 1G. Referring to the steady-state waveforms shown in Figure 2.17b, it can be seen that the peak of inductor current Δi_L equals $\Delta i_{C,dc}$. Therefore, $\Delta i_{C,dc}$ can be derived from the sampling signals of i_L. Furthermore, using (2.20), R_{ESR} and C of C_{dc} can be estimated. In addition, according to (2.13), reference [34] uses the mid-frequency signal of v_{dc} and i_L to simplify the calculation of R_{ESR} of C_{dc}, i.e.

$$R_{ESR,dc} \approx \frac{\Delta v_{dc}}{\Delta i_{C,dc}} = \frac{\Delta v_{dc}}{\Delta i_L}. \tag{2.21}$$

The schemes presented in [34] can realize the CM of C_{dc}. However, a resistive load is assumed to connect with boost converters. Usually, an inverter load is connected at the rear-end of boost converters, as shown in Figure 2.18a. The

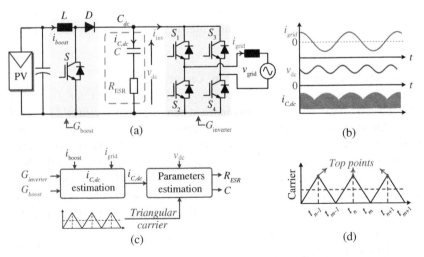

Figure 2.18 System operation model-based scheme (1C and 1G). (a) Implementation scheme. (b) Steady-state waveforms. (c) Parameter estimation procedure. (d) The Triangular Carrier.

typical waveforms of grid current i_{grid}, capacitor voltage v_{dc}, and capacitor current $i_{C,dc}$ are shown in Figure 2.18b. In [35], a system operation model-based scheme (i.e. 1C and 1G) is presented to estimate R_{ESR} and C, whose estimation procedure is shown in Figure 2.18c. Referring to Figure 2.18a, the capacitor current can be derived using

$$i_{C,dc} = \overline{G}_{boost} i_{boost} - i_{inv} = \overline{G}_{boost} i_{boost} - f(G_1, G_2, G_3, G_4) i_{grid} \qquad (2.22)$$

where G_{boost}, G_1, G_2, G_3, and G_4 represent the status of switches S, S_1, S_2, S_3, and S_4, respectively. $\overline{G}_{boost} i_{boost}$ represents the average inductor current i_{boost}. i_{inv} is the input current of the inverter. Generally, the dc-link voltage v_{dc} equals the capacitor voltage when the capacitor current $i_{C,dc} = 0$. Therefore, C can be calculated using v_{dc} in two different instants in which $i_{C,dc} = 0$, and the integral of $i_{C,dc}$ between these two instants. The typical instants t_{n-1} and t_n are shown in Figure 2.18d. Using Eq. (2.12), C can be estimated as

$$C = \int_{t_{n-1}}^{t_n} i_{C,dc} dt / [v_{dc}(t_n) - v_{dc}(t_{n-1})] \qquad (2.23)$$

where $v_{dc}(t_n)$ and $v_{dc}(t_{n-1})$ represent v_{dc} at t_n and t_{n-1}, respectively. Furthermore, R_{ESR} can be calculated using v_{dc} and $i_{C,dc}$ when $i_{C,dc} \neq 0$, i.e.

$$R_{ESR} = \frac{v_{dc}(t_m) - \left(\frac{1}{C} \int_{t_n}^{t_m} i_{C,dc} dt + v_C(t_n) \right)}{i_{C,dc}(t_m)} \qquad (2.24)$$

where t_m represents the instant that $i_{C,dc} \neq 0$.

In [35], sampled signals at particular instants are used to estimate the parameters of C_{dc}. Reference [4] presents a similar scheme for CM of C_{pv}, as shown in Figure 2.19a. Figure 2.19b shows the steady-state waveforms of i_L, $i_{C,pv}$, v_{ESR}, and v_C when the boost converter operates in a continuous conduction mode (CCM). It can be seen that the difference between v_{pv} sampled at $t = 0$ (i.e. point a) and $t = DT_s$ (i.e. point c) is only due to R_{ESR}. Because C does not affect this difference, no net charge is absorbed by the capacitor in this duration. Similarly, the difference between v_{pv} sampled at $t = DT_s/2$ (i.e. point b) and $t = (1 + D)T_s/2$ (i.e. point d) is only due to C. Using (2.9), the capacitor parameters can be estimated as

$$\begin{cases} R_{ESR} = L[v_{pv}(0) - v_{pv}(DT_s)]/(DT_s v_{pv}) \\ C = \dfrac{v_{pv} DT_s^2}{8L\{v_{pv}(DT_s/2) - v_{pv}[(1 + D)T_s/2]\}} \end{cases} \qquad (2.25)$$

Similarly, C and R_{ESR} for converters operating in a discontinuous conduction mode (DCM) can be estimated using the same way.

In addition, reference [36] presents an external signal injection-based method (i.e. 1D and 1H) in order to realize the CM of C_{pv} in PV boost converters (cf.,

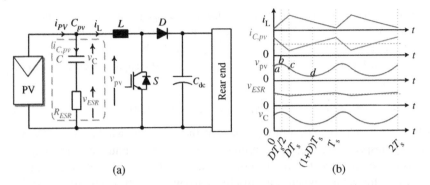

Figure 2.19 System operation model-based scheme (1C and 1G). (a) PV boost converter. (b) Steady-state waveforms.

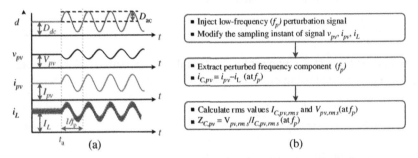

Figure 2.20 External signal injection-based method (1D and 1H). (a) CM procedure. (b) Key waveforms.

Figure 2.19a). Referring to Figure 2.20a, the system works in a steady state before the instant t_a and the duty cycle of the converter is D_{dc}. The amplitude of PV voltage v_{pv} and PV current i_{pv} are V_{pv} and I_{pv}, respectively. The reference value of the inductor current i_L is I_L. At t_a, a small perturbation signal d_{ac} is added to the original duty cycle D_{dc}. The peak amplitude and frequency of d_{ac} are D_{ac} and f_p, respectively. The injected low-frequency perturbation results in the appearance of a low-frequency ripple as shown in Figure 2.20a. Then, according to the sampling of low-frequency signals, the low-frequency impedance of C_{pv} can be calculated as

$$Z_{C,pv}(f_p) = V_{pv,rms}^{f_p} / I_{C,pv,rms}^{f_p} \tag{2.26}$$

where $V_{pv,rms}^{f_p}$ and $I_{C,pv,rms}^{f_p}$ are the RMS values of v_{pv} and $i_{C,pv}$ at f_p. Here, the capacitor current $i_{C,pv}$ is calculated as $i_{C,pv} = i_{pv} - i_L.$, as shown in the CM procedure in Figure 2.20b.

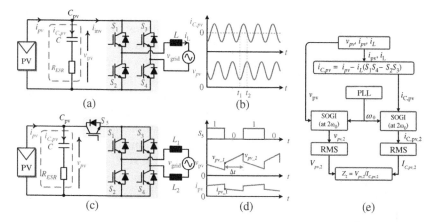

Figure 2.21 Single-phase grid-connected PV H4 inverter and H5 inverter. (a) Structure of H4 inverter. (b) Average capacitor current and voltage waveforms. (c) Structure of H5 inverter. (d) Typical waveform of PV voltage and PV current in active and zero states. (e) Implementation scheme presented in [38]. Source: Adapted from [38].

2.3.3.2 CM for C_{pv} in Single-Stage Inverter

Besides the double-stage configuration, single-stage structures are widely used in PV applications. In [37], a derived scheme-based method 1C is proposed to estimate C of C_{pv} in a single-phase grid-connected PV H4 inverter, as shown in Figure 2.21a. Here, the average capacitor current $i_{C,pv}$ is calculated as $i_{C,pv} = i_{pv} - i_{inv} = i_{pv} - i_L(S_1S_4 - S_2S_3)$, where $S_1 - S_4$ are the state of the switches. The estimated capacitor current and capacitor voltage waveform are shown in Figure 2.21b. C_{pv} is sampled at two special instants t_1 and t_2, where average capacitor current equals to 0, i.e. $i_{C,pv}(t_1) = i_{C,pv}(t_2) = 0$. Thus, the capacitance is calculated using (2.12), i.e.

$$C = \frac{\int_{t_1}^{t_2} i_{C,pv}(t)dt}{v_{pv}(t_2) - v_{pv}(t_1)}. \tag{2.27}$$

Using the same capacitor current reconstruction method, reference [38] estimates the capacitor impedance Z_2 at second harmonic frequency of a PV inverter as $Z_2 = V_{pv,2}/I_{C,pv,2}$, where $V_{pv,2}$ and $I_{pv,2}$ represent the RMS values at second harmonic frequency, as shown in Figure 2.21e. Here, a second-order generalized integrator (SOGI) extracts components of twice the grid frequency ($2\omega_0$).

Moreover, a CM scheme is proposed for a PV H5 inverter, as shown in Figure 2.21c [39]. The capacitance is calculated during the zero states, i.e. the power switch S_5 is turned off, as shown in Figure 2.21d. Using (2.12), C is calculated as $C = (i_{pv\_1} \times \Delta t)/(v_{pv\_2} - v_{pv\_1})$.

Besides 1C- and 1G-based schemes, a current injection scheme is also proposed for a grid-connected PV inverter [30], as shown in Figure 2.13e. Here, the CM of capacitors is taken during the night. At night, there is no voltage on the PV panel/string. Here, a current at hth harmonic frequency is injected into the grid, which causes $(h-1)$th and $(h+1)$th voltage and current ripples to appear on the dc-link capacitors. Using (2.9), R_{ESR} and C can be estimated using the voltage and current ripples at $(h-1)$th or $(h+1)$th based on the LMS algorithm.

Furthermore, based on the idea of power losses (i.e. method 1J), references [40, 41] proposed a power extraction efficiency (PEE) method to monitor the health state of capacitors. The PEE is defined as PEE $= P_{av}/P_{max}$, where P_{av} and P_{max} represent the average PV power and maximum PV power, which are calculated as

$$\begin{cases} P_{av} = \sum_{k=1}^{N} p_{pv}[k]/N \\ P_{max} = \sqrt{2}P_{ripp,rms} + P_{av} \end{cases} \tag{2.28}$$

where N is the total number of samples in a period of π/ω_0. $p_{pv}[k]$ is the kth sample of instantaneous PV power. It is calculated from $v_{pv}[k]$ and $i_{pv}[k]$, i.e. $p_{pv}[k] = v_{pv}[k] \times i_{pv}[k]$. $P_{ripp,rms}$ is the power of ripples, which is calculated as

$$P_{ripp,rms} = \sqrt{\sum_{k=1}^{N} (p_{pv}[k])^2/N - P_{av}^2}. \tag{2.29}$$

Then, by sampling the PV voltage and current, the PEE is calculated to estimate the health state of capacitors.

2.4 Data-Driven-based Condition Monitoring

2.4.1 Concept of Data-Driven-based CM

Recently, artificial intelligence (AI) techniques have been widely used in power electronic systems to realize the health status assessment of systems and components. Figure 2.22 shows the procedure to apply the data and AI for RUL prediction, anomaly detection, and degradation state identification [42]. Generally, two steps exist to apply AI techniques in CM, i.e. health indicators/features acquisition (Step I) and health status prediction (Step II).

For Step I, there are two categories. The first one is to directly use the electrical signals, such as voltage and current, as the input of Step II. The second one is to identify the health indicators using the neural network and AI-assisted tools. As discussed in Section 2.3.1, the nonelectrical indicators (e.g. temperature, volume, and weight) can be translated to electrical indicators (e.g. R_{ESR} and C) using

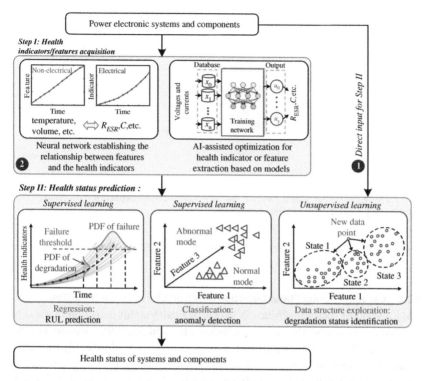

Figure 2.22 Different functions of AI in condition monitoring of power electronic systems.

the neural network. Moreover, the electrical indicators can be identified using the electrical signals (e.g. voltage and current) of power electronic systems based on AI-assisted tools.

The health status indicators obtained from Step I can be used as the input information for Step II to realize the RUL prediction and anomaly detection. Notice that the RUL prediction and anomaly detection can use the known relationship between health indicators and RUL (or anomaly status), called supervised learning. For the cases without labeled data, unsupervised learning methods (e.g. data clustering and data compression) can be used to determine the degradation states based on similarities in historical data.

2.4.2 Case Studies

Generally, the physical model-based CM schemes are dependent on the circuit topologies. Therefore, the existing CM schemes discussed in Section 2.3 are specially designed for PV converters. Unlike that, the data-driven-based techniques

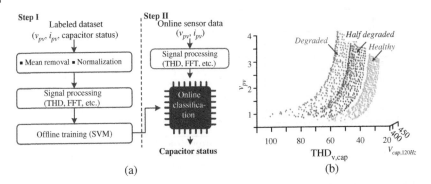

(a) (b)

Figure 2.23 Example of support vector machine (SVM)-based capacitor monitoring. (a) Monitoring procedure. (b) Labeled datasets.

are independent of the topologies of converters. Therefore, the following case studies are not specifically for PV applications.

2.4.2.1 Capacitor Health Status Detection Based on Classification Algorithms

In [43], a support vector machine (SVM)-based health status detection method is presented to distinguish the healthy, half-degraded and degraded capacitors in PV inverters (cf., Figure 2.16b). Referring to the monitoring procedure in Figure 2.23a, it mainly includes two steps, i.e. offline training (Step I) and online identification (Step II). In Step I, the labeled training and testing datasets were generated based on a simulation model. Here, three tuples of $\{C, R_{ESR}\}$ values are used to represent the healthy, half-degraded, and fully degraded capacitors. The relationship between capacitor status and electrical signals can be obtained using the SVM training algorithm, as shown in Figure 2.23b. In Step II, the obtained training results are imported to an MCU. Then, online monitoring can be implemented.

Similarly, reference [44] presents a convolutional neural network (CNN)-based classification method to realize the health status monitoring of dc-link capacitors in three-phase inverters, as shown in Figure 2.24a. Firstly, the sampled time-domain current signal is transformed into time-frequency domain images using a continuous wavelet transform (CWT), as shown in Figure 2.24b. Then, these images are classified into one of five capacitor health categories using CNNs. Finally, a hidden Markov model (HMM) algorithm improves the accuracy of classification results, as shown in Figure 2.24c.

2.4.2.2 Capacitance Estimation Based on Intelligent Algorithms

Besides anomaly monitoring, intelligent algorithms, such as artificial neural networks (ANN) and PSO, are widely used in parameter estimation. Figure 2.25 gives the typical structure of an intelligent algorithms-based capacitor monitoring

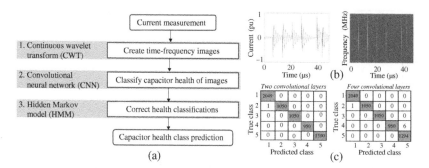

Figure 2.24 Example of convolutional neural network (CNN)-based capacitor monitoring. (a) Monitoring procedure. (b) Time-frequency signal. (c) Classification results.

Figure 2.25 Intelligent algorithm-based CM of capacitors.

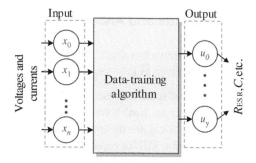

scheme. Usually, the input signals are the existing signals of a power electronic converter, such as input voltage, input current, output voltage, and output current. The output is the target parameters, such as R_{ESR} and C.

In [45], ANN algorithm-based schemes are proposed to estimate the capacitor parameters of the adjustable speed drive (ASD) system. The implementation example is shown in Figure 2.25. The key issue of these schemes is to obtain and define a suitable training data set. To accurately obtain the relationship between capacitor parameters and input training data (such as input current, input voltage, dc-link voltage, output current, and output voltage), different capacitance and load conditions must be considered in the training process. Similarly, an adaptive neuro-fuzzy inference system (ANFIS) algorithm and a PSO-based support vector regression (PSO-SVR) are employed to train data to monitor the capacitor health state in [46, 47].

2.4.2.3 RUL Prediction Based on Intelligent Algorithms

In [48], the ANFIS algorithm predicts the RUL of capacitors. The structure is similar to that in Figure 2.25. However, the input signals are the dataset of R_{ESR} and C. The output signal is the predicted RUL of capacitors. Taking six tuples of $\{C, R_{ESR}\}$

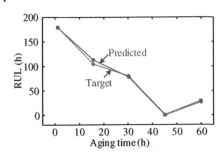

Figure 2.26 Comparison of the Predicted Results and Target Results.

values as input of the training network, Figure 2.26 shows the predicted results. It can be seen that the predicted results are consistent with the target results.

2.5 Results and Analysis

Table 2.5 summarizes the CM schemes for capacitors in PV applications. The remarks are given as follows:

1) The derived schemes based on 1A and 1E (i.e. capacitor voltage-based schemes) can predict the health status of capacitors without capacitor current sensors. However, there are no specific values of R_{ESR} and C, which is challenging to calculate the RUL of capacitors. Therefore, these schemes are not considered in the actual PV applications.

2) The derived schemes based on 1B and 1F in [33, 34] can realize the CM of C_{pv} and C_{dc} in PV boost converters. However, capacitor current sensors and high-frequency sampling devices are required.

3) The converter's operation model schemes (based on 1C and 1G) can realize the CM using the relationship between voltage ripple and current ripple. Although additional capacitor current sensors are not required, additional ripple extraction circuits are usually needed to sample the high-frequency ripples of boost converters [35–37]. For single-stage inverters [38–40], the capacitor current estimation schemes (based on 1C, 1G) are real-online schemes without additional hardware circuits and current sensors. However, [39] is only suitable for PV H5 inverter, which has a low applicability

4) For PV boost converters, the calculation model of the signal injection scheme (based on 1D, 1H) in [36] is simple and the sampling frequency is low. However, a simple ripple extraction circuit is required, and there are no specific values of R_{ESR} and C. For single-stage inverters, the signal injection scheme in [30] (based on 1D, 1H) is a quasi-online scheme with relatively high estimation accuracy. However, the CM is implemented during the night and an additional power diode is required.

Table 2.5 Summary of CM schemes for PV applications

Capacitors	Methods	Exp. error	Advantages and limitations
C_{pv}, C_{dc} (PV boost)	1B, 1F	R_{ESR}: <4.9% [33] C: <5.4% [33]	• Real-online monitor • Additional capacitor current sensor, high-frequency sampling, not suitable for converter loads
C_{dc} (double-stage configuration)	1C, 1G	N/A [35]	• Real-online monitor, no capacitor current sensor • Estimated i_C, additional ripple extraction circuit in actual applications
C_{pv} (PV boost)	1C, 1G	R_{ESR}: <6.2% [4]	• Real-online monitor, low-frequency sampling, no capacitor current sensor • Addition hardware circuit
C_{pv} (PV boost)	1D, 1H	Z: <0.64% [36]	• Real-online monitor, no capacitor current sensor • External signal injection, additional circuits, no specific values of R_{ESR} and C
C_{pv} (PV inverter)	1C, 1G	C: <2.56% [37]	• Real-online estimation, no additional hardware circuits and current sensors. • Estimated i_C.
C_{pv} (PV inverter)	1C, 1G	Z: <17.2% [38]	• Real-online estimation, no additional hardware circuits and current sensors • Estimated i_C, low estimation accuracy, no specific values of R_{ESR} and C, specially defined failure criterion.
C_{pv} (PV inverter)	1C, 1G	N/A [39]	• No additional sensors, real-online estimation, no additional hardware circuits and current sensors • Additional ripple extraction circuit in actual applications, only suitable for H5 inverter
C_{pv} (PV inverter)	1D, 1H	R_{ESR}: <3.65% [30] C: <1.88%	• Quasi-online estimation, no input source (at night), additional diode • External signal injection
C_{pv} (PV inverter)	1J	N/A [40, 41]	• Real-online estimation, no additional hardware circuits and current sensors • No specific values of R_{ESR} and C, specially defined failure criterion.
C_{pv} (PV inverter)	Data-driven	N/A [43]	• Real-online estimation, no additional hardware circuit and current sensors • Need a large amount of training data and complexly algorithms.

5) The power losses schemes in [40, 41] (based on 1J) can real-online monitor the health status of capacitors without additional hardware circuits and current sensors. However, specially defined failure criteria are required.
6) In the data-driven-based schemes, additional sensors and hardware circuits are not required. However, complexly training algorithms and a large amount of training data are required, which is the maximum challenge.

2.6 Summary

The aging detection of dc-link capacitors has great significance in enhancing the reliability of the power electronic converters in PV applications. This chapter summarizes the wear-out failure characteristics of capacitors, including degradation models and end-of-life criteria. Based on this, the physical-model and data-driven-based CM schemes for PV applications are summarized. Future research opportunities are summarized from the authors' point of view are given in the following:

1) Further research in wear-out mechanisms of capacitors to obtain an accurate degradation model and end-of-life criteria based on nonelectrical parameters.
2) Designing accurate CM schemes for real industrial applications (working in harsh EMI environments) without additional hardware cost and reliability risk. Future studies need to improve estimation accuracy while reducing the complexity of sampling data and algorithms.
3) Monitoring capacitors' status while monitoring other key components in converters, such as semiconductor switches.
4) Designing low-cost and high-accuracy CM schemes suitable for all types of dc-link applications.
5) Designing emerging capacitors with built-in monitoring components such as internal thermal and pressure sensors.
6) Applying the state-of-the-art computation-light AI tools to improve the computation efficiency.

References

1 Yang, Y., Ma, K., Wang, H., and Blaabjerg, F. (2014). Mission profile translation to capacitor stresses in grid-connected photovoltaic systems. *2014 IEEE Energy Conversion Congress and Exposition (ECCE)*, Pittsburgh, PA, USA (14–18 September 2014), pp. 5479–5486.

2 Gupta, Y., Ahmad, M.W., Narale, S., and Anand, S. (2019). Health estimation of individual capacitors in a bank with reduced sensor requirements. *IEEE Transactions on Industrial Electronics* 66 (9): 7250–7259.

3 Araujo, S., Zacharias, P., and Mallwitz, R. (2010). Highly efficient single phase transformerless inverters for grid-connected photovoltaic systems. *IEEE Transactions on Industrial Electronics* 57 (9): 3118–3128.

4 Ahmad, M.W., Agarwal, N., and Anand, S. (2016). Online monitoring technique for aluminum electrolytic capacitor in solar PV-based DC system. *IEEE Transactions on Industrial Electronics* 63 (11): 7059–7066.

5 Zhao, Z., Davari, P., Lu, W., and Blaabjerg, F. (2022). Online dc-link capacitance monitoring for digital-controlled boost PFC converters without additional sampling devices. *IEEE Transactions on Industrial Electronics* 70 (1): 907–920.

6 Wang, H. and Blaabjerg, F. (2014). Reliability of capacitors for dc-link applications in power electronic converters—an Overview. *IEEE Transactions on Industry Applications* 50 (5): 3569–3578.

7 Rubycon Corporation (2021). Film capacitor technical notes, November 2021. http://www.rubycon.co.jp/en/products/film/technote.html (accessed 10 August 2022)

8 TDK Corporation (2021). Capacitors selection guide, November 2021. https://product.tdk.com/en/products/selectionguide/capacitor.html (accessed 10 August 2022).

9 CDE Cornell Dubilier (2020). Type SLPX 85 °C snap-in Aluminum electrolytic, May 2020. http://www.cde.com/resources/catalogs/SLPX.pdf (accessed 10 August 2022)

10 TDK Corporation (2020). Film capacitors—metallized polypropylene film capacitors, May 2020. https://www.tdk-electronics.tdk.com/inf/20/20/db/fc_2009/MKP_B32774_778.pdf (accessed 10 August 2022)

11 Zhao, Z., Davari, P., Lu, W., Wang, H., and F. Blaabjerg (2021). An overview of condition monitoring techniques for capacitors in dc-link applications. *IEEE Transactions on Power Electronics* 36 (4): 3692–3716.

12 TDK Corporation (2020). Film capacitors-general technical information, March 2020. https://www.tdk-electronics.tdk.com/download/530754/480aeb04c789e45ef5bb9681513474ba/pdf-generaltechnicalinformation.pdf (accessed 10 August 2022).

13 Nichicon Corporation (2021). General description of aluminum electrolytic capacitors, November 2021. https://www.nichicon.co.jp/english/products/pdf/aluminum.pdf (accessed 10 August 2022).

14 Jianghai Corporation (2021). Film capacitors-technical notes, November 2021. http://www.jianghai.com/file/upload/2020/03/09/1583731313744.pdf (accessed 10 August 2022).

15 Gupta, A., Yadav, O. P., DeVoto, D., and Major, J. (2018). A review of degradation behavior and modeling of capacitors. *Proceedings of InterPACK 2018,* Hilton San Francisco, California, USA (28–30 August 2018).

16 R. Celaya, C. Kulkarni, G. Biswas, S. Saha, and, K. Goebel, "A model-based prognostics methodology for electrolytic capacitors based on electrical overstress accelerated aging", *Proceedings of Annual Conference of the PHM Society,* vol. 3, no. 1, 2011.

17 Sun, B., Fan, X., Qian, C., and Zhang, G. (2016). PoF-simulation-assisted reliability prediction for electrolytic capacitor in LED Drivers. *IEEE Transactions on Industrial Electronics* 63 (11): 6726–6735.

18 Wang, H., Nielsen, D., and Blaabjerg, F. (2015). Degradation testing and failure analysis of DC film capacitors under high humidity conditions. *Microelectronics Reliability* 55 (9–10).

19 Gulbrandsen, S., Arnold, J., Kirsch, N., and Caswell, G. (2014). A new method for testing electrolytic capacitors to compare life expectancy. *Proceedings Additional Conference (Device Package, HiTEC, HiTEN, & CICMT)* (January 2014), pp. 1759–1786.

20 Kulkarni, C., Celaya, J., Goebel, K., and Biswas, G. (2012). Physics based electrolytic capacitor degradation models for prognostic studies under thermal overstress. *Proceedings of the European 1st Conference of the Prognostics and Health Management Society,* Dresden, Germany (3–5 July 2012).

21 Dou, Z., Rong, X., Alfonso, B., Javaid, Q., and Cynthia, P. (2010). "Performance of aluminum electrolytic capacitors and influence of aluminum cathode foils", *CARTS Europe,* Munich, Germany (10–11 November 2010).

22 Shrivastava, A., Azarian, M.H., Morillo, C. et al. (2014). Detection and reliability risks of counterfeit electrolytic capacitors. *IEEE Transactions on Reliability* 63 (2): 468–479.

23 Li, Z., Li, H., Qiu, T., Lin, F. and Wang, Y., Moisture induced weight gain relation to capacitance loss characteristic of film capacitors under harsh environment. *2021 IEEE 4th International Electrical and Energy Conference (CIEEC),* Wuhan, China (28–30 May 2021), pp. 1–4.

24 Vogelsberger, M.A., Wiesinger, T., and Ertl, H. (2011). Life-cycle monitoring and voltage-managing unit for dc-link electrolytic capacitors in PWM converters. *IEEE Transactions on Power Electronics* 26 (2): 493–503.

25 Chen, Y.M., Wu, H.C., Chou, M.W., and Lee, K.Y. (2011). Online failure prediction of the electrolytic capacitor for *LC* filter of switching-mode power converters. *IEEE Transactions on Industrial Electronics* 55 (1): 400–406.

26 Imam, A., Habetler, T., Harley, R., and Divan, D. (2005). LMS based condition monitoring of electrolytic capacitor. *31st Annual Conference of IEEE Industrial Electronics Society,* Raleigh, NC (6–10 November 2005), pp. 848–853.

27 Sundararajan, P., Sathik, M.H.M., Sasongko, F. et al. (2020). Condition monitoring of dc-link capacitors using goertzel algorithm for failure precursor parameter and temperature estimation. *IEEE Transactions on Power Electronics* 35 (6): 6386–6396.

28 Yao, K., Tang, W., Hu, W., and Lyu, J. (2015). A current-sensorless online ESR and C identification method for output capacitor of buck converter. *IEEE Transactions on Power Electronics* 30 (12): 6993–7005.

29 Amaral, A. and Cardoso, A. (2009). A simple offline technique for evaluating the condition of aluminum electrolytic capacitors. *IEEE Transactions on Industrial Electronics* 56 (8): 3230–3237.

30 Agarwal, N., Ahmad, M.W., and Anand, S. (2018). Quasi-online technique for health monitoring of capacitor in single-phase solar inverter. *IEEE Transactions on Power Electronics* 33 (6): 5283–5291.

31 Zhao, Z., Davari, P., Wang, Y., and Blaabjerg, F. (2021). Online capacitance monitoring for dc/dc boost converters based on low-sampling-rate approach. *IEEE Journal of Emerging and Selected Topics in Power Electronics*, online published (27 August 2021), doi: 10.1109/JESTPE.2021.3108420.

32 Zhao, Z., Lu, W., Davari, P. et al. (2021). An online parameters monitoring method for output capacitor of buck converter based on large-signal load transient trajectory analysis. *IEEE Journal of Emerging and Selected Topics in Power Electronics* 9 (4): 4004–4015.

33 Miao, W., Lam, K.H., and Pong, P.W.T. (2020). Online monitoring of aluminum electrolytic capacitors in photovoltaic systems by magnetoresistive sensors. *IEEE Sensors Journal* 20 (2): 767–777.

34 Miao, W., Liu, X., Lam, K.H., and Pong, P.W.T. (2019). Condition monitoring of electrolytic capacitors in boost converters by magnetic sensors. *IEEE Sensors Journal* 19 (22): 10393–10402.

35 Sepehr, A., Saradarzadeh, M. and Farhangi, S. (2016). A noninvasive on-line failure prediction technique for aluminum electrolytic capacitors in photovoltaic grid-connected inverters. *7th Power Electronics and Drive Systems Technologies Conference (PEDSTC)*, Tehran, Iran (16–18 Februay 2016), pp. 356–361.

36 Ahmad, M.W., Agarwal, N., Kumar, P.N., and Anand, S. (2017). Low-frequency impedance monitoring and corresponding failure criteria for aluminum electrolytic capacitors. *IEEE Transactions on Industrial Electronics* 64 (7): 5657–5666.

37 Ahmad, M.W., Kumar, P.N., Arya, A., and Anand, S. (2018). Noninvasive technique for dc-link capacitance estimation in single-phase inverters. *IEEE Transactions on Power Electronics* 33 (5): 3693–3696.

38 Arya, A., Ahmad, M.W., Agarwal, N., and Anand, S. (2017). Capacitor impedance estimation utilizing dc-link voltage oscillations in single phase inverter. *IET Power Electronics* 10 (9): 1046–1053.

39 Agarwal, N., Ahmad, M. W., and Anand, S. (2016). Condition monitoring of dc-link capacitor utilizing zero state of solar PV H5 inverter. *10th International Conference on Compatibility, Power Electronics and Power Engineering (CPE-POWERENG)*, Bydgoszcz, Poland (29 June–1 July 2016), pp. 174–179.

40 A. Arya, Md. W. Ahmad, and S. Anand, "Online monitoring of power extraction efficiency for minimizing payback period of solar PV system," in *IEEE International Conference on Industrial Technology (ICIT)*, Seville, Spain, March 2015, pp. 2863–2868.

41 Agarwal, N., Arya, A., Ahmad, M.W., and Anand, S. (2016). Lifetime monitoring of electrolytic capacitor to maximize earnings from grid-feeding PV system. *IEEE Transactions on Industrial Electronics* 63 (11): 7049–7058.

42 Zhao, S. and Wang, H. (2021). Enabling data-driven condition monitoring of power electronic systems with artificial intelligence: concepts, tools, and developments. *IEEE Power Electronics Magazine* 8 (1): 18–27.

43 Seferian, V., Bazzi,m A., and Hajj, H. (2020). Condition monitoring of dc-link capacitors in grid-tied solar inverters using data-driven techniques. *IEEE Energy Conversion Congress and Exposition (ECCE)*, Detroit, Michigan, USA (11–15 October 2020), pp. 5318–5323.

44 McGrew, T., Sysoeva, V., Cheng, C.-H., and Scott, M. (2021). Condition monitoring of DC-link capacitors using hidden Markov model supported-convolutional neural network. *IEEE Applied Power Electronics Conference and Exposition (APEC), APEC 2021 Virtual Conference + Exposition* (14–17 June 2021), pp. 2323–2330.

45 Soliman, H., Abdelsalam, I., Wang, H., and Blaabjerg, F. (2017). Artificial neural network based DC-link capacitance estimation in a diode-bridge front-end inverter system. *IEEE 3rd International Future Energy Electronics Conference and ECCE Asia (IFEEC 2017 – ECCE Asia)*, Kaohsiung, Taiwan (3–7 June 2017), pp. 196–201.

46 Kamel, T., Biletskiy, Y., and Chang L. (2015). Capacitor aging detection for the DC filters in the power electronic converters using ANFIS algorithm. *IEEE 28th Canadian Conference on Electrical and Computer Engineering (CCECE)*, Halifax, Nova Scotia, Canada (3–6 May 2015), pp. 663–668.

47 Abo-Khalil, A.G., Alyami, S., Alhejji, A., and Awan, A.B. (2019). Real-time reliability monitoring of dc-link capacitors in back-to-back converters. *Energies* 12 (12): 2369.

48 Jamshidi, M. B. and Alibeigi, N. (2017). Neuro-fuzzy system identification for remaining useful life of electrolytic capacitors. *2nd International Conference on System Reliability and Safety (ICSRS)*, Milan, Italy (20–22 December 2017), pp. 227–231.

3

Photovoltaic Module Fault. Part 1: Detection with Image Processing Approaches

V S Bharath Kurukuru and Ahteshamul Haque

Advance Power Electronics Research Lab, Department of Electrical Engineering, Jamia Millia Islamia (A Central University), New Delhi, India

3.1 Overview

This chapter presents an efficient fault classification technique for monitoring the condition of photovoltaic (PV) modules. The proposed approach aims at early and efficient detection of fault to achieve reliable operation for solar PV modules. Generally, infrared thermography inspections can give meaningful support to identify the faults and assess the quality and performance of PV modules. However, the defect detection principle operates on the basis of heat radiated from the PV modules, and the impact of other heat-emitting objects that are nearby the panels is considered a major source of misinterpretation and false identification of the faults. Therefore, there is a need to eliminate the unwanted elements or any external noises from the captured thermal image. In this paper, a module fault classification algorithm is developed by adapting a neural network classifier. Initially, the thermal images of different module faults are captured and then preprocessed to train with the classifier. Further, in the testing stage or while performing real-time monitoring, an image processing algorithm developed using edge detection and Hough transform techniques is adapted. This eliminates the background information in the thermal image ad aids for efficient fault classification. The processed images are further subjected to feature extraction and passed through the trained classifier for localization and identification of the type of fault. The experiment results depicted 96.2% testing accuracy when tested with three different fault samples.

Fault Analysis and its Impact on Grid-connected Photovoltaic Systems Performance, First Edition.
Edited by Ahteshamul Haque and Saad Mekhilef.

3.2 Background Information

There are many types of faults that can be identified through a thermal survey inspection and each of them has a unique indication in the thermal image. In the case there is a faulty PV module, the warm areas will show up clearly in the thermal image. Depending on the shape and location, these hot spots and areas can indicate different types of faults. The main thermal abnormalities in PV power plants that can be detected by thermal diagnosis are [1]:

a. Heated bypass diodes inside the PV module junction box.
b. Defective PV module.
c. Hot cells formally named as hot spots by
   - Breakage of front glazing
   - External shading
   - Internal cell problems
d. Entirely heated string made of several PV modules.

In general, the emissivity of the surface of a material depends on the energy emitted as thermal radiation. Different materials have different thermal emissivities, for example, the emissivity value for the front cover (glass) of a PV module is typically 0.85 [2]. Hence, some important characteristics regarding thermography for PV applications need to be considered as follows: The image resolution is an important characteristic as it deals with pixels, which contain the data acquisition points for thermal measurements, and these data are used to create a thermal image from the thermal profile. The more pixels and data points per investigated area, the more accurate the thermal interpretation will be. In addition, the noise equivalent temperature difference or thermal sensitivity is considered to be of great importance. Furthermore, there is a significant problem with the glass module cover in PV module thermographic analysis due to the fact that the surface is extremely reflective and behaves differently for different wavelengths [3]. In fact, there are different types of glass with different opacity for various wavelengths and most PV modules are made with a glass cover, which minimalizes the opacity in the infrared range, enabling heat to escape. Due to the aforementioned, it is suggested that the viewing angle of the PV module by the thermal camera is between 5° and 60°, where 0° is perpendicular to the PV module plane [3]. But this process is difficult to adapt while dealing with large PV plants or while using aerial thermography. Hence more efficient methods are required for efficient classification and localization of module faults.

Infrared thermography inspections proved to be a meaningful support to assess the quality and performance of PV modules [4]. Aside from that, other technical inspection methods are electroluminescence (EL) imaging [5] and I–V curve tracing [6]. Some common defects, e.g. potential induced degradation (PID) [7], inactive cells, hot spots, and microcracks, can be found by applying these mentioned methods [8]. However, the implementation of a cost-effective method to scan and check huge PV plants represents different challenges, such as the cost and time of detecting PV module defects with their classification and exact localization within the solar plant. Some PV plants are too large and time-consuming to be inspected by manned technologies, especially if they are located on a building roof or in difficult access areas. Furthermore, current methods of PV inspections are not able to provide online information about failures or defects in the monitored plants and most of them use a lot of time only for the data acquisition task, without taking into account the subsequent important analysis steps [9].

Besides, different power electronics converters are used for utilizing the generated PV power with domestic loads and integrating them with grid-connected systems. But, due to the effect of the module degradation and hidden faults, the module area changes resulting in mismatch losses, where a PV array's available maximum power output is lower than the sum of each module's theoretical maximum power output [10, 11]. These mismatch losses have a huge impact on the operation of power electronics converters resulting in power loss, frequent load switching, and failure of power electronic converters [12]. Hence, it is important to identify the change in the area of the module due to degradation and faults and evaluate their effect on the performance of the system.

Generally, the fault detection and degradation analysis process is based on manual visual inspections [13], voltage-current characteristics (V–I curve)-based field measurements [14, 15], thermal evaluations using infrared imaging [16], and a combination of IV curves and thermal behaviors of selected individual modules from an array [17]. As most of these inspection processes are practiced manually, the chances of identifying all the anomalies are very low. This resulted in enhanced risks for failure and degradation of the modules.

Further, automatic inspection and monitoring systems, which deal with vision-based, and infrared thermography aided with different image processing techniques are performed to achieve efficient inspection of the modules [18, 19]. Both of these methods lacked in specific areas with visual inspection failing to identify the internal changes and faults in a module and thermography failing due to calibration and ambient metrological conditions [16, 20]. In [21], the unmanned aerial vehicle-based thermal imaging for degradation and fault analysis of PV modules is achieved at reduced monitoring cost and time. From the results, it is observed that the developed research does not identify the effect of degraded panel area on the operation of the array or the string connected to the inverter. Further, in [22], the EL degradation analysis is performed to estimate the performance of PV modules. The developed approach estimated the relative power loss caused due to PID effect by calculating the logarithmic ratio of EL images captured at different applied currents. In all the image-based degradation analysis methods, the effect of background information while capturing images is considered a major drawback. Apart from the image-based approaches, model-based degradation analysis methods are discussed in [23–25]. These methods employ series resistance estimation based on a single diode model [23], a data-based approach considering the statistical model [24], and comprehensive modeling through dry deposition [25] to analyze the degradation of PV modules. The main disadvantage of these approaches is that their performances are not validated or analyzed rigorously with different degradation datasets.

In light of these drawbacks and requirements, this work initially develops a fault classification algorithm using a large dataset of images divided into five different fault classes. The classification algorithm involves image preprocessing through the padding process and feature extraction through texture features for training the classifier. A multilayer perceptron neural network (MLPNN) classifier is used as the training classifier. Further, in the monitoring process, the acquired images are preprocessed using edge detection [26] and Hough transform [27]-based image-processing techniques to extract the required features and test them with the trained algorithm. Thereafter, a fusion of visual imaging and thermal imaging information is proposed for the monitoring of PV panels. The visual imaging system analyzes the irregular irradiation surface while the thermal imaging system analyzes the irregular temperature distribution on a PV panel. As the image acquisition devices capture the PV modules and some background information [28], the region of interest is specified with the help of a polygon selection. This forms a localized PV panel in the acquired image and helps in extracting the histogram-based statistical features. Further, an optimal threshold value to differentiate the foreground and background [29, 30] of the captured image is calculated and combined with the segmented image information to find out the PV panel surface. Besides, the features extracted from the localized panel surface are analyzed to correlate with performance degradation. Further, by reviewing the advantages of correlating image-based approaches and electrical characteristics of PV modules for degradation analysis in [17, 31, 32], mathematical relations are established between image feature variation and performance degradation of electrical characteristics of the module. These electrical characteristics estimate the mismatch losses due to degrading areas that will affect the operation of the PV system by correlating the module area and power output using linear regression analysis. The novelty of the proposed lies in the following aspects.

3.3 Fault Classification Approach

To develop the classifier, firstly, thermal images must be sorted and matched according to the fault type. For a non-defective PV module, its temperature should be relatively uniform, with no areas of significant temperature difference.

Although, the PV module will be warmer around the junction box in comparison to the rest, since there is more current concentration in this area and heat does not dissipate so well to the surrounding environment. A thermographic image with abnormalities should show at minimum one whole module, pointing out the position of the junction box and the lower edge within the installation [33]. Further, different fault types have different patterns and it is important to identify these differences for developing a fault classifier. Hence, a feature extraction algorithm is adapted to identify the fault patterns in a thermal image. The textural feature extraction aims to capture the granularity and pictorial patterns within an image by transforming image data into quantifiable information. This can be done through different methods which can be categorized mainly into four groups [34, 35]: statistical – co-occurrence and autocorrelation features, geometrical – structural and polygonal features, model – random field and fractal parameters, and signal processing – spatial frequency and filter banks and wavelet transform. Aside from the geometrical approach, the three other categories were applied for this study. The statistical parameters such as mean gray-level intensity, standard deviation, entropy, and histogram are basic features that can easily be extracted from an image. In general, computation of a gray-level co-occurrence matrix (GLCM) is standard practice for the extraction of textural characterization [36]. Generally, these gray-level occurrences are calculated at a given displacement and angle. The algorithm for GLCM feature extraction is discussed in [37] and the feature extraction process for thermal images has been discussed in [36, 38]. These features form a feature vector, which is further trained with their corresponding classes to develop a classifier [39].

3.3.1 Algorithm Development

The fault classification approach for PV modules is developed by considering six different faults and operating conditions. One condition is directed toward a normal operating module, and the other five conditions corresponds to five different faults. In this research, the different fault types (bypass diode, cell, string, and hot spots) were identified and their corresponding thermal images were captured from the experiment site. The inspection of PV plants is performed under a thermal steady-state condition [40]. Table 3.1 shows the required environmental conditions for inspections to be performed.

The cloud coverage should be given in okta, which means part of 8 of cloud coverage, with "8 okta = full cloud coverage" and "1 okta = no cloud coverage at all" [13].

Total 21 image samples corresponding to 6 different faults are captured for processing and training to achieve the fault classification mechanism. The resolution of each captured image is 280 * 500 pixels. Further, these captured images are

Table 3.1 Required inspection conditions.

Parameter	Limits
Irradiance	Minimum 600 W/m^2 in the plane of the PV module
Wind speed	Maximum 8 m/s
Cloud coverage	Maximum 2 okta

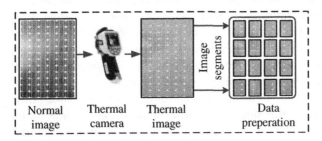

Normal image   Thermal camera   Thermal image   Image segments   Data preperation

Figure 3.1 Acquired thermal images are divided into multiple subregions. Source: tanokm/123RF.

divided into equal subregions as shown in Figure 3.1 to isolate the fault area in a module from the healthy area. This process is achieved by inheriting for loop and padding process.

Provided an image, the loop and padding process is applied to get 14×14 sub-regions with each subregion having a resolution of 25×35 pixels. This resolution is identified to provide all the image characteristics without any depreciation of feature quality and also reduces the time taken. This provides 196 subregions of an image and the training process is achieved with 4116 such subregions. Immediately after the subregions are achieved, the GLCM feature extraction is applied to extract seven different features. The energy, entropy, contrast, homogeneity, correlation, shade, and prominence features are extracted for all the image subregions to form a feature vector of size 7×4116. The obtained feature vector is tabulated along with the corresponding fault and normal operating classes to train with the classifier.

3.3.2 Training of Classifier

A feature matrix of size 7×4116 with 7 features and 4116 samples for each feature corresponding to 6 different fault types and operating conditions is considered for training with the MLPNN classifier. Further, the neural network is formed with 12 neurons in a hidden layer, which form a structure of 7–12–6, where the input layer has 7 neurons, hidden layer has 12 neurons, and output layer has 6 neurons. The scaled conjugate gradient algorithm [41] is applied to achieve the training and the performance is estimated by the means of cross-entropy, mean square error, and error percentage. The training session is scheduled for 1000 epochs, but based

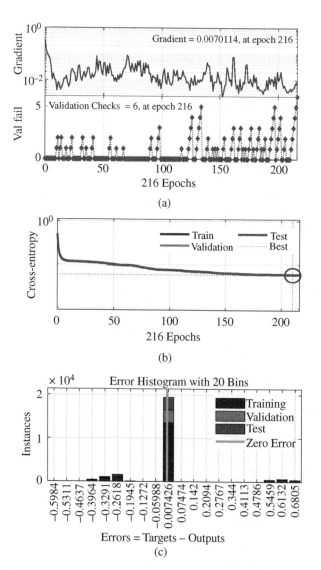

Figure 3.2 Results with classifier training. (a) Validation of training states. (b) Convergence of the training process. (c) Error histogram for different training instances

on the convergence rate and validation checks performed by the algorithm, the training is terminating 210 epochs. The training results of the classifier are shown in Figure 3.2.

After training for 210 epochs, the cross-entropy for best validation performance is observed to be 0.058849. The results in Figure 3.2a plot the gradients and validation checks during the training process. On a logarithmic scale, the gradient value at 216 epochs is observed to be 0.0070114. Thereafter, the performance of the training approach along with testing and validation performance and best performance is shown in Figure 3.2b. From the results, it is identified that after 166 epochs, the cross-entropy error converges gradually around 0.058849. Furthermore, the error histogram of the training process is plotted in Figure 3.2c. The plot indicates that the training error lies in a range of −0.5984 and 0.6805. In addition to the above results, an overall confusion matrix for the training, testing, and validation process is also plotted as shown in Figure 3.3.

Along the diagonal of the truly classified samples, the classification accuracy is observed to be 95.9%. This accuracy is estimated as an average of the training accuracy (95.7%), the validation accuracy (97.8%), and the testing accuracy (95%). Further, a brief comparison of the training accuracies of the trained algorithm with other conventional approaches is depicted in Table 3.2.

3.3.3 Algorithm Testing

The proposed testing process is depicted in Figure 3.4. In order to test the trained classifier, three different thermal images corresponding to different faults in

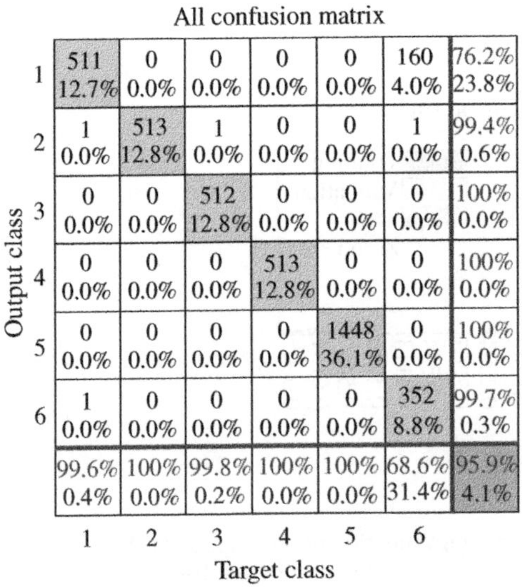

All confusion matrix

Output class	1	2	3	4	5	6	
1	511 / 12.7%	0 / 0.0%	0 / 0.0%	0 / 0.0%	0 / 0.0%	160 / 4.0%	76.2% / 23.8%
2	1 / 0.0%	513 / 12.8%	1 / 0.0%	0 / 0.0%	0 / 0.0%	1 / 0.0%	99.4% / 0.6%
3	0 / 0.0%	0 / 0.0%	512 / 12.8%	0 / 0.0%	0 / 0.0%	0 / 0.0%	100% / 0.0%
4	0 / 0.0%	0 / 0.0%	0 / 0.0%	513 / 12.8%	0 / 0.0%	0 / 0.0%	100% / 0.0%
5	0 / 0.0%	0 / 0.0%	0 / 0.0%	0 / 0.0%	1448 / 36.1%	0 / 0.0%	100% / 0.0%
6	1 / 0.0%	0 / 0.0%	0 / 0.0%	0 / 0.0%	0 / 0.0%	352 / 8.8%	99.7% / 0.3%
	99.6% / 0.4%	100% / 0.0%	99.8% / 0.2%	100% / 0.0%	100% / 0.0%	68.6% / 31.4%	95.9% / 4.1%

1 2 3 4 5 6
Target class

Figure 3.3 Analysis of training accuracy.

Table 3.2 Training results of different classifiers for same sample data.

Classifier	Training time (s)	Training accuracy (%)	Test time (s)	Test accuracy (%)
K-nearest neighbor	1.36	67.92	0.0196	61.12
Support vector machine	0.93	81.35	0.0167	89.63
Neural network	0.32	95.7	0.0013	95

multiple PV strings are captured as shown in Figure 3.5. The captured images represent a discrete palette, where the fault is identified by a brighter color. Further, it can be observed that there are other heat-emitting objects in the image, especially the frames which exhibit the same color as a fault. This results in misclassification while testing with the trained classifier. Hence, the captured images need to be preprocessed for better classification.

The first step of preprocessing involves converting the captured images into grayscale as depicted in Figure 3.6. This helps in identifying the important features and edges in the thermal image. Further, the lower complexity of gray-level image highlights the contours, texture, shadows and so on without addressing the color. Figure 3.6a–c shows the gray-scale representation of thermal images in Figure 3.5a–c.

Once the grayscale of the thermal image is obtained, the background information is eliminated by performing edge detection. The Canny edge detection method discussed in the above section is used to perform the process. Initially, the Gaussian filtration is applied using Canny edge detection [42, 43] to remove the unwanted details and reduce the noise. This eliminates the unwanted textures and provides the edge details of the grayscale images in Figure 3.6. Further, the strength of the edges is identified by computing the gradient of the filtered image. In order to prevent major loss of information during the filtration process, a threshold is applied. The threshold secures all the edge elements while the noise is being suppressed. In the final step, the non-maxima pixels are suppressed in order to maintain thin edge ridges. The edge images for grayscale images in Figure 3.6 are shown in Figure 3.7a–c. In the second step of preprocessing, the Hough transform is created for the edge-detected images in Figure 3.7. As it can be identified from Figure 3.7 that the edge-detected image does not possess any information about the quantity and identity of the features within the edge boundary. Hence, the HT identifies the lines and peaks in the edge-detected image for detecting the geometric structure of the faulty panel as depicted in Figure 3.8a–c.

Further, the geometric coordinates of the faulty panel are identified by using the Hough peaks and lines. The Hough peaks and lines extract the segments in

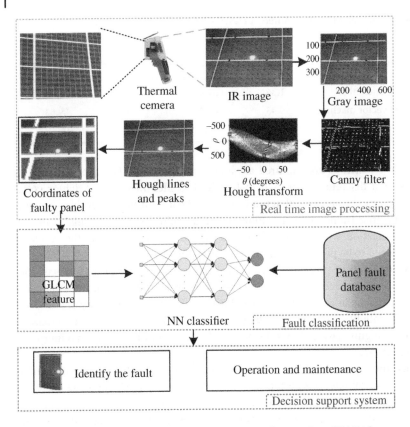

Figure 3.4 Fault classification process. Source: Fluke Corporation; CAN·UAS.

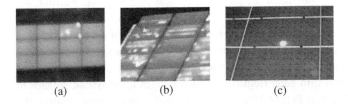

    (a)   (b)   (c)

Figure 3.5 Acquired thermal image for testing. Source: (c) CAN·UAS.

the binary image and depict the orientation of modules as shown in Figure 3.9a–c. The symbol X in yellow color determines the starting point of the panel area and the symbol X in red color depicts the ending point of the panel area. These points are formulated by the Hough peaks, whereas the Hough lines define the boundary around the panel area.

Figure 3.6 Grayscale of the captured thermal image.

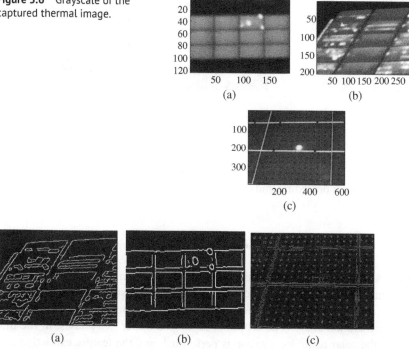

(a)

(b)

(c)

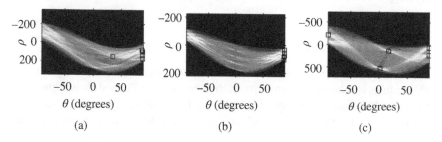

(a)   (b)   (c)

Figure 3.7 Canny-filtered image (binary image).

(a)   (b)   (c)

Figure 3.8 Hough transform applied to the filtered image.

Further, based on the coordinates identified, the image is segmented into sub-regions in such a way that each subregion has a pixel value of 25×35. These sub-regions are subjected to GLCM feature extraction and the extracted features are tabulated to test with the classifier developed earlier. The results identified faults in Figure 3.5a as a panel with multiple hotspots, the fault in Figure 3.5b is module failure and the fault in Figure 3.5c is identified as a warm cell. The results

(a)   (b)   (c)

Figure 3.9 Coordinates identification with the filtered image.

classified the module fault with 96.2% efficiency when tested with the developed image detection procedure and classification process.

3.4 Panel Area Degradation Analysis

A block diagram of the proposed solar panel health-monitoring system is shown in Figure 3.10. The prior knowledge of solar panel dimensions and location are the basic elements for monitoring the modules. Therefore, the proposed panel surface area degradation analysis algorithm is developed under two phases. In the first phase, the solar panel localization is performed, and the feature extraction and analysis are developed. Further, in the second phase, the effect of PV panel surface area degradation is analyzed on the power output of the PV system.

3.4.1 Data Preparation

Generally, the data preparation with the thermal image digital representation of PV modules is achieved in two steps. In the first step, the PV module is segmented from the PV string using the digital image processing approach discussed in Section 3.3.2. Further, the second step adapts the localization process of the

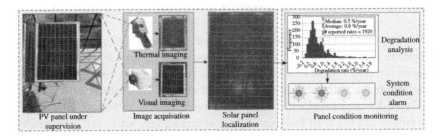

Figure 3.10 The proposed health monitoring of solar PV system.

segmented module images to localize the final solar panel surface for proceeding with the feature extraction process.

3.4.1.1 Segmentation Process

The PV module segmentation process aids in the efficient implementation of the feature extraction process and analyzing the surface area degradation of the module. For a sample of visual images of the PV modules, the grayscale transformation and the digital representation of a thermal image are shown in Figure 3.11. Further, this representation is subjected to various digital image processing techniques for achieving the module segmentation.

To begin with, consider I as a function of the visual image for a thermal image,

$$I : u \rightarrow [0, 1]^c \tag{3.1}$$

where u represents the pixel density information in a range of $\{0, ..., m-1\} \times \{0, ..., n-1\}$ with $m, n \in \mathbb{N}$ as the number of rows and columns respectively, and $c \in \{1, 3\}$ corresponds to the number of channels. Further, $I(i, j, c)$ provides the

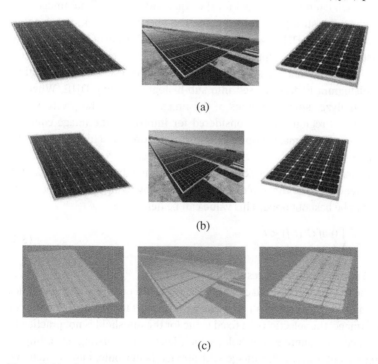

(a)

(b)

(c)

Figure 3.11 Digital representation of solar panel in visual, grayscale, and thermal forms. (a) Sample of captured solar panel visual images. (b) Sample of grayscale transformation of the captured solar panel visual images. (c) Sample of captured solar panel thermal images. Source: (a) Ethan Miller/Getty Images News/Getty Images; INDESOLAR LLP (b) Ethan Miller/Getty Images News/Getty Images; INDESOLAR LLP.

image intensity at pixel coordinates (i, j, c). For any given condition, if the value of coordinate $c = 3$, then the representation of a color image in the RGB space is given by

$$I(i, j, c) = (I_r(i, j), I_g(i, j), I_b(i, j)) \tag{3.2}$$

where $I_r(i, j) = I(i, j, 1)$ corresponds to the red channel, $I_g(i, j) = I(i, j, 2)$ represents the green channel, and $I_b(i, j) = I(i, j, 3)$ refer to the blue channel.

Further, a color space change is performed to segment the PV module which transforms the image function I into a grayscale function G. This image transformation process is given as

$$G(i, j) = 0.299 I_r(i, j, 1) + 0.587 I_g(i, j, 2) + 0.114 I_b(i, j, 3). \tag{3.3}$$

Typically, the contrast value on the thermography is very low. Hence, it is important to adopt methods that can conveniently improve it. Conventionally, the global histogram equalization approach is used for improving the contrast, but this approach cannot be adapted for the equalization of thermal image local brightness characteristics [44]. To overcome the above drawback, the dynamic histogram equalization (DHE) [45] along with the equalization of adaptive limited contrast histogram (CLAHE) approach [46, 47] is implemented. In this approach, based on the local maxima and specifically assigned gray level for each partition, the global histogram is decomposed into sub-histograms using DHE. Whereas, the CLAHE analyzes smaller regions of the image such that the pixels, which correspond to this region can be considered for improving the image contrast. As the contrast of the image improves, noise reduction is achieved using the Gaussian filter.

Further, a thresholding technique is used for separating the PV module from the backgroundGenerally, the threshold value is selected by fixing t_h such that a new pixel value can be best obtained. This value can be defined as

$$B(i, j) = \begin{cases} 0 & if \, G^*(i, j) < t_h \\ 1 & if \, G^*(i, j) \geq t_h \end{cases} \tag{3.4}$$

where $B(i, j)$ corresponds to the resulting binary image.

Usually, the choice of threshold can be achieved based on image characteristics and/or based on the desired result. But due to variations in the characteristics of the thermal image, the selection of a fixed value for the threshold is not practically possible. Hence, an adaptive threshold, which keeps on adjusting according to different regions of a particular image will provide better outcomes. To achieve this, Otsu's method is implemented to perform automatic thresholding of the image [48, 49]. In parallel with the automatic thresholding approach, the opening morphological operation [50] is used to remove the imperfections in the image

Figure 3.12 Segmented solar PV module after applying adaptive threshold process.

structure during the segmentation process. This operation is mathematically defined as

$$B \circ K = (B \ominus K) \oplus K \tag{3.5}$$

where B indicates the previously obtained binary image, $K \in \mathbb{R}^2$ corresponds to the structuring element which defines the neighborhood relation of pixels with reference to the task of shape analysis, $\ominus$ is the erosion operator, and $\oplus$ is a dilatation operator. The erosion and dilation operators are defined by Eqs. (3.6) and (3.7), respectively, as

$$(B \ominus K)(i,j) = \inf\{B(x + x', y + y') \mid (x', y') \in K\} \tag{3.6}$$

$$(B \oplus K)(i,j) = \sup\{B(x - x', y - y') \mid (x', y') \in K\} \tag{3.7}$$

The segmented image after applying the optimal threshold is shown in Figure 3.12.

3.4.1.2 Localization of Solar Panel

From Figure 3.12, it is observed that even after performing segmentation and thresholding, there exist some parts of background information in the image. Hence, to refine the segmented image, and achieve good degradation estimation, the Hough transform is used to localize the PV modules. The Hough transform identifies the lines and peaks of the segmented image and determines the geometric structure of the PV module. Generally, these Hough lines are generated by specifying the numerical matrix of the image in a polar space (θ, ρ). Here θ angle of line rotation in radians and ρ represents the distance between coordinates and the origin of the space. Further, the Hough peaks identify the minimum and maximum values of coordinates in rows and columns. The segmented image after applying the Hough transform is shown in Figure 3.13.

From the figure, the red line indicates the minimum length of the Hough line, whereas the green lines indicate the Hough lines with maximum length. Similarly, the yellow peaks indicate the minimum peak in the coordinate and the red peaks indicate the maximum peaks in the coordinates. This approach efficiently localizes the PV module around its coordinates in the polar space. Further, the feature extraction process is carried out to achieve the panel area degradation analysis.

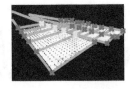

Figure 3.13 Segmented solar PV module after the localization approach.

The flow chart for the proposed PV panel area degradation approach is shown in Figure 3.14.

3.4.2 Feature Extraction

Features are extracted from the acquired images to reduce the dimension and complexity of useful information. Extracted features are divided into statistical feature [51] and geometrical features [52]. The statistical features are utilized for localization of the solar panel region for visual inspection, while geometrical features are utilized for establishing a relation between performance degradation and change in image features for real-time monitoring of the system. A detailed description of the features is given as follows:

Statistical features: The histogram-based features used in this work are first-order statistics that include mean, variance, and standard deviation. Let Z be a random variable denoting image gray levels and $p(Z_i)$, where, $i = 0, 1, 2, 3, ...,$ $L-1$, be the corresponding histogram, here L represents the number of distinct gray levels. Further, the features are calculated using the histogram $p(Z_i)$ as given in Eqs. (3.8)–(3.10).

The mean feature gives the average gray level of each region and it is useful only as a rough idea of intensity, does not provide any information about the texture. This is mathematically represented as

$$M = \sum_{i=0}^{L-1} Z_i \times p(Z_i) \tag{3.8}$$

Similarly, the variance feature gives the number of gray level fluctuations from the mean gray level value and can be calculated as

$$V = \sum_{i=0}^{L-1} (Z_i - m)^2 \times p(Z_i) \tag{3.9}$$

Further, the standard deviation is estimated by calculating the square root of the variance feature. This measures the information about how features are spread out for a given image region.

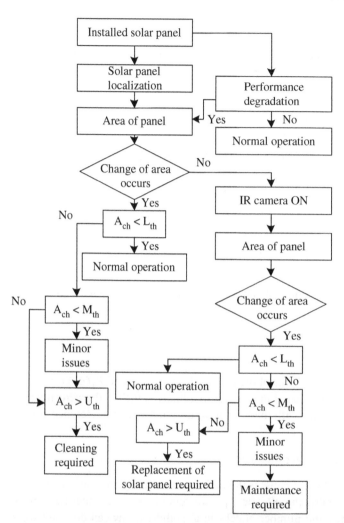

Figure 3.14 Flow chart of the proposed photovoltaic panel area degradation approach.

From this discussion, a set of feature samples obtained for a set of coordinates on the solar panel surface are shown in Table 3.3.

Geometrical features: The geometrical features of image regions for a detected solar panel correspond to the area and perimeters of major and minor axes [52]. The area, change of area, and percentage change of area features are utilized to establish the relationship between performance degradation and extracted features. An area of the region represents the size of the PV module. The

Table 3.3 Sample of statistical features.

S. No.	Coordinates of polygon	Mean	Variance	Standard deviation
1	[(10, 24), (114, 11), (63, 214), (211, 73)]	75.16	3.9×10^3	62.47
2	[(14, 23), (144, 40), (97, 158), (270, 67)]	97.90	2.2×10^3	47.06
3	[(10, 24), (114, 11), (63, 214), (211, 73)]	144.05	1.24×10^3	35.28
4	[(10, 24), (114, 11), (63, 214), (211, 73)]	96.11	4.9×10^3	70.67
5	[(10, 24), (114, 11), (63, 214), (211, 73)]	102.69	2.2×10^3	47.13

mathematical expression for calculating the area from the detected solar panel is given as

$$A_D = \sum_{i=0}^{m} \sum_{j=0}^{n} F_T(i,j) \tag{3.10}$$

where, A_D is the detected area of the localized solar panel and $F_T(i, j)$ represents the final localized solar panel in the acquired image. Further, the percentage of area change is calculated with the help of the detected area and reference area.

3.4.3 Degradation Effect Analysis

The surface area of the PV panel is the main asset for analyzing the performance degradation. Generally, the electrical parameters of the solar modules depend upon the total area and number of cells in a module. If any cell does not work properly, then it means that a particular area of the system is not generating power and cause degradation in system performance [53]. If the area of the PV panel is covered by any external object, then that particular region does not irradiate properly. A similar scenario occurs when any part of the solar array depicts a hotspot due to cell damage. This indicates that the cell area does not play a proper role in identifying the degradation.

3.4.3.1 Effect of Degradation Rate on Module Characteristics

The degradation of PV modules results in a gradual change in area with respect to time. This phenomenon affects the efficiency of the module, which results in the degradation of the electrical parameters of the module. The major affected parameters are open circuit voltage (V_{oc}), short circuit current (I_{sc}), maximum power (P_{max}), and the form factor (*FF*). Mathematical expression for degradation rate of the electrical parameters is given by

$$R_D(X)\% = \left(1 - \frac{X}{X_0}\right) \times 100 \tag{3.11}$$

where X depicts the electrical parameters effected due to degradation, and X_0 depicts the reference value of electrical parameters obtained under STC or as depicted by the manufacturer. The yearly degradation rate of X is evaluated as:

$$R_{Dt}(X)\% = \left(\frac{RD(X)\%}{\Delta t}\right) \times 100 \tag{3.12}$$

where Δt = Operating time of PV modules in years from the time of commencement of operation.

The major advantage of the proposed system is that it creates an alarm according to the condition of the system. The lower threshold area (A_{TL}) is chosen around 1–10%, medium threshold area (A_{TM}) is chosen 10–30% and upper threshold area (A_{TU}) is chosen 30% and above.

3.4.4 Effect of Degradation Rate on System Output

In this section, a theoretical prediction of the mismatch losses of the string inverter system is calculated. Generally, the prediction can be performed using the flash test data that is shipped along every module or with the characteristics of a healthy operating system. A bivariate regression analysis is used to present a relation between the power output, which is a dependent variable determined by the module area and solar irradiance. In order to determine the relationship between the two variables, the bivariant analysis uses a linear regression model, which fits the variation of the variables along a straight line. The coefficient of variation in maximum operating current and voltage is the most important input value, although the variation in voltage is only in effect if a system has multiple strings connected in parallel. The model can be derived from [54] as

$$\Delta P = \frac{(C+2)}{2}\left[\sigma_{I_{mpp}}^2\left(1 - \frac{1}{L}\right) + \frac{\sigma_{V_{mpp}}^2}{L}\left(1 - \frac{1}{M}\right)\right] \tag{3.13}$$

where ΔP (%) is the fractional power loss due to mismatch, C is the cell characteristic factor, $\sigma_{I_{mpp}}^2$ is variance in maximum power current, $\sigma_{V_{mpp}}^2$ is variance in maximum power voltage, L is the number of modules in string, and M is the number of strings. For one string, the relation is simplified as

$$\Delta P = \frac{(C+2)}{2}\left[\sigma_{I_{mpp}}^2\left(1 - \frac{1}{L}\right)\right] \tag{3.14}$$

Further, the cell characteristic factor (C), which can be expressed in terms of the fill factor, is usually defined as

$$FF = \frac{I_{m_{pp}} \cdot V_{m_{pp}}}{I_{sc} \cdot V_{oc}} \tag{3.15}$$

where I_{sc} is short circuit current, and V_{oc} is open circuit voltage. Once the fill factor is known, the following equation for C can be solved using a numerical solver function.

$$FF = \frac{C^2}{(1+C)(C + \ln(1+C))} \tag{3.16}$$

In addition to C, the variance in maximum power current and maximum power voltage are given as

$$\sigma^2_{I_{mpp}} = \left(\frac{\sigma_{I_{mpp}}}{\overline{I_{mpp}}} \right)^2 \tag{3.17}$$

$$\sigma^2_{V_{mpp}} = \left(\frac{\sigma_{V_{mpp}}}{\overline{V_{mpp}}} \right)^2 \tag{3.18}$$

The variance can be computed by knowing the standard deviation:

$$\sigma_{I_{mpp}} = \sqrt{\frac{\sum (I_{mpn} - \overline{I_{m_p}})^2}{N - 1}} \tag{3.19}$$

$$\sigma_{V_{mpp}} = \sqrt{\frac{\sum (V_{mpn} - \overline{V_{m_p}})^2}{N - 1}} \tag{3.20}$$

where N is the total number of modules in the array.

Finally, the calculation of ΔP for a string of modules can be defined as a percentage of fractional power loss that can be expected due to variations between the power-generation capabilities of modules within a string. This prediction of fractional power loss can be used to identify the change in performance of the system due to the effect of degrading modules. Further, a sensitivity analysis of the model is used for estimating the mismatch losses. This focuses on how the changes in coefficient of variation, number of parallel strings, and reduction of the fill factor affect the estimation of fractional power loss in a system.

3.4.5 Experimental Testing for Degradation Analysis

3.4.5.1 Experimental Setup
This section outlines and highlights the methods used in the collection and the analysis of the data for measuring the degradation of solar modules. The experimental setup for monitoring of solar PV system is shown in Figure 3.15. An Eldora

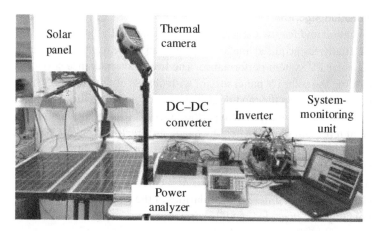

Figure 3.15 Experimental setup for degradation analysis.

Table 3.4 Image acquisition device specifications.

Attributes	Value
Model number	Fluke TIS45 9 Hz
Temperature measurement range	−20 to + 350 °C
Thermal sensitivity	≤0.09 °C @ 30
Best temperature measurement accuracy	±2 °C
Detector resolution	160 × 120 pixels
Display resolution	320 × 240 pixels

40 polycrystalline module is considered for testing the developed degradation estimation algorithm. The experimental analysis is carried out considering STC (Air Mass 1.5 G where G is global and includes both diffused and direct radiation, 1000 W/m² Irradiance, and 25 °C temperature). An Amprobe solar-100 solar power meter is used to constantly monitor the change in irradiance. The operating characteristics of the PV module are observed by operating it under full load. The output characteristics of the PV module are measured using Yokogawa WT310E digital power meter. In order to measure the degradation rate of the module, a portion of the PV module is covered and the performance of the system is measured. To analyze the change in area due to degradation and constantly monitor the effect on cell temperatures, a Fluke TiS 45 thermal camera [55] is installed at an angle prescribed by [56]. The specifications of the thermal imager are given in Table 3.4.

3.4.5.2 Degradation Algorithm

PV panel localization and feature extraction are performed with the help of MAT-LAB R2017a. Initially, the acquired images are processed and features are extracted to monitor the solar PV system performance. The irregular irradiance surface is created by covering the solar PV panel surface with objects of different sizes. Further, different sizes of area on the module surface are covered to analyze the effect of change in the irradiance on the system performance. The images in Figure 3.16 shows the visual representation of a healthy panel, an irregular irradiance panel, and an irregular temperature distribution. Further, the corresponding thermal and grayscale representations are shown in Figures 3.17 and 3.18 respectively for analyzing the change on the surface area. Similar to the above conditions, a total of 500 images of PV panel with different irregularities of the irradiance and temperature on the healthy and degraded module surface area are analyzed and related

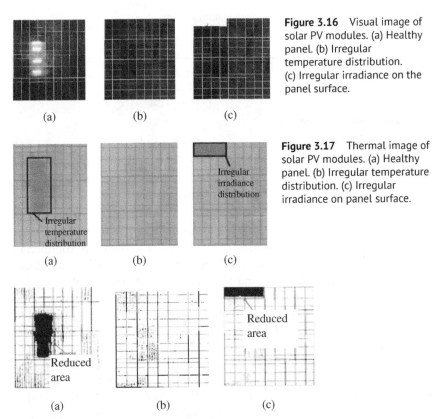

Figure 3.16 Visual image of solar PV modules. (a) Healthy panel. (b) Irregular temperature distribution. (c) Irregular irradiance on the panel surface.

Irregular irradiance distribution

Figure 3.17 Thermal image of solar PV modules. (a) Healthy panel. (b) Irregular temperature distribution. (c) Irregular irradiance on panel surface.

Irregular temperature distribution

Reduced area

Reduced area

Figure 3.18 Grayscale image of Solar PV modules. (a) Area of healthy PV panel. (b) Area changes due to Irregular temperature distribution. (c) Area changes due to Irregular irradiation surface

Table 3.5 Sample of image feature from PV panel cell.

S. No.	A_{REF} (Pixel^2)	A_D (Pixel^2)	A_{CH} (Pixel^2)	A_{CHP} (%)
SPC1	17896860	17896860	0	0
SPC2		16914696	982164	5.48
SPC3		14131898	3764962	21.03
SPC4		10366936	7529924	42
SPC5		7584138	10312722	57.62
SPC6		6274586	11622274	64.9
SPC7		2509624	15387236	85.9

with the performance degradation. The area, change in the area, and percentage change of area with respect to the reference area of the solar PV module is calculated. The sample data of extracted features for the images obtained are shown in Table 3.5.

3.4.5.3 Relation Between Image Features and Performance Degradation

Performance degradation of the system is analyzed with the change in image features. Degradation in voltage and current is noted down and related to the image features. It is analyzed from the extracted features and recorded performance data that the voltage and current of the system decrease linearly with a change in the irradiation surface area. Samples of the feature extracted, such as reference voltage, reference current, the percentage change of surface area, change in voltage and change in current, are shown in Table 3.6.

Table 3.6 Sample of voltage and current degradation with feature change.

S. No.	V_{ref}	I_{ref}	A_{chp}	Voltage change	Current change
SPC1	15.8	0.046	0	0	0
SPC2			5.48	15.34	0.043
SPC3			21.03	13.39	0.035
SPC4			42.00	10.75	0.033
SPC5			57.62	8.8	0.027
SPC6			64.9	7.88	0.02
SPC7			85.9	5.24	0.014

Table 3.7 Comparison of the proposed monitoring method with the conventional approaches.

S. no.	Parameters	Accuracy	Environmental factor	Fault misclassification
1.	Hog feature extraction approach	81.7%	14%	Yes
2.	Hog feature extraction approach with proposed image processing	93.4%	11%	No
3.	Developed feature extraction	88.6%	12%	Yes
4.	Developed feature extraction with proposed image processing	98.23%	1%	No

Further, Table 3.7 shows a comparison with the existing feature extraction-based degradation analysis approaches. The comparison analysis is identified based on various parameters like accuracy [57] of degradation analysis, environmental factor effects on monitoring [58], and fault misclassification aspect. Further, the accuracy of the image processing approach is estimated as given in Eq. (3.21) [59].

$$Accuracy = \frac{Matched(R, S)}{|S|} \tag{3.21}$$

where R = Ground truth of PV panel region, and S is the segmented or detected PV panel.

From this analysis, it is observed that the developed approach has better accuracy in training and identifying the degradation of PV modules. Further, the fault misclassification rate indicates that the segmentation and localization process can effectively identify the panel in the system and aid with the feature extraction approach. Besides, the threshold limit set for change in the area below the lower limit value means the system has no issues. Whereas change in an area above the lower limit of the threshold value and below the medium limit indicates that the system has fewer issues. Further, any change in an area greater than the upper limit of the threshold value indicates that the system has major issues.

3.4.5.4 Relation Between Module Performance Degradation and PV System Output

The results and the sensitivity analysis for two systems (System A: degrading effect implemented system, and System B: healthy operating system) consisting of 16 multi-crystalline Si modules with a rated maximum power of 210 W per module are tested for evaluating the effect of degradation, irradiance on the power output of the system. System A is tested with an even degradation effect on all the module areas in order to perform the module area degradation effect analysis. There is no

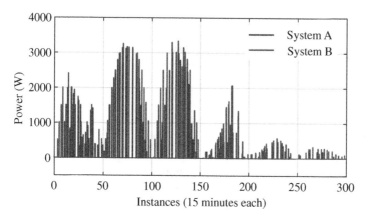

Figure 3.19 Measured power output.

Table 3.8 Measured power output.

System	Total energy output	Average energy output
A	69.1 kWh	0.79 kWh
B	76.7 kWh	0.87 kWh

significant difference in the local shading of the systems. Power output data has been collected from two systems under different hourly irradiance values as per the STRÅNG model [60], and the instantaneous power output measurements are made for every 15 minutes between 6:00 to 18:00 hours from 3 April 2020 to 8 April 2020. The data obtained is plotted as shown in Figure 3.19.

After having performed a comparison of the two systems, A and B, it can be concluded that system B is outperforming system A. The curves of the two systems are very similar. However, system B has a higher power output at almost every peak of the curve. In Table 3.8, the total energy output for all the intervals is presented as well as the average power output for the same period. It shows a higher total power output for system B as well as higher average power output, which complies with the healthy operation of the modules.

The theoretical value of the fractional power loss between both systems is listed in Table 3.9.

The fractional power loss of the 16 modules connected in series is estimated to be 2.6% based on the test data of healthy operating modules. This indicates that the degradation of cells is affecting the system's mismatch losses at a faster rate. The estimation of mismatch was not performed on the array with module-level

Table 3.9 Result from an estimation of mismatch losses.

Parameter	Result
Form factor	0.76
Cell characteristic factor	12.02
Number of modules in a string	16
Number of strings	1
$\sigma_{I_{mpp}}$	0.032
$\sigma^2_{I_{mpp}}$	0.0041
ΔP	0.026

Table 3.10 Parameters of the regression analysis.

System	Intercept of x-axis (W)	Coefficient of the slope	Standard error of estimate (W)	R-squared
A	885	0.305	71	0.809
B	808	0.308	73	0.804

inverters. In addition to this test, when the mismatch losses of system A are compared with the flash test data available with the modules at the time of installation, the measured energy output resulted in a 5.6% mismatch, which is very high considering that the modules are only two years old.

Further, a linear regression analysis has been made in order to determine a relationship between the module area and power output for the two systems. To be able to determine the correlation, the function *linear model fit* which performs a linear regression analysis is performed using MATLAB software. Table 3.10 and Figure 3.20 present the results from the regression analysis. System A is displayed as the red line and system B as the blue line in Figure 3.21. The intercept of the x-axis is of interest, as it will provide an indication of the capability of producing power under varying irradiation conditions.

The predicted values for each system show that system A has a higher intercept value and a lesser standard error of the estimate. However, this means that system A needs more solar irradiance in order to produce the same power as system B. The coefficient of determination, R-squared, indicates how well the amount of disparity in the response variable power output is explained by the independent variable solar irradiance in the linear regression model. The larger the *R*-squared is, the more disparity is explained by the linear regression model. The model

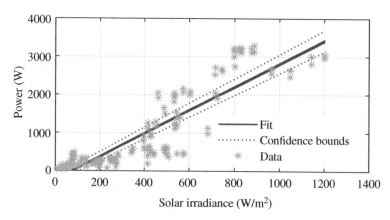

Figure 3.20 Relationship between solar irradiance and power output for system A and system B.

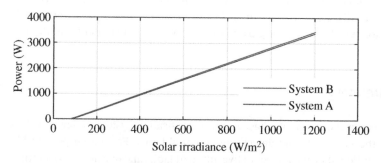

Figure 3.21 Results from the regression analysis for system A, system B.

used adequately explains both systems and it shows that system B outperforms system A.

3.4.5.5 Sensitivity Analysis
This section will discuss the precision of the result, focusing on the prediction of mismatch in system A and the measurements of both systems. The uneven degradation of modules has proven to increase mismatch in a PV system. As time passes, uneven degradation becomes a factor that increases mismatch in the system. The increase in a mismatch is caused by a greater variation between the modules in the array and it is therefore of interest to know how the coefficient of variation is affecting the prediction of mismatch. Table 3.11 shows fractional power loss at different values for the coefficient of variation of maximum power current.

The investigated mismatch losses due to area degradation could potentially rise to 12% for series-connected modules, which implies a coefficient of variation in the

Table 3.11 Fractional power loss.

$\sigma^2_{I_{mpp}}$	0.000027	0.0036	0.0072	0.001	0.014	0.01	0.02
$\Delta P\%$	0.02	2.29	4.57	6.86	9.14	11.4	13.9

Table 3.12 Cell characteristic factor vs. Fill factor.

Form factor	0.75	0.7	0.65	0.6	0.55	0.5	0.45
Cell characteristic factor	11.17	8.13	6.09	4.66	3.61	2.81	2.2

maximum operating current of 2%. The cell characteristic factor is derived from its relation to the fill factor, displayed in Eq. (3.15). Table 3.12 shows how C varies with the fill factor, which can decrease by up to 38% operating under field conditions.

3.5 Summary

In this paper, a thermography-based fault classification system is developed to recognize the PV panels and detect any potential faults. For the purpose of PV panel recognition, thermal images of five different faults and one normal operating panel were gathered. The acquired thermal images are subjected to image preprocessing where each captured image is divided into multiple subregions using the padding process. Further, the subregions are subjected to feature extraction using GLCM to obtain the texture features. Seven different features are extracted for each subregion and the corresponding feature vector is subjected to training with the MLPNN classifier. The developed classifier depicted 95.7% training accuracy and 95% testing accuracy. Further, while adapting the trained classifier for fault classification or monitoring purposes, the acquired samples should go through a special preprocessing as they may contain background information, which needs to be suppressed. In this context, the edge detection and Hough Transform were applied to remove the unwanted information from the panel, identify the exact module area and avoid misclassification of fault. Further, the processed image is subjected to feature extraction and passed through a classification algorithm for localization and identification of the type of fault. The developed technique overcomes the drawbacks of conventional techniques by providing advantages with efficient remote monitoring, least detection time, and high classification efficiency. The complete testing procedure is tested on three different image samples and the results depict an efficient classification with 96.2% efficiency.

Further, the relation between module performance degradation and PV system output is estimated for two PV systems using linear regression analysis. The results predicted a mismatch of 2.6 % and 5.6% for the string inverter system based on healthy operating module system data and flash test data, respectively. Further, this work can be extended by analyzing the degradation caused due to irregular faults on the panel surface.

References

1 Haque, A., Bharath, K.V.S., Khan, M.A. et al. (2019). Fault diagnosis of photovoltaic modules. *Energy Sci. Eng.* September: ese3.255. https://doi.org/10.1002/ese3.255.

2 Hammami, M., Torretti, S., Grimaccia, F., and Grandi, G. (2017). Thermal and performance analysis of a photovoltaic module with an integrated energy storage system. *Appl. Sci.* 7 (11): 1107. https://doi.org/10.3390/app7111107.

3 Glavas, H., Vukobratovic, M., Primorac, M., and Mustran, D. (2017). Infrared thermography in inspection of photovoltaic panels, *2017 International Conference on Smart Systems and Technologies (SST)*, Osijek, Croatia (18–20 October), pp. 63–68. doi: https://doi.org/10.1109/SST.2017.8188671.

4 Schuss, C., Remes, K., Leppänen, K. et al. (2018). Detecting defects in photovoltaic panels with the help of synchronized thermography. *IEEE Trans. Instrum. Meas.* 67 (5): 1178–1186. https://doi.org/10.1109/TIM.2018.2809078.

5 Hu, X., Chen, T., Xue, J. et al. (2017). Absolute electroluminescence imaging diagnosis of GaAs thin-film solar cells. *IEEE Photonics J.* 9 (5): 1–9. https://doi.org/10.1109/JPHOT.2017.2731800.

6 dos Santos, R.L., Ferreira, J.S., Martins, G.E. et al. (2017). Low cost educational tool to trace the curves PV modules. *IEEE Lat. Am. Trans.* 15 (8): 1392–1399. https://doi.org/10.1109/TLA.2017.7994784.

7 Kaden, T., Lammers, K., and Möller, H.J. (2015). Power loss prognosis from thermographic images of PID affected silicon solar modules. *Sol. Energy Mater. Sol. Cells* 142: 24–28. https://doi.org/10.1016/j.solmat.2015.05.028.

8 Koch, S., Weber, T., Sobottka, C. et al. (2016). Outdoor electroluminescence imaging of crystalline photovoltaic modules: comparative study between Manual Ground-Level Inspections and Drone-Based Aerial Surveys. *32nd European Photovoltaic Solar Energy Conference and Exhibition*, Munich, Germany (20–24 June 2016), vol. 53, no. 9, pp. 1736–1740. doi: https://doi.org/10.4229/EUPVSEC20162016-5DO.12.2.

9 Daliento, S., Chouder, A., Guerriero, P. et al. (2017). Monitoring, diagnosis, and power forecasting for photovoltaic fields: a review. *Int. J. Photoenergy* 2017: 1–13. https://doi.org/10.1155/2017/1356851.

10 Niazi, K.A.K., Yang, Y., and Sera, D. (2019). Review of mismatch mitigation techniques for PV modules. *IET Renew. Power Gener.* 13 (12): 2035–2050. https://doi.org/10.1049/iet-rpg.2019.0153.

11 Manganiello, P., Balato, M., and Vitelli, M. (2015). A survey on mismatching and aging of PV modules: the closed loop. *IEEE Trans. Ind. Electron.* 62 (11): 7276–7286. https://doi.org/10.1109/TIE.2015.2418731.

12 Pendem, S.R. and Mikkili, S. (2018). Modeling, simulation and performance analysis of solar PV array configurations (Series, Series–Parallel and Honey-Comb) to extract maximum power under Partial Shading Conditions. *Energy Reports* 4: 274–287. https://doi.org/10.1016/j.egyr.2018.03.003.

13 Djordjevic, S., Parlevliet, D., and Jennings, P. (2014). Detectable faults on recently installed solar modules in Western Australia. *Renew. Energy* 67: 215–221. https://doi.org/10.1016/j.renene.2013.11.036.

14 van Dyk, E.E., Gxasheka, A.R., and Meyer, E.L. (2005). Monitoring current–voltage characteristics and energy output of silicon photovoltaic modules. *Renew. Energy* 30 (3): 399–411. https://doi.org/10.1016/j.renene.2004.04.016.

15 G.-H. Kang Kim, K.-S., Park, C.-H. et al. (2006). Current-voltage characteristics with degradation of field-aged silicon photovoltaic modules. *21st European Photovoltaic Solar Energy Conference*, Dresden, Germany (4–8 September 2006).

16 Muttillo, M., Nardi, I., Stornelli, V. et al. (2020). On field infrared thermography sensing for PV system efficiency assessment: results and comparison with electrical models. *Sensors* 20 (4): 1055. https://doi.org/10.3390/s20041055.

17 Chattopadhyay, S., Dubey, R., Kuthanazhi, V. et al. (2014). Visual degradation in field-aged crystalline silicon PV modules in India and correlation with electrical degradation. *IEEE J. Photovoltaics* 4 (6): 1470–1476. https://doi.org/10.1109/JPHOTOV.2014.2356717.

18 Tsanakas, J.A. and Botsaris, P.N. (2012). An infrared thermographic approach as a hot-spot detection tool for photovoltaic modules using image histogram and line profile analysis. *Int. J. Cond. Monit.* 2 (1): 22–30. https://doi.org/10.1784/204764212800028842.

19 Tsanakas, J.A., Chrysostomou, D., Botsaris, P.N., and Gasteratos, A. (2015). Fault diagnosis of photovoltaic modules through image processing and Canny edge detection on field thermographic measurements. *Int. J. Sustain. Energy* 34 (6): 351–372. https://doi.org/10.1080/14786451.2013.826223.

20 Tsanakas, J.A., Ha, L., and Buerhop, C. (2016). Faults and infrared thermographic diagnosis in operating c-Si photovoltaic modules: a review of research and future challenges. *Renew. Sustain. Energy Rev.* 62: 695–709. https://doi.org/10.1016/j.rser.2016.04.079.

21 Grimaccia, F., Leva, S., and Niccolai, A. (2017). PV plant digital mapping for modules' defects detection by unmanned aerial vehicles. *IET Renew. Power Gener.* 11 (10): 1221–1228. https://doi.org/10.1049/iet-rpg.2016.1041.

22 Bedrich, K.G. et al. (2018). Quantitative electroluminescence imaging analysis for performance estimation of PID-influenced PV modules. *IEEE J. Photovoltaics* 8 (5): 1281–1288. https://doi.org/10.1109/JPHOTOV.2018.2846665.

23 Bastidas-Rodriguez, J.D., Franco, E., Petrone, G. et al. (2015). Model-based degradation analysis of photovoltaic modules through series resistance estimation. *IEEE Trans. Ind. Electron.* 62 (11): 7256–7265. https://doi.org/10.1109/TIE.2015.2459380.

24 Bala Subramaniyan, A., Pan, R., Kuitche, J., and TamizhMani, G. (2018). Quantification of environmental effects on PV module degradation: a physics-based data-driven modeling method. *IEEE J. Photovoltaics* 8 (5): 1289–1296. https://doi.org/10.1109/JPHOTOV.2018.2850527.

25 Sengupta, S., Sengupta, S., and Saha, H. (2020). Comprehensive modeling of dust accumulation on PV modules through dry deposition processes. *IEEE J. Photovoltaics* 10 (4): 1148–1157. https://doi.org/10.1109/JPHOTOV.2020.2992352.

26 Canny, J. (1986). A computational approach to edge detection. *IEEE Trans. Pattern Anal. Mach. Intell* 8 (6): 679–698. https://doi.org/10.1109/TPAMI.1986.4767851.

27 Duda, R.O. and Hart, P.E. (2002). Use of the Hough transformation to detect lines and curves in pictures. *Commun. ACM* 15 (1): 11–15. https://doi.org/10.1145/361237.361242.

28 Havens, K.J. and Sharp, E.J. (2016). Remote Sensing. In: Thermal Imaging Techniques to Survey and Monitor Animals in the Wild, 35–62. Elsevier.

29 Li, Z., Liu, C., Liu, G. et al. (2010). A novel statistical image thresholding method. *AEU – Int. J. Electron. Commun.* 64 (12): 1137–1147. https://doi.org/10.1016/j.aeue.2009.11.011.

30 Liu, X. and Wang, D. (2006). Image and texture segmentation using local spectral histograms. *IEEE Trans. Image Process.* 15 (10): 3066–3077. https://doi.org/10.1109/TIP.2006.877511.

31 Hu, Y., Cao, W., Ma, J. et al. (2014). Identifying PV module mismatch faults by a thermography-based temperature distribution analysis. *IEEE Trans. Device Mater. Reliab.* 14 (4): 951–960. https://doi.org/10.1109/TDMR.2014.2348195.

32 Chattopadhyay, S. et al. (2018). Correlating infrared thermography with electrical degradation of PV modules inspected in all-india survey of photovoltaic module reliability 2016. *IEEE J. Photovoltaics* 8 (6): 1800–1808. https://doi.org/10.1109/JPHOTOV.2018.2859780.

33 Coşgun, A.E. and Uzun, Y. (2017). Thermal fault detection system for PV solar modules. *Electr. Electron. Eng. An Int. J.* 6 (3): 9–15. https://doi.org/10.14810/elelij.2017.6302.

34 Randen, T. (1997). Filter and filter bank design for image texture recognition. *Sci. Technol.* 138. [Online]. http://www.ux.uis.no/~tranden/thesis/thesis.pdf.

35 Tuceryan, M. and Jain, A.K. (1993). Texture analysis. In: Handbook of Pattern Recognition and Computer Vision, 235–276. World Scientific.

36 Jarc, A., Perš, J., Rogelj, P. et al. (2007). Texture features for affine registration of thermal (FLIR) and visible images. *Comput. Vis. WinterWorkshop* 1–7.

37 GLCM Texture Feature (2020). https://support.echoview.com/WebHelp/Windows_and_Dialog_Boxes/Dialog_Boxes/Variable_properties_dialog_box/Operator_pages/GLCM_Texture_Features.htm (accessed 29 January 2020).

38 Kurukuru, V.S.B., Haque, A., Khan, M.A., and Tripathy, A.K. (2019). Fault classification for photovoltaic modules using thermography and machine learning techniques. *2019 International Conference on Computer and Information Sciences (ICCIS)*, Sakaka, Saudi Arabia (3–4 April 2019), pp. 1–6. doi: https://doi.org/10.1109/ICCISci.2019.8716442.

39 Kurukuru, V.S.B., Haque, A., Tripathy, A.K., and Khan, M.A. (2022). Machine learning framework for photovoltaic module defect detection with infrared images. *Int. J. Syst. Assur. Eng. Manag.* https://doi.org/10.1007/s13198-021-01544-7.

40 Woyte, A., Richter, M., Moser, D. et al. (2014). Analytical monitoring of grid-connected photovoltaic systems good practices for monitoring and performance analysis, IEA-PVPS T13-03: 2014, IEA PVPS Task 13, Subtask 2. doi: 10.13140/2.1.1133.6481.

41 Baghirli, O. (2015). Comparison of Lavenberg-Marquardt, scaled conjugate gradient and bayesian regularization backpropagation algorithms for multistep ahead wind speed forecasting using multilayer perceptron feedforward neural network. Dissertation, Uppsala University. doi: diva2:828170.

42 Kurukuru, V.S.B., Haque, A., Khan, M.A., and Tripathy, A.K. (2019). Fault classification for photovoltaic modules using thermography and machine learning techniques. *2019 International Conference on Computer and Information Sciences (ICCIS)*, Sakaka, Saudi Arabia (3–4 April 2019), pp. 1–6. doi: 10.1109/ICCISci.2019.8716442.

43 Niu, S., Yang, J., Wang, S., and Chen, G. (2011). Improvement and parallel implementation of canny edge detection algorithm based on GPU. *Proc. Int. Conf. ASIC* 6: 641–644. https://doi.org/10.1109/ASICON.2011.6157287.

44 Chien, S.-C., Chang, F.-C., Hua, K.-L., Chen, I.-Y., and Chen, Y.-Y. (2017). Contrast enhancement by using global and local histogram information jointly. *2017 International Conference on Advanced Robotics and Intelligent Systems (ARIS)*, Taipei, Taiwan (68 September 2017), pp. 75–75. doi: 10.1109/ARIS.2017.8297188.

45 Ooi, C. and Mat Isa, N. (Nov. 2010). Quadrants dynamic histogram equalization for contrast enhancement. *IEEE Trans. Consum. Electron.* 56 (4): 2552–2559. https://doi.org/10.1109/TCE.2010.5681140.

46 Momeni Pour, A., Seyedarabi, H., Abbasi Jahromi, S.H., and Javadzadeh, A. (2020). Automatic detection and monitoring of diabetic retinopathy using efficient convolutional neural networks and contrast limited adaptive histogram equalization. *IEEE Access* 8: 136668–136673. https://doi.org/10.1109/ACCESS .2020.3005044.

47 Majeed, S.H. and Isa, N.A.M. (2021). Adaptive entropy index histogram equalization for poor contrast images. *IEEE Access* 9: 6402–6437. https://doi.org/10 .1109/ACCESS.2020.3048148.

48 Otsu, N. (1979). A threshold selection method from gray-level histograms. *IEEE Trans. Syst. Man. Cybern.* 9 (1): 62–66. https://doi.org/10.1109/TSMC .1979.4310076.

49 Khambampati, A.K., Liu, D., Konki, S.K., and Kim, K.Y. (2018). An automatic detection of the ROI using otsu thresholding in nonlinear difference EIT imaging. *IEEE Sens. J.* 18 (12): 5133–5142. https://doi.org/10.1109/JSEN.2018 .2828312.

50 De Natale, F.G.B. and Boato, G. (2017). Detecting morphological filtering of binary images. *IEEE Trans. Inf. Forensics Secur.* 12 (5): 1207–1217. https://doi .org/10.1109/TIFS.2017.2656472.

51 Aggarwal, N. and Agrawal, R.K. (2012). First and second order statistics features for classification of magnetic resonance brain images. *J. Signal Inf. Process.* 03 (02): 146–153. https://doi.org/10.4236/jsip.2012.32019.

52 Teelen, K. and Veelaert, P. (2009). Computing regions of interest for geometric features in digital images. *Discret. Appl. Math.* 157 (16): 3457–3472. https://doi .org/10.1016/j.dam.2009.02.002.

53 Jaffery, Z.A. and Dubey, A.K. (2014). Design of early fault detection technique for electrical assets using infrared thermograms. *Int. J. Electr. Power Energy Syst.* 63: 753–759. https://doi.org/10.1016/j.ijepes.2014.06.049.

54 Webber, J. and Riley, E. (2013). Mismatch loss reduction in photovoltaic arrays as a result of sorting photovoltaic modules by max-power parameters. *ISRN Renew. Energy* 2013: 1–9. https://doi.org/10.1155/2013/327835.

55 Fluke Corporation (2019). Fluke TiS45 Infrared Camera | Fluke. https://www .fluke.com/en-us/product/thermal-cameras/tis45 (accessed 23 November 2019).

56 Zefri, Y., ElKettani, A., Sebari, I., and Lamallam, S.A. (2018). Thermal infrared and visual inspection of photovoltaic installations by UAV

photogrammetry—application case: Morocco. *Drones* 2 (4): 41. https://doi
.org/10.3390/drones2040041.

57 Lizák, F. and Kolcun, M. (2008). Improving reliability and decreasing
losses of electrical system with infrared thermography. *Acta Electrotech.
Inform* 8 (1): 60–63. [Online]. https://pdfs.semanticscholar.org/987f/
b53a7596a661c922f62df31862f2fcd248d4.pdf.

58 Masson, V., Bonhomme, M., Salagnac, J.-L. et al. (2014). Solar panels reduce
both global warming and urban heat island. *Front. Environ. Sci.* 2: https://doi
.org/10.3389/fenvs.2014.00014.

59 Jiang, X. (2005). Performance Evaluation of Image Segmentation Algorithms.
Handbook of Pattern Recognition and Computer Vision. In: Handbook of
Pattern Recognition and Computer Vision, 525–542. World Scientific.

60 SMHI (2017). STRÅNG – a mesoscale model for solar radiation, 2014. http://
strang.smhi.se/ (accessed 23 April 2020).

4

Photovoltaic Module Fault. Part 2: Detection with Quantitative-Model Approach

V S Bharath Kurukuru and Ahteshamul Haque

Advance Power Electronics Research Lab, Department of Electrical Engineering, Jamia Millia Islamia (A Central University), New Delhi, India

4.1 Introduction

Photovoltaic (PV) systems, such as PV power plants or smaller-scale PV applications, rely on continuous operations and maintenance (O&M) routines to ensure long-term up-time, higher system efficiencies, and economic viability. The continuous growth of this industry enhanced the importance of O&M activities. Augmented challenges are found in solar PV plants located in remote places, with difficult access and poor communication infrastructures [1, 2]. One main O&M issue in the PV industry is the number of components that need to be inspected in large PV plants, especially solar panels. A study done for grid-connected systems in Germany in the 1990s revealed that solar panels, or PV modules, accounted for 15% of the total system failures, whereas inverters contributed 63% and other system components contributed 22%. Despite being only 15% of the total system failures, PV module failures affect the overall system's efficiency and can jeopardize energy production. Well-maintained PV systems present on average 6% higher performance than poorly maintained ones [3, 4].

Currently, plenty of PV module failures are known, but their rapid detection is not an easy task [5–8]. The sooner a faulty module is identified and replaced, the lower is the energy loss it causes. Hence, it is essential to develop robust and accurate monitoring systems and manage how faults impact energy production.

There are three kinds of monitoring systems for the PV system:

- Quantitative model-based solutions: Offer compact representations of the system dynamics, where a difference or differential equation may describe the system for a broad set of input functions and initial states [9].

Fault Analysis and its Impact on Grid-connected Photovoltaic Systems Performance, First Edition.
Edited by Ahteshamul Haque and Saad Mekhilef.
© 2023 The Institute of Electrical and Electronics Engineers, Inc. Published 2023 by John Wiley & Sons, Inc.

- Process history-based solutions: Collect, store, and replay historical and continuous system data, using it to analyze and understand system performance.
- Signal processing-based solutions: Focus on fault detection through signal analyzing and processing.

All these methods are applied to employ scheduled routines; some of them even demand the disconnection of the solar panels. Thus, they are called off-line methods. On the other hand, an online method – capable of continually monitoring the PV system's status – offers apparent advantages in terms of response time. Online methods are typically based on measurements of the electric characteristics, e.g. current, voltage, and power, at some point in the circuit [10–12].

In this chapter, the problems regarding current and voltage measurements in online monitoring will be studied as well as a set of recommendations on quantitative model-based monitoring systems which use electrical measurements of current, voltage, and power. It will focus on characterizing the error in voltage and current measurements of PV cells and study their impact on the temperature on the cell output power voltage and current. The main objectives of this work are the following:

- Define the errors associated with current and voltage measurements.
- Evaluate how these errors and temperature influence fault detection.
- Recommend practices to improve the accuracy of monitoring and fault detection.

4.2 Photovoltaic System Characteristics

PV systems are energy-conversion systems that are designed to generate electricity by converting sunlight into electricity, the physical phenomenon that occurs is the PV effect, and it is the basis of this conversion. This effect is a specific case of the photoelectric effect, where electrons are excited to a higher energy level when exposed to sunlight [13–15]. What distinguishes these effects is that in the PV effect, excited carriers, or free electrons, are still contained within the material and are not emitted to the exterior. This effect happens in PV cells and can be described simply as follows: light, pure energy, enters a PV cell and transfers enough energy to release some electrons. The PV cell consists of semiconductor material typically in the form of a p–n junction, this junction is the interface between two semiconductors made of the same material: the "n" (negative) side contains electrons in excess, and the "p" (positive) side contains a shortage of electrons, also known as holes. Because the p–n junction is composed of inversely polarized materials, an electric field is formed. This electric field causes the electrons to flow in one direction (n side) and the holes in the other direction (p side). The movement of

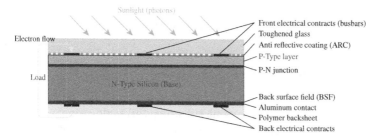

Figure 4.1 PN-junction of a PV cell.

the electrons in one direction is what constitutes the electric current in the cell. The characteristics of the electric field depend on the semiconductor material. The potential difference associated with this electric field is typically 0.6 to 0.7 V for silicon (Si) semiconductors and 0.3 to 0.35 V for germanium (Ge) semiconductors (Figure 4.1) [16, 17].

Most of the energy that reaches the cell in the form of sunlight is lost before it is converted into electricity. Light-to-electricity conversion reaches peak values in the order of 30% (much higher values are achieved for other more complex cell design types), but typically panel efficiencies are 15–20%. Certain physical phenomena limit the efficiency of solar cells; some are inherent and cannot be avoided; others can be improved by changes in design and the production processes [16].

The phenomena that most limit the efficiency of the PV cell are the following [18–21]:

- Reflection of the cell surface
- The light that does not contain enough energy to separate the electrons from their atomic bonds
- The light that has too much energy, beyond what is needed, to separate the electrons from their bonds
- Electrons and holes (empty links) that randomly meet and recombine before they can contribute to cell performance
- Electrons and holes that recombine in the surface and defects of the cell material
- Resistance to current flow
- Performance degradation due to nonoptimal operating temperatures

Further, the materials and structures used in the PV panel-manufacturing process are very similar independently from different types of solutions. That is why the manufacturing process plays a fundamental role, experience, and research to achieve the best PV module design. The assembly process of a PV panel is related to integrating each raw material by adopting all the optimizations required to improve the quality of the final product [22, 23]. The currently used

PV panels are commonly formed with the following structure: PV solar cell, Front glass, Back sheet, Encapsulant material, Frame, and Junction Box.

4.2.1 Topologies of PV Systems

The connection of PV inverters with PV panels, converters, and transformers considers four basic topologies: (i) central, (ii) string, (iii) multi-string, and (iv) module-integrated. The generation of each of these topologies is affected by location, availability of solar irradiance, temperature, and shading effect. Hence, it becomes very important that the correct choice of topology is selected during the design process according to the power output, location, reliability, cost, and efficiency [24]. A brief overview of the interconnection between the arrangement of PV panels and the converters is shown in Figure 4.2.

The central topology in Figure 4.2a interconnects multiple PV panels to one inverter. This disposition is clustered into PV arrays, with each array accumulating a series of PV strings [24–27]. Further, the string topology connects one PV string with one DC/AC converter as shown in Figure 4.2b. In Figure 4.2c, the multi-string topology is depicted as it connects each PV string with a DC/DC converter, and the multiple configurations are connected to one inverter which may or may not be close to the DC-DC converter. Finally, the module-integrated topology as shown in Figure 4.2d has one inverter per PV panel [27]. The electrical characteristics of these topologies are given in Table 4.1.

Further, based on the performance, power losses, power quality, and cost, these topologies are differentiated as shown in Table 4.2.

4.2.2 Faults in PV Systems

A fault in a PV panel origins one of two events, it causes the power degradation of the module [28–32], which is irreversible through standard procedures, or it creates safety problems. A purely aesthetic problem that does not give rise to one of the two events mentioned earlier is not considered a fault. A failure in a module

Table 4.1 Electrical characteristics of PV inverter topologies.

Inverter topology	P (kW)	V_{in} MPPT DC (V)	V_{out} AC (V)	f (Hz)
Central	100–1500	400–1000	270–400	50,60
String	0.4–5	200–500	110–230	50,60
Multi-string	2–30	200–800	270–400	50,60
Module-integrated	0.06–0.4	20–100	110–230	50,60

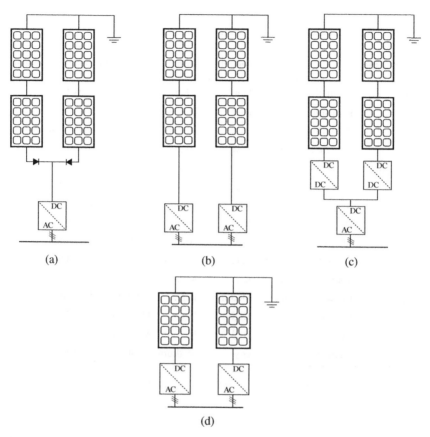

Figure 4.2 Different PV inverter topologies. (a) Central. (b) String. (c) Multi-string. (d) Module integrated.

is relevant regarding insurance when the fault occurs under normal and expected operating conditions [33]. Some defects in the modules occur during their production. These defects may be the reason why some modules do not function as well as expected. As long as these defects are not relevant to the safety or the labeled power level, taking into account the loss of power rating for imperfections due to production, these defects are not considered a module failure [34]. A PV system can be subject to a wide variety of faults, including those of the PV array, those of the inverter and those of the electricity grid. This chapter focuses on PV arrays, which have nonlinear characteristics of limited current and voltage, (I–V) curves. Failures in PV systems require special attention and careful evaluation. A brief overview of various faults at different components in a PV system is shown in Figure 4.3 [35].

Table 4.2 Main characteristics of PV inverter topologies.

Characteristics		Central	String	Multi-string	Module-integrated
Performance	Reliability	Low	High	Medium	Very high
	Robustness	High	Low	Medium	Very low
	Flexibility	Low	High	Medium	Very high
	MPPT efficiency	Low	High	Medium	Very high
Power losses	Mismatching	High	Low	Low	Very low
	Switching	High	Low	Medium	Very low
	AC power loss	Low	Medium	Medium	High
	DC power loss	High	Low	Medium	Very low
	AC voltage variation	Low	High	Medium	Very high
Power quality	DC voltage variation	Very High	Medium	High	Very low
	Voltage balance	High	Medium	Low	Low
Cost	Installation cost	Medium	High	Medium	Very high
	DC cables	High	Low	Medium	Very low
	AC cables	High	Medium	Medium	High
	Maintenance	Low	Medium	High	Very high

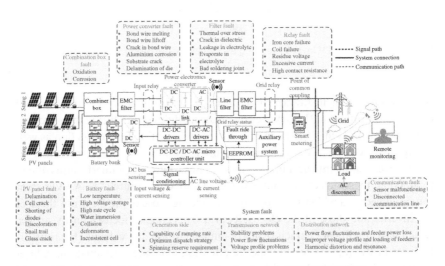

Figure 4.3 Faults in PV Systems. Source: Kurukuru et al. [35]/MDPI/CC BY.

A wide variety of faults are usually found within PV arrays. In this chapter, it is considered that the PV array is the only source of fault current since most of the inverters are endowed with galvanic isolation between the PV arrays and the electric power grid. Among these faults, ground faults and line-to-line faults are those that have the greatest potential to cause high fault currents [36–38]. Without proper protection and fault detection, these defects can cause serious problems in the PV array, such as DC arcs and even fire hazards. Also, parallel or serial DC arcs can fall into the categories of ground faults, line-line fault, open-circuit faults, and mismatch faults. Further, the defects in PV modules can include discoloration, cracking, snail tracks, antireflection coating damage, bubbles, soiling, busbar oxidation, corrosion, split encapsulation over cells interconnections, and back sheet adhesion loss, etc., A detailed understanding of these faults is provided in [7, 39–41].

4.2.3 Monitoring of PV Systems

Monitoring systems are a crucial part of any electrical installation; it allows to collect, analyze, and use the information to manage the system and track faults. Monitoring is conducted throughout the operation period of the PV system [35, 42]. As mentioned in Section 4.1, the monitoring approaches are categorized as quantitative model-based solutions, process history-based solutions, and signal processing-based solutions. Further, each of these approaches is classified depending on the measuring characteristics, algorithms used, and indices adapted.

Generally, the quantitative analysis method is classified as current-voltage (I–V), performance comparison, performance ratio (PR), and capture loss [43, 44]. The I–V curve analysis monitors the PV operation based on the operating points of PV modules in a string or array. The impacts of various faults in the IV curve of a PV module are shown in Figure 4.4.

Based on the shape characteristics of the I–V curve, the PV module faults are classified and detected in this analysis. This approach mostly allows detecting series losses, mismatch losses, losses in the metallic parts of the system, and other losses causing reduced current and voltage [42]. Further, the performance comparison approach is adapted for monitoring the PV system performance for energy loss by studying the fault patterns and by developing the fault detection approaches [43, 44]. Generally, this approach estimated the expected performance of the system and compares it with the actual performance by considering the weather information for fault detection [43, 44]. The third type of quantitative analysis method is the PR method. Here PR estimated as normalized parameter of PV system energy yield to identify the performance of the system. This approach evaluates the system performance irrespective of the orientation and inclination

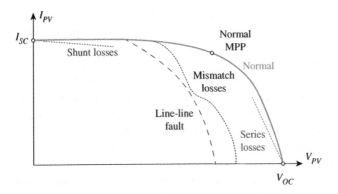

Figure 4.4 I–V curve with fault impact representation.

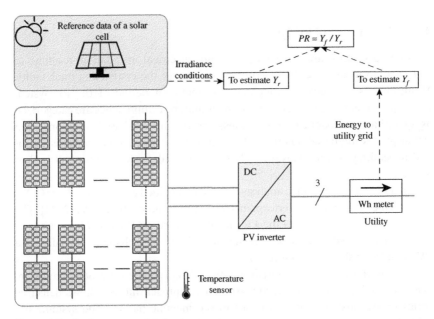

Figure 4.5 Performance ratio diagram.

of the panel. PR considers the overall effects of the system losses so that it can be used for fault detection [43, 44].

The estimation process with the PR approach is shown in Figure 4.5. The estimation of PR is defined by the ratio of final yield (Y_f) to reference yield (Y_r) as

$$PR = \frac{Y_f}{Y_r} \ (dimensionless) \tag{4.1}$$

Figure 4.5 illustrates how to obtain Y_f and Y_r in a PV system. The Y_f identifies the normalized AC energy output to the grid as given in Equation (4.2)

$$Y_f = \frac{E}{P_0} (\text{kWh/kW}) \; or \; (\text{hours}) \tag{4.2}$$

where E is the net AC power output, and P_0 is the nominal PV power [44].

Further, Y_r identifies the conditions under normalized solar irradiation as given in (4.3)

$$Y_r = \frac{H}{G_{STC}} (\text{hours}) \tag{4.3}$$

where H is the total in-plane solar irradiation (kWh/m^2) and G_{stc} is 1 kW/m^2 [43].

The final classification of the quantitative analysis method is the capture loss analysis. In this approach, the system capture loss (L_c) is estimated to analyze the abnormality at the system level as given in Equation (4.4)

$$L_c = Y_r - Y_A \tag{4.4}$$

Here, Y_A is the daily array energy output per kW of the installed PV array. Besides, Lc can be formulated based on two types of losses, thermal capture loss (L_{ct}) and miscellaneous capture loss (L_{cm}) are given in Equation (4.5).

$$L_c = L_{ct} - L_{cm} \tag{4.5}$$

Here, the measured capture loss ($L_{c_{mes}}$) is constrained to a theoretical if the PV array is normal. Therefore, fault detection is achieved using Equation (4.6) as

$$[L_{c_{sim}} - 2\delta] < L_{c_{mes}} < [L_{c_{sim}} + 2\delta] \tag{4.6}$$

where δ is the standard deviation of the simulated capture losses $L_{c_{sim}}$ [43].

As can be seen in Figure 4.6, module parameters, plant configuration, irradiance, and module temperature are inputs for the simulation model. This simulation has as output the instantaneous simulated energy produced, which is subtracted to the reference yield, obtaining the corresponding losses. The simulated losses are then compared with the measured losses obtained through electrical measurements of voltage and current. This comparison allows for deciding whether a fault does or does not exist. It is then possible to classify faults based on ratios of voltage and current between their simulated and measured values.

Further, the process history-based analysis methods are categorized based on two different approaches – statistical methods and machine learning (ML) [45]. The statistical methods adapt the inferential and descriptive statistics of the power generation to detect the abnormalities in the system [44, 45]. Whereas, ML analyzes the characteristics of the information obtained from the given PV dataset to predict the operating state of the system. Based on how the data is processed for the purpose of learning, these ML algorithms are categorized as supervised, unsupervised, and semi-supervised approaches.

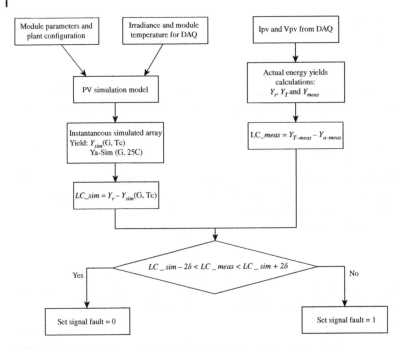

Figure 4.6 Capture loss analysis decision tree.

The last classification of the monitoring methods is signal-processing solutions. Monitoring with these methods is achieved through different approaches like, DC arc-fault circuit interrupter [46, 47], insulation resistance monitor [48], residual current detector [49–51], and time-domain reflectometry [52, 53]. The DC arc-fault circuit interrupter generates a significant amount of noise in the system to develop a fault signature such that the series-parallel arcing faults in the sources and output are detected and interrupted. This approach fills the protection shortcomings of conventional ground fault detection interrupters and open circuit protection devices. Similarly, the insulation resistance monitor is used as ground-fault protection in ungrounded PV systems when the system is de-energized. Generally, the insulation resistance of the PV array ranges from $k\Omega$ to $M\Omega$, and a significant decrease is observed when the insulation or direct contact faults occur. This reduced insulation resistance indicates the occurrence of a ground fault in the system [44]. Further, the residual current detector in a power system acts as a differential relay for monitoring the current difference between input and output terminals of the protected zone as shown in Figure 4.7. The primary currents and corresponding currents coming in and leaving out from the protected zone are indicated with black and red arrows respectively and were labelled as I_{1a} and I_{1b}. The difference between these currents is given by I_{op} as shown in the figure [22].

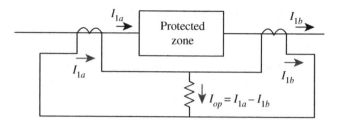

Figure 4.7 Residual current detector for internal fault detection.

Here, if this monitored difference exceeds the threshold, a fault alarm is raised. However, the leakage current is caused by the parasitic capacitance of the PV array for the systems with transformerless inverters, which results in nuisance tripping on a residual current detector.

In the time-domain reflectometry approach characteristics of the electrical cables are determined by injecting step signals or impulses to observe the reflected waveforms. Further, the measured reflections of an unknown system are compared with the reflections obtained from a known or reference system. The schematic for setting up time-domain reflectometry in a PV field is given in Figure 4.8.

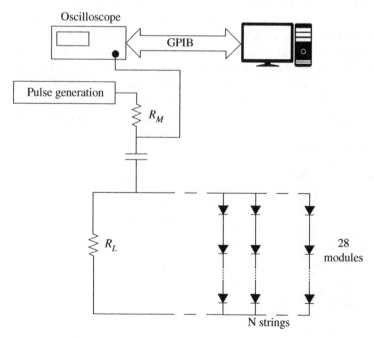

Figure 4.8 Setup for time-domain reflectometry in the PV field.

4.3 Solar Cell Characterization and Modelling

In order to test the possible detection of the most usual faults, gain sensibility with this subject, learn how to draw an I–V curve, and make an analysis of how voltage and current vary from healthy to faulty cells, a set of experimental measurements and simulations were carried out in the laboratory.

4.3.1 Experimental Analysis of a Solar Cell

A series of laboratory tests were made in order to obtain voltage and current curves for different irradiance values. For this analysis, two SUNTECH PV panels were used, one in perfect condition and another one damaged with the glass cracked and with a broken cell. This allowed us to perform this experiment. The corresponding pictures are shown in Figure 4.9.

The panel used the experiments has the technical characteristics listed in Table 4.3.

In order to have access to the cell terminals, the EVA coating was melted, allowing to perform the desired voltage and current measurements. The full I–V curve could not be observed at first this was due to high resistance in the sensors and cabling. Taking into account that the cell electrical characteristics at STC conditions are $V_{oc} = 0.63$ V, $I_{sc} = 5.62$ A, and that the maximum observable current value was around 0.5 A, knowing that $V = R.I$, this means that the resistance of the measuring system was around 1 Ω. Hence it had to reduced 10 times to around 0.1 Ω, in order to observe the full, I–V curve until the short circuit current value I_{sc} was reached. To accommodate the losses associated with the cabling connectors and sensors, different gain values are introduced into the system. This causes resistance in the system, which leads to a decrease in the accuracy of fault

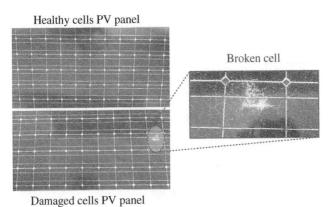

Healthy cells PV panel

Broken cell

Damaged cells PV panel

Figure 4.9 Healthy and damaged cell PV panels.

Table 4.3 Characteristics of a PV panel.

Characteristics	
Maximum power voltage (V_{mp})	36.6 V
Maximum power current (I_{mp})	5.20 A
Open-circuit voltage (V_{oc})	45.2 V
Short-circuit current (I_{sc})	5.62 A
Maximum power for STC (P_{max})	190 W
Module efficiency	14.9%
Normal operating condition temperature (T_{noct})	45 °C ± 2 °C
Weight	15.5 kg
Size	1580 mm × 808 mm × 35 mm
Maximum system voltage	1000 V
Cell technology	mono-Si
Number of cells	72 (6 × 12)

Data presented in standard test conditions (STCs), air mass (AM) = 1.5 G = 1000 W/m², T_c = 25 ° C.

detection. In order to correct this phenomenon, measurements were made as given in Table 4.4. Here G is the incident irradiance, I_{cc} is the current when the cell terminals are short-circuited, V_m is the voltage at the terminals of the voltage sensor, $VR1$ is the voltage between the positive terminal of the cell and the positive output of the voltage sensor, $VR2$ is the voltage between the negative terminal of the cell and the negative output of the voltage sensor, VR is the sum of $VR1$ and $VR2$, T_{cell} is the temperature of the solar cell at the time of measurement and R is the resulting resistance value of the cabling, connectors and sensor obtained in the measuring system.

Table 4.4 Healthy cell measurements for resistance correction.

Healthy cell							
G(W/m²)	I_{cc}(A)	V_m(V)	$VR1$(V)	$VR2$(V)	VR(V)	T_{cell}(deg)	R(Ω)
1000	4.37	0.21	0.0484	0.0685	0.1169	64	0.026751
800	3.25	0.1425	0.0384	0.0562	0.0946	55	0.029108
600	2.46	0.0946	0.0285	0.0382	0.0667	47	0.027114
400	1.595	0.0701	0.0192	0.026	0.0452	39	0.028339
230	0.903	0.0373	0.0109	0.0136	0.0245	32	0.027132

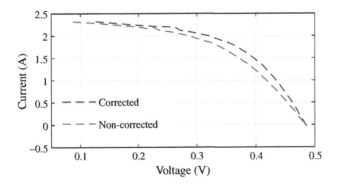

Figure 4.10 Broken cell corrected and non-corrected I–V curve for $G = 1000\,\text{W/m}^2$.

These unwanted voltage drops in the monitoring system can eventually bring problems in fault detection, e.g. if 0.05 V in the positive side of the cabling sums up with 0.05 V on the negative side this will lead to an error voltage of 0.1 V. Knowing that the output voltage of the cell is 0.21 V the voltage drop in the measuring system accounts for around 50% of the output. These values reduce as the irradiance decreases since the current is smaller and so is the voltage drop in the measuring components. This experiment was made for just one cell, if this is applied to an entire PV array or string, much more cabling will be involved and thus the error will be higher. Further, a short-circuited cell is equivalent to losing 0.5 V, which also means that if 5 cells present the above-mentioned error a false alarm of a SC fault could appear. The values of the voltage drop in all these components were used to correct the plots of the I–V curves. The voltage drops on the measuring system were added to the voltage output of the cell, correcting the obtained I–V curves. The example given was for the healthy cell situation; these same corrections were performed for the other two cases corresponding to the damaged panel. A corrected and non-corrected I–V curve for a $G = 1000\,\text{W/m}^2$ for the Broken cell is shown in Figure 4.10. As can be seen in the figure, the series losses have decreased significantly with the correction.

4.3.2 Error Computation

An error propagation analysis along the measuring chain is performed to achieve the resistance correction. In order to compute the error associated with the resistance correction the following rules that can be derived from the Gauss equation for errors with a normal distribution were used. Where a, b, c, … are measured quantities and δa, δb, δc, … are the uncertainties associated with those measurements [54]. Computing the error in sums and subtractions: If Q is a combination

Table 4.5 Resistance correction associated error.

Calculating the error				
VR1 error	VR2 error	I_{cc} error	VR error	R error
0.000345	0.000406	0.11925	0.000533	0.00074
0.000315	0.000369	0.09125	0.000485	0.000831
0.000286	0.000315	0.07150	0.000425	0.000807
0.000258	0.000278	0.049875	0.000379	0.000917
0.000233	0.000241	0.032575	0.000335	0.001047

of sums and subtractions, i.e.

$$Q = a + b + \cdots + c - (x + y + \cdots + z) \tag{4.7}$$

So,

$$\delta Q = \sqrt{(\delta a)^2 + (\delta b)^2 + \cdots + (\delta c)^2 + (\delta x)^2 + (\delta y)^2 + \cdots + (\delta z)^2} \tag{4.8}$$

Computing the error in multiplication and division:
If,

$$Q = \frac{ab \ldots c}{xy \ldots z} \tag{4.9}$$

So,

$$\delta Q = \sqrt{\left(\frac{\delta a}{a}\right)^2 + \left(\frac{\delta b}{b}\right)^2 + \cdots + \left(\frac{\delta c}{c}\right)^2 + \left(\frac{\delta x}{x}\right)^2 + \left(\frac{\delta y}{y}\right)^2 + \cdots + \left(\frac{\delta z}{z}\right)^2} \cdot |Q| \tag{4.10}$$

The error propagation in the correction is computed as given in Table 4.5.

These values represent the error associated with measuring equipment and process. These instruments were used to measure the values that led to the correction of the voltage drop across the connections, they are significant and must be accounted for. As it is possible to conclude there are many errors associated with measuring and they appear across all stages of monitoring. They can have a serious impact on fault detection and on the accuracy of monitoring systems.

4.3.3 Current-Voltage Characteristics

Solar cell I–V characteristic curves are graphs of output voltage versus current for different levels of insolation and temperature and can tell a lot about a PV cell or panel's ability to convert sunlight into electricity. The most fundamental values for calculating a particular panel power rating are the voltage and current at maximum power rating.

4.3.3.1 Healthy Cell I–V Curves

The I–V and Power–Voltage (P–V) curves for the healthy cell under all the proposed irradiances are given in Figures 4.11 and 4.12, respectively.

It can be observed that the I–V curves correspond to a healthy cell. The power-produced scales with the irradiance available on the slopes of the curve are small, which indicates small shunt and series resistance losses as shown in Figure 4.12. The I_{SC} observed for the irradiance of 1000 W/m^2 is of around 4 A, this is smaller when compared with the one given at STC of 5.2 A, this is mainly because the spotlight light quality (radiation spectrum amplitude) is not the same as the sun. The voltage V_{oc} is also far from the one at STC conditions; this is

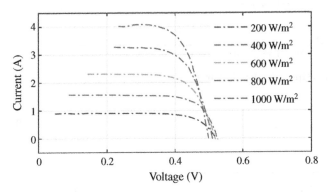

Figure 4.11 Healthy cell I–V curves.

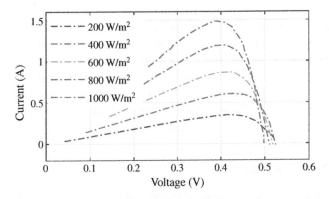

Figure 4.12 Power-voltage curves for a healthy cell.

mainly related to the temperature increase since in this experiment, the module temperature was 64 °C and not 25 °C. The V_{oc} value is 0.63 V at STC conditions, as was expected the value closest to this one corresponds to the irradiation of 230 W/m² since the module temperature was at the lowest among the others for higher irradiations. In this situation the temperature of the cell was roughly above 32 °C, which leads to an open-circuit voltage loss of −0.018 V and so a V_{oc} of 0.61 V, this value is still far from the one observed in Figure 4.11, this might be due to other reasons like performance degradation and aging. The slope of the I–V curve is affected by the amount of shunt resistance in the electrical circuit. Reduced shunt resistance results in a steeper slope in the I–V curve near I_{sc} and a reduced fill factor. A decrease in shunt resistance may be due to changes within the PV cells or modules. Potential causes are shunt paths exist in PV cells, shunt paths exist in the PV cell interconnects, and module I_{sc} mismatch. The slope of the I–V curve between V_{mp} and V_{oc} is affected by the amount of series resistance internal to the PV modules and in the array wiring. Increased resistance reduces the steepness of the slope and decreases the fill factor. The potential causes are that PV wiring has excess resistance, or it is insufficiently sized, electrical interconnections in the array are resistive, and the series resistance of PV modules has increased.

4.3.3.2 Broken Glass but Healthy Cell I–V Curves

The I–V and Power–Voltage curves for the healthy cell with broken front glass under all the proposed irradiances are given in Figures 4.13 and 4.14, respectively.

Here it is observed that there is a slight reduction in the current since the irradiance that reaches the cell is reduced due to reflections caused by the cracked glass. No significant difference can be observed comparing the curves in Figure 4.11 with

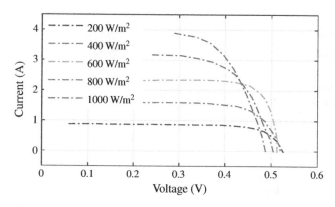

Figure 4.13 Healthy cell with a front broken glass I–V curves.

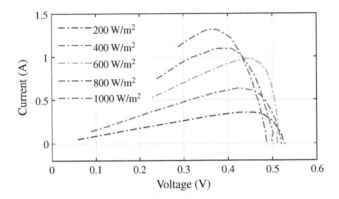

Figure 4.14 Power–voltage curves for healthy cell with a front broken glass.

the previous one in Figure 4.13 of the healthy cell. The same expected V_{oc} loss due to temperature increase with higher irradiance is visible.

4.3.3.3 Broken Glass and Damaged Cell I–V Curves

The I–V and Power–Voltage curves for the damaged cell with broken front glass under all the proposed irradiances are given in Figures 4.15 and 4.16.

In this case, the I–V curve in Figure 4.15 curve presents a reduction in the power produced by the cell, as can be seen in Figure 4.16, with a much flatter slope. The fill factor is reduced tremendously, thus indicating that this cell has some defect. The cell can be broken in an infinite number of ways, and so it could not be characterized for simulation purposes. In order to evaluate the three previous scenarios, the maximum power points were plotted and compared with the reference values from the datasheet for the irradiance of 1000 W/m^2.

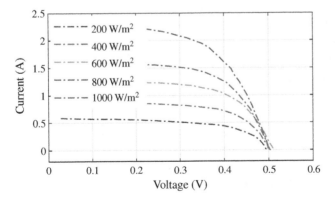

Figure 4.15 Broken glass with damaged cell I–V curves.

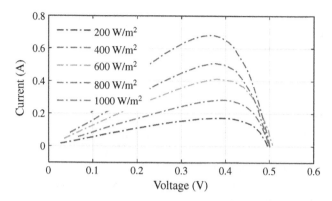

Figure 4.16 Broken glass with a damaged cell power–voltage curves.

Table 4.6 Maximum power points for the three scenarios.

MPP	Healthy	BGHC	BGBC	Reference
Power (W)	1.4849	1.3153	0.6843	2.6389
Voltage (V)	0.3863	0.3615	0.3670	0.5083
Current (I)	3.8438	3.6383	1.9647	5.6200

Further, the maximum power points for all three previous scenarios of, healthy cell (H), broken glass with a healthy cell (BGHC) and the broken front glass with a broken cell (BGBC) for 1000 W/m² are presented. Despite the experiments being conducted at the same irradiance of 1000 W/m², they differed in module temperature (Table 4.4) as seen before the increase in temperature leads to a decrease in V_{oc}. The maximum power point values for power, voltage, and current are listed in Table 4.6.

Comparing the MPP values with the reference ones, given in the datasheet for STC conditions, the variations are tabulated as shown in Table 4.7:

Table 4.7 Maximum Power Point Variations.

δMPP	Healthy	BGHC	BGBC
δP (W)	−1.1540	−1.3236	−1.9546
δV (V)	−0.1220	−0.1468	−0.1413
δI (A)	−1.7762	−1.9817	−3.7553

As can be concluded from the data obtained, the broken glass condition affects the voltage mainly due to the broken glass whilst the current decreases significantly in the broken cell scenario.

4.3.4 Simulation of I–V curves

4.3.4.1 PV Cell Simulation Model

In this chapter, the solar cell electric model and the introductions made by the system advisor model (SAM) for changes in temperature and irradiance are briefly explained. The electric current generated by the photoelectric effect is represented by the source current (I_L). The semiconductor p-n junction is represented by the diode that has a current given by (4.11)

$$I_D = I_0 \left[\exp \left(\frac{q.V_{pv}}{n.k_B.T_{cell}.N_S} \right) - 1 \right]$$ (4.11)

Applying Kirchhoff's law, the electric current supplied by the solar cell is given by:

$$I = I_L - I_0 \left[\exp \left(\frac{q.V_{pv}}{n.k_B.T_{cell}.N_S} \right) - 1 \right]$$ (4.12)

Equation (4.12) represents the ideal solar cell. It is said to be an ideal model because it does not take into consideration the power losses due to imperfections of the semiconductor and the nonzero resistance of the metallic cell interconnections. Those losses are modeled by adding a parallel resistance R_{sh} and a series resistance R_s. This model with the added resistances is called the five-parameter model (I_L, I_0, R_s, R_{sh}, n) and its output current expression is given by Equation (4.13) [54].

$$I = I_L - I_0 \left(\exp \left[\frac{V_{pv} + I_{pv}.R_s}{a} \right] - 1 \right) - \frac{V_{pv} + I_{Rs}}{R_{sh}}$$ (4.13)

The modified nonideality factor a defined in (4.14) encapsulates the diode thermal voltage V_T, the number of cells in series in the PV module N_s and the diode nonideality factor n [55]. Given that the thermal voltage is defined as $V_T = k . T_{cell}/q$, where k is the Boltzmann constant, T_{cell} the cell temperature and q the elementary charge, the modified nonideality factor is:

$$a = N_s \frac{k_B.T_{cell}}{q}.n$$ (4.14)

At STC, $T_{cell} = 298$ K, resulting in $a \approx 0.025. n. N_s$ [55].

Normally, values such as I_{sc}, V_{oc}, I_{mp}, and V_{mp} can be found in the datasheet of the solar panel. These values depend on temperature (T) and irradiance (G), and can be found on the panel datasheet under STCs. The test conditions of irradiance, cell temperature, and AM, defined by this international industrial-wide standard to indicate the performance of PV modules, are: $G = 1000\ Wm^2/T_{cell} = 25\,°C$ or $298.25\ K$, and $AM = 1.5$. In order to accurately compute the R_s and R_{sh} parameters, reliable measurements of the I–V curve under controlled irradiance and temperature conditions are necessary.

As mentioned, Equation (4.13) depends on the temperature and irradiance, and the five parameters are valid in STC conditions. It is then necessary to also model the dependency of these parameters with temperature and irradiance changes. The SAM takes into consideration these variations and adds a sixth parameter to the five-parameter model, the adjust parameter. The purpose of this new parameter is to adjust the short-circuit current and open-circuit voltage temperature coefficients (α_{sc} and β_{oc}) that are also in the solar panel datasheet. The idea is to adjust α_{sc} and β_{oc}, down and up, respectively, by the same percentage, so that the maximum power point temperature coefficient (γ), also in the datasheet, matches the one predicted by this new six-parameter model. Below are the equations that translate the temperature coefficients adjustment and the variation of the MPP with temperature [56].

$$\alpha'_{sc} = \alpha_{sc}\left(1 - \frac{Adjust}{100}\right) \tag{4.15}$$

$$\beta'_{oc} = \beta_{oc}\left(1 + \frac{Adjust}{100}\right) \tag{4.16}$$

$$\gamma_{spec} = \frac{\partial P}{\partial T}\Big|_{mp} \approx \frac{P|_{T_{cell}} - P|_{T_{STC}}}{P_{mp,stc}(T_{cell} - T_{STC})} \tag{4.17}$$

To solve Equation 4.17, it is necessary to know how the MPP varies with temperature. By knowing the temperature dependencies of the remaining parameters, one can calculate different MPPs for different cell temperatures and irradiances. The current generated by the incident radiation (source current in the five-parameter model) varies almost linearly with the irradiation available but also depends on the temperature as can be seen in Equation (4.18).

$$I_L = \frac{G}{G_{STC}} \cdot \frac{AM}{AM_{STC}}[I_{L_{STC}} + \alpha_{sc}(T_{cell} - T_{cell,STC})] \tag{4.18}$$

where G is the irradiance and AM is the air mass.

On the other hand, the saturation current of the diode representing the p–n junction only depends on the temperature:

$$I_0 = I_{0_{ref}}\left[\frac{T_{cell}}{T_{STC}}\right]^3 \exp\left[\frac{1}{k_B}\left(\frac{E_g}{T}\Big|_{T_{STC}} - \frac{E_g}{T}\Big|_{T_{cell}}\right)\right] \tag{4.19}$$

The energy bandgap (E_g) depends on the semiconductor used (1.12 eV for silicon), and has also its dependency on the temperature:

$$E_g = E_g T_{STC}[1 - 0.0002677(T_{cell} - T_{cellSTC})] \tag{4.20}$$

Parameter R_{sh} depends only on the incident irradiation:

$$R_{sh} = R_s \frac{G_{STC}}{G} \tag{4.21}$$

Parameter R_s is considered constant no matter the temperature and irradiance conditions [56]. The ideality factor will be contained inside the parameter called modified nonideality factor (a), which also encapsulates the diode thermal voltage and the number of cells in series (N_s) in the module. Its equation and temperature dependence are described in Equations (4.22) and (4.23).

$$a = N_s \frac{k_T}{q_n} \tag{4.22}$$

$$a = a_{STC} \frac{T_{cell}}{T_{STC}} \tag{4.23}$$

Given all these temperature dependencies and the datasheet parameters, it is possible to formulate constraints to solve the module I–V equation and determine the six parameters (a, R_s, R_{sh}, I_L, I_0, $Adjust$). The first constraint derives from the short circuit condition where $V = 0$ and $I = I_{sc}$:

$$I_{sc} = I_L - I_0 \left[\exp\left(\frac{I_{sc}}{R_{sa}} \right) - 1 \right] - \frac{I_{sc}R_{sa}}{a} \tag{4.24}$$

The second constraint derives from the open circuit condition where $V = V_{oc}$ and $I = 0$:

$$I_L - I_0 \left[\exp\left(\frac{V_{oc}}{a} \right) - 1 \right] - \frac{V_{oc}}{R_{sh}} = 0 \tag{4.25}$$

The third constraint derives from the maximum power point, where $I = I_{mp}$ and $V = V_{mp}$:

$$I_{mp} = I_L - I_0 \left[\exp\left(\frac{V_{mp} + I_{mp}R_s}{a} \right) - 1 \right] - \frac{V_{mp} + I_{mp}R_s}{R_{sh}} \tag{4.26}$$

The fourth constraint requires that the slope of the PV curve be zero around the MPP: Since $P = I. V$ [56].

$$\frac{d(IV)}{dV} \big|_{mp} = I_{mp} - V_{mp} \frac{d(I)}{dV} \big|_{mp} = 0 \tag{4.27}$$

The fifth constraint ensures that the manufacturer specified β_{oc} matches the value determined by the six-parameter model. V_{oc} temperature dependence is as follows:

$$V_{oc_{T'}} = \beta_{oc} \left(1 + \frac{Adjust}{100} \right) \Delta T + V_{oc_{STC}} \tag{4.28}$$

where $T' = \Delta T + T_{STC}$.

Substituting the open circuit constraint equation (4.25) with a, I_L and I_0 and their temperature adjustments stay:

$$\left[\alpha_{sc}\left(1 - \frac{Adjust}{100}\right)\Delta T + I_L\right] - I_{0_{T'}}\left[\exp\left(\frac{V_{oc_T}}{a_{T'}}\right) - 1\right] - \frac{V_{oc_{T'}}}{R_{sh}} = 0 \qquad (4.29)$$

The sixth and final constraint forces the manufacturer to maximum power point temperature coefficient (γ_{spec}) to equal the maximum power point temperature coefficient calculated by the six-parameter model (γ_{model}):

$$\gamma_{spec} - \gamma_{model} = 0 \qquad (4.30)$$

These six nonlinear constraint equations are solved simultaneously using a globally convergent variant of Newton's method. Through this methodology, a database of already pre-calculated six parameters for several solar panels of different manufacturers is available online in the MATLAB toolbox.

4.3.4.2 Simulation Model with SAM Parameters

After performing the previous experiments and learning about each component's influence on the behavior of the system and its impact on the I–V curve, simulations of short-circuit and open-circuit cells were made. Using a graphical programming environment for modelling, simulating, and analyzing multidomain dynamical systems, containing the SAM parameter values for STC conditions, a five-parameter model of the PV cell was created. According to the model, the values for this solar panel cell are $R_{sh} = 258.20\ \Omega$, $R_s = 0.6\ \Omega$, and $I_L = 5.633\ A$. There is more than one diode model available in the block library. The diode that correctly models the PV cell p–n junction is the exponential diode model. This diode model has SAM model parameters that can be seen in Equation (4.11). Creating a block out of this PV panel, connecting it with a load, an electrical ground, and sensors for voltage and current measurements, the PV monitoring system is obtained.

The load was defined as the MPP load, $R_{mp} = V_{mp}/I_{mp}$, in order to set the operating point of the system in MPP conditions since the input parameters for the model are the SAM parameters in STC conditions. We can observe in Figure 4.17 that the voltage reading correctly corresponds to the expected maximum power voltage, $V_{mp} = 36.69\ V$, for STC conditions, given in the datasheet of the panel. Using this model, faults were induced in the system.

As an example, 10 short-circuited cells were introduced in Panel 1, as can be seen in the image. As it is possible to observe the voltage in Panel 1 reduced from 36.69 V to 31.97 V, this reduction of 4.99 V matches the expected since each cell has a voltage drop of around 0.5, i.e. 5 V for ten cells in short-circuit. In Figure 4.18, the I–V curve of the simulated panel can be seen, the voltage and current values from the datasheet of the panel approximately match the ones obtained.

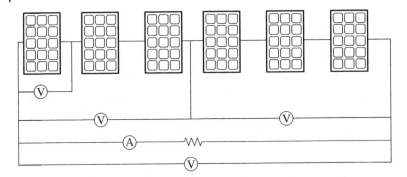

Figure 4.17 Six-panel PV system with voltage and current monitoring.

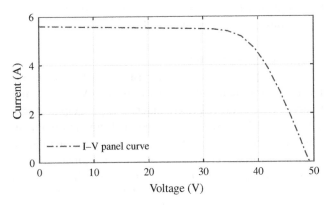

Figure 4.18 PV panel I–V curve.

4.4 Power Variations

In this chapter, power loss induced by temperature variation along the year and its influence on fault detection in PV systems will be presented.

4.4.1 Power Variation with Temperature

The voltage, current, and, consequently, the power generated in a PV cell vary depending on irradiance and cell temperature values which affect the short-circuit and open-circuit voltage temperature coefficients, α_{sc} and β_{oc}, correspondingly. Through the power data access viewer, sets of meteorological data were obtained [57]. Firstly, a connection to the NASA Surface meteorology and Solar Energy database for a particular location is needed. This is made by selecting the "power single point solar access" for taking data for a specific point on the map [57]. Following, keeping the default selection "SSE-Renewable energy selection, a temporal average had to be chosen. The inter-annual option was picked since it presents the monthly average radiation for the chosen years. The monthly option was picked since it presented a significant difference in the values obtained between months, while in the daily average temperature sampling, the values did not present a significant difference between them. At last, the selection of the desired output file format and the selection of the desired meteorological parameters to be presented are asked. The output format of ASCII and the following meteorological parameters were selected, as shown in Table 4.8. Since α_{sc} and β_{oc} change with irradiance and temperature, thus affecting the power produced, the fault sensibility of any monitoring system needs to vary along the year in order to differentiate a fault, from a voltage deviation or a power loss originated by the temperature increase. A monthly temperature reference calibration is proposed. As previously said, a significant difference in temperatures and irradiance between months is observed, e.g. even if the irradiance is much higher in August as can be seen in Table 4.8. The temperature is also much higher, and this will have, as a consequence, an associated power loss and voltage deviation. The smooth performance of a solar PV module is strongly geared to the factor temperature. Higher than standard conditions temperatures can mean losses in maximum output power which is why we would usually aim at optimally cooling the modules. Each solar cell technology comes with a set of unique temperature coefficients. These temperature coefficients are crucial; the temperature of the

Table 4.8 Average characteristics of the PV system.

2021	T_{max} (deg)	T_{min} (deg)	IRR_{avg} (W/m^2)	IRR^*_{avg} Monthly	T^*_{avg} Monthly	$\Delta\theta^*$ at 25 °C
Jan	15.0	9.49	230	452.7	28.66	3.66
Feb	14.5	8.21	326	589.0	32.22	7.22
Mar	14.93	10.4	400	672.3	35.10	10.10
Apr	17.0	11.5	517	843.9	42.32	17.32
May	19.5	12.7	640	962.9	48.48	23.48
Jun	21.8	15.5	637	922.8	49.56	24.56
Jul	23.3	17.0	682	948.3	51.81	26.81
Aug	26.94	17.69	664	912.1	54.30	29.30
Sep	25.9	17.5	508	772.7	49.14	24.14
Oct	21.8	14.8	389	609.0	40.07	15.07
Nov	16.9	12.2	225	442.8	30.23	5.23
Dec	16.2	10.4	208	434.3	29.26	4.26
Ann	19.5	13.1	451			

solar cell has a direct influence on all electrical quantities being produced by it. Once the temperature at which a solar PV cell operates increases, its voltage and power output will decrease significantly, and its current will slightly increase. Crystalline solar cells are the leading cell technology and usually come with a temperature coefficient of the maximum output power of $-0.5\%/°C$. For the panel used this coefficient presents a value of $-0.45\%/°C$ which is better than the average value. The open-circuit voltage temperature coefficient (β_{oc}) for this panel is $-0.34\%/°C$. The rated power as generally indicated on the module's label is measured at $25°C$, STC conditions, with any temperature increase above it; there is an associated maximum power loss of 0.9% and an open-circuit voltage decrease of 0.68%/2 °C increase [58]. Most installed solar modules in sunny countries easily reach higher temperatures than $25°C$. Cell temperatures of $50°C$ and above are easily reached [57]. There is a simplified module temperature model that consists of assuming that the module temperature concerning the ambient temperature is directly proportional to the incident irradiance multiplied by a constant [59–61].

$$\theta_m = \theta_a + kG \tag{4.31}$$

where θ_m is the module temperature in (°C), θ_a is the ambient temperature in (°C), G is the incident irradiance in (mW/cm^2), and k is the Ross constant with a value of 0.3 in (cm$^2 \cdot °C \cdot$ mW^{-1}) [60]. Further, the averaging of the maximum incident irradiances in (W/m^2) for all the months is shown in Table 4.8. Using the maximum and minimum average temperatures obtained with the database and averaging the maximum and minimum incident irradiances obtained, the calculation of the average maximum power loss for each month is possible. This allowed the observation of the power and voltage variation with temperature and irradiance throughout the year at the chosen location. The average module maximum temperature in (°C), computed with (4.31) is shown in Table 4.8. The minimum irradiance is 0, and so for $G = 0$ W/m^2, using (4.31), Equation (4.32) is obtained.

$$\theta_m = \theta_a \tag{4.32}$$

Since the lower temperature values correspond to the time of the day with small to nonexistent radiation, these are not considered. Having obtained the module average monthly temperatures, and knowing the maximum output power and open-circuit voltage temperature coefficients, the computation of the associated average maximum power and open-circuit voltage decrease was carried out. First, it is needed to compute the difference in (°C) between the module temperature θ_m and the $25°C$ of the STC reference conditions:

$$\theta_m - \theta_{STC} = \Delta\theta \tag{4.33}$$

Further, multiplying the monthly temperature variation $\Delta\theta$ by the maximum output power loss (Eq. (4.34)) and by the open-circuit voltage (Eq. (4.35))

Table 4.9 Loss characteristics of the PV system.

2021	ΔP (%)	ΔV (%)	ΔP(W)	ΔV(V)	MP(W)	OCV(V)
Jan	−1.65	−1.25	−3.13	−0.56	186.8	44.64
Feb	−3.25	−2.45	−6.17	−1.11	183.8	44.09
Mar	−4.55	−3.43	−8.64	−1.55	181.8	43.65
Apr	−7.79	−5.89	−14.81	−2.66	175.1	42.54
May	−10.57	−7.98	−20.07	−3.61	169.9	41.59
Jun	−11.05	−8.35	−21.00	−3.78	169.0	41.42
Jul	−12.06	−9.12	−22.92	−4.12	167.0	41.08
Aug	−13.19	−9.96	−25.05	−4.50	164.9	40.70
Sep	−10.86	−8.21	−20.64	−3.71	169.3	41.49
Oct	−6.78	−5.12	−12.88	−2.32	177.1	42.88
Nov	−2.36	−1.78	−4.48	−0.80	185.5	44.40
Dec	−1.92	−1.45	−3.64	−0.65	186.3	44.55

temperature coefficients, the power and voltage decrease percentage due to the temperature factor is obtained, listed in Table 4.9.

$$\Delta P(\%) = \Delta\theta.\gamma \qquad (4.34)$$

$$\Delta V(\%) = \Delta\theta.\beta_{oc} \qquad (4.35)$$

For the solar panel with 72 cells, with maximum power at STC of 190 W and an open-circuit voltage of 45.2 V the associated maximum average power loss in (W), (Table 4.9), and maximum open-circuit voltage decrease in (V), (Table 4.9), for each month of 2018 is given by Equations (4.36) and (4.37).

$$\Delta P(W) = Power\ Loss(\%) \times P_{max_{STC}} \qquad (4.36)$$

$$\Delta V(V) = Voltage\ Decrease(\%) \times V_{oc_{STC}} \qquad (4.37)$$

This means that the average maximum power produced (4.38) and open-circuit voltage) (Eq. (4.39)) in the hours of maximum solar irradiation are given by:

$$MP(W) = P_{max_{STC}} + \Delta P(W) \qquad (4.38)$$

$$OCV(V) = V_{oc_{STC}} + \Delta V(W) \qquad (4.39)$$

When monitoring the PV system, these power and voltage losses need to be considered as these variations could be seen as a fault by the monitoring system. The system should be calibrated considering this variation so that the fault detection system differentiates a faulty panel from a power or voltage loss associated with a temperature increase. If the detection method is calibrated monthly considering these new values as a reference, the uncertainty window will decrease. With the calibration, a new threshold for fault detection would be set, and faults could be detected more accurately.

4.4.2 Manufacturing Process Power Variation

In the datasheet of the PV panel, it is said by the manufacturer that a positive power deviation can exist and that it can vary from 0 to +5%. The monitoring system must be calibrated for the power deviation measured in each PV panel. This power deviation changes accordingly to the equipment used. For this panel, there are 72 cells, and the maximum power at STC conditions of 190 W, which using (4.40) corresponds to a power of 2.64 W per cell.

$$P_{cell} = \frac{P_{STC}}{N_{cell}} \qquad (4.40)$$

Knowing that the maximum power deviation indicated by the manufacturer is +5%, the power produced by a cell can reach at least 2.772 W, which multiplied by the number of cells of the panel corresponds to a PV panel output power that varies between 190 and 199.5 W, this corresponds according to (4.41) to a maximum power deviation of +9.5 W.

$$\Delta P = P_{max} - P_{STC} \qquad (4.41)$$

There is a necessity for calibration; this power deviation should be considered. If the monitoring system is calibrated for the STC reference and there are faults in the PV system that correspond to that power value interval, the monitoring system will not be able to detect them.

4.5 Fault Detection Zone Evaluation

In this section, some recommendations for a fault detection method based on uncertainty induced by temperature variation are proposed, being uncertainty the threshold that separates a clear fault from a power loss of unknown origin. This could lead to a simple and comfortable method to implement because just voltage sensors and one current sensor are necessary per string. The consequences from an electrical point of view of any defect in a PV panel will eventually lead to a short-circuited cell or an open-circuit. Taking this as a premise, these are faults being analyzed in this work specifically the short-circuited cell since open-circuit causes the activation of the bypass diode which corresponds to all the cells in the string being short-circuited and consequently losing all the power generated by the string.

Assuming that N_s is the number of panels in a string and that N_d is the number of panels monitored by a single voltage sensor, the power monitored by this sensor P_{sensor} is given by (4.42), and the total power in the string P_{string} is given by (4.43).

$$P_{sensor} = N_d . V_p . I_d \tag{4.42}$$

$$P_{string} = N_s . V_p . I_d \tag{4.43}$$

where V_p is the output voltage of the panel and I_d is the current in the string, the following relation is obtained:

$$P_{weight} = \frac{P_{sensor}}{P_{string}} = \frac{N_d . V_p . I_d}{N_s . V_p . I_d} = N_d N_s \tag{4.44}$$

where P_{weight} is the weight of the power in (%) being monitored by the voltage sensor in the string. The value of P_{weight} will be defined after the operator decides the power loss uncertainty zone. The benefit of having a wider uncertainty zone is that fewer voltage sensors are necessary, which makes the monitoring system cheaper and less complex. The downside is that it increases the uncertainty of the monitoring system. The accuracy of the decision made on whether the maintenance team should intervene or not will depend on the size of the uncertainty area. Allowing more uncertainty, the operator loses the accuracy of the monitoring system. Two different scenarios become possible inside the uncertainty area; the power loss could be caused by a temperature variation or by one or more faults. P_{weight} decreases with the reduction of the uncertainty zone. Taking as an example the worst-case scenario, in the month of August an average voltage deviation for the maximum irradiance hour of -4.5 V (Table 4.9) can be expected per PV panel, if compared with the voltage drop loss associated with SC cells seen in Table 4.10, this corresponds to 9 SC cells and a power loss of 23.75 W. If two PV panels are covered by a voltage sensor these values will be doubled, and a voltage drop loss of 9 V with a power loss of 47.5 W becomes the new uncertainty zone range. The owner or operator of the installation should choose the number of panels in series being monitored by each voltage sensor, caring for this uncertain fault detection zone size caused by power variation induced by temperature increase.

The power and voltage loss uncertainty percentage remain the same with the increase in the number of panels covered by a voltage sensor, what will change is the power on which this percentage is applied which will have implications on the choice of the number of panels being monitored by each sensor. As an example, if eight PV panels were to be connected in series a total power of 1520 W would be produced at STC conditions. Covering this group composed of eight panels with a

Table 4.10 Short-circuit cells associated power and voltage losses.

Short-circuit cells	Voltage drop loss (V)	Voltage drop loss (%)	Power loss (W)
1	0.50	1.4	2.64
2	1.0	2.8	5.28
3	1.50	4.2	7.92
4	2.0	5.6	10.56
5	2.5	6.9	13.19
6	3.0	8.3	15.83
7	3.50	9.7	18.47
8	4.0	11.1	21.11
9	4.50	12.5	23.75
10	5.0	13.9	26.39
11	5.50	15.3	29.03
12	6.0	16.7	31.67

Table 4.11 Uncertainty of power and voltage range for the PV panel.

PV panels	Voltage loss (V)	Power loss (W)
1	4.5	23.75
2	9	47.5
3	13.5	71.25
4	18	95
5	22.5	118.75
6	27	142.5
7	31.5	166.25
8	36	190

voltage sensor would result in an uncertainty zone of 190 W (Table 4.11) resulting from the temperature variation in the month of August. This means that an uncertainty region with the range of power equivalent to a PV panel would be admitted therefore the owner of the plant would have to decide on how much power loss he can tolerate as an uncertainty window.

If the system is calibrated for power and voltage reference values that account for temperature variations (in this case monthly) the false alarm window diminishes.

Multiple arrangements for the position of the voltage sensors can be made as long as they measure voltage drops between nodes of the same string. The size of the uncertainty zone is a decision of the PV plant owner. The certainty of fault only exists when the maximum power loss due to temperature variation is overcome. If the power loss value falls inside the power variation due to the temperature increase variation window, the outcome on whether a fault causes it or not will be inconclusive. The no calibration leads to a power loss grey zone that increases with the number of panels in series being covered by a sensor, hence a fault or a false alarm can happen. In this case, the deployment of a maintenance team to perform tests and substitute a faulty panel has a cost that has to be justified. Because of this, fault detection must have a significant level of certainty. Setting new monthly references, the uncertainty zone decreases and so do the maintenance costs associated with false alarms on healthy panels. In case of fault detection, in order to pinpoint the exact faulty panel, a team or technician is deployed on-site. The maintenance team has to carry measuring equipment to the field in order to pinpoint the faulty module inside the faulty zone. This method would require more time spent on the field by the maintenance team in comparison with other more immediate and precise methods since tests need to be carried out on-site.

The precision of this method is inversely proportional to the number of panels being monitored by the same voltage sensor. Some problems may arise when using this approach. If there is a power variation in some of the panels in a given string and the sum corresponds to the power deviation defined as a fault the system would still see it has a fault. The supposed detected fault might be the sum of the power loss due to the regular PV panel performance decrease with aging. In case where the PV panels are covered by warranty and present an abnormal performance decrease that does not correspond to the expected specified by the manufacturers, the panels can be changed for new ones. As explained before, the voltage and consequently the power produced varies with irradiance and temperature throughout the year. For this reason, for each month, the uncertainty zone can acquire different sizes, as shown in Figure 4.19.

Observing Figure 4.19, if the power loss value corresponds to the orange zone a fault does or doesn't exist, while if the power loss falls inside the yellow zone, a fault exists with certainty. If the power loss value falls into the orange zone tit can be due to a fault or to the temperature increase and lead to a false fault alarm. If a new reference that starts in the yellow zone is set, the monitoring system will be able to detect faults much more accurately. The problem of calibrating the system to new references values is that they are for the case of average maximum power loss of a given month. If the panel produces more power than the reference, then some fault might occur without notice. In this case, the power deviation is positive in relation to the reference if the fault power loss falls in that window the monitoring system will not detect it. Applying this methodology with one panel per sensor

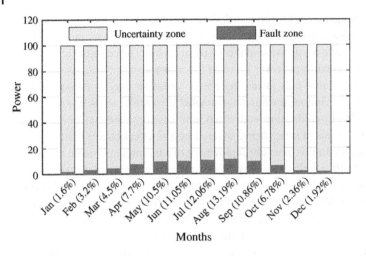

Figure 4.19 Fault decision zones.

Table 4.12 Number of faults necessary for power loss to fall in the fault certainty zone.

Month	P_{loss} Zone (W)	No of fault	P_{loss} (W)	No of fault	P_{loss} (W)	No of fault	P_{loss} (W)	No of fault	P_{loss} (W)
Jan	3.13	2	5.98	1	63.3	3	3.97	2	3.91
Feb	6.17	3	7.92	1	63.3	5	6.62	4	7.82
Mar	8.64	4	10.56	1	63.3	7	9.27	5	9.77
Apr	14.81	6	15.83	1	63.3	12	15.8	8	15.6
May	20.07	8	21.11	1	63.3	16	21.1	11	21.5
Jun	21.00	8	21.11	1	63.3	16	21.1	11	21.5
Jul	22.92	9	23.75	1	63.3	18	23.8	12	23.4
Aug	25.05	10	26.39	1	63.3	19	25.1	13	25.4
Sep	20.64	8	21.11	1	63.3	16	21.1	11	21.5
Oct	12.88	5	13.19	1	63.3	10	13.2	7	13.6
Nov	4.48	2	5.28	1	63.3	4	5.29	3	5.89
Dec	3.64	2	5.28	1	63.3	3	3.97	2	3.91

and observing Table 4.12, the fault from the SC fault simulated in (Section 4.3.4.2) could be detected with certainty for all the months of the year. With less than ten cells in short-circuit, fault detection could not be guaranteed, and its possible detection would depend on the corresponding month, e.g. nine short-circuited cell faults with a corresponding power loss of 23.75 W could not be guaranteed for the month of August since the uncertainty power loss window is 25.05 W.

For a power loss to fall inside the fault zone the following number of faults, as listed in Table 4.12, must happen. If less than that number of faults occur the power loss will fall into the uncertainty zone. This case corresponds to one panel per sensor. The number of faults that can fall into the uncertainty zone increases with the number of panels being monitored by each voltage sensor.

As it can be observed in Table 4.12, the worst-case scenario happens in August where for fault detection to occur with certainty, 10 short-circuits, 1 open-circuit, 19 BGHC, and 13 BGBC have to exist, otherwise, the origin of the power loss cannot be accurately pinpointed.

4.6 Summary

In this chapter, a series of PV module failure detection methods were studied, and some possible fault detection method recommendations were made. Most of the existing fault detection methods can be complex and require expensive equipment and significant computational power. It was seen that the module temperature and solar irradiance have a significant impact on the power produced by a PV panel, hence the importance of developing a low-cost and straightforward fault detection method that accounts for the variations of these physical quantities.

References

1 Shapiro, D., Robbins, C., and Ross, P. (2014). Solar PV operation & maintenance issues. *Desert Research Institute* 1–13.

2 Sanchez-Miralles, A., Calvillo, C., Martín, F., and Villar, J. (2014). *Use, Operation and Maintenance of Renewable Energy Systems: Experiences and Future Approaches*, Green Energy and Technology, 1ee, vol. 2014, XII, 385 (ed. M.A. Sanz-Bobi). Springer Cham https://doi.org/10.1007/978-3-319-03224-5.

3 International Energy Agency (2020). *Global Energy Review. 2020*. International Energy Agency https://doi.org/10.1787/a60abbf2-en.

4 Köntges, M., Oreski, G., Jahn, U. et al. (2017). Assessment of photovoltaic module failures in the fiel. *Report IEA PVPS Task* 13-ST3: 1–123.

5 Ferrara, C. and Philipp, D. (2012). Why do PV modules fail? *Energy Procedia* 15: 379–387. https://doi.org/10.1016/j.egypro.2012.02.046.

6 Köntges, M., Kurtz, S., Packard, C.E. et al. (2014). *Review of Failures of Photovoltaic Modules*. International Energy Agency, Photovoltaic Power System Programme https://doi.org/978-3-906042-16-9.

7 Haque, A., Bharath, K.V.S., Khan, M.A. et al. (2019). Fault diagnosis of photovoltaic modules. *Energy Sci Eng* 7: https://doi.org/10.1002/ese3.255.

8 Jaffery, Z.A., Dubey, A.K., and Irshad, H.A. (2017). Scheme for predictive fault diagnosis in photo-voltaic modules using thermal imaging. *Infrared Phys Technol* 83: 182–187. https://doi.org/10.1016/j.infrared.2017.04.015.

9 Patton, R.J., Chen, J., and Nielsen, S.B. (1995). Model-based methods for fault diagnosis: some guide-lines. *Trans Inst Meas Control* 17: 73–83. https://doi.org/10.1177/014233129501700203.

10 Naseri, F., Farjah, E., Allahbakhshi, M., and Kazemi, Z. (2017). Online condition monitoring and fault detection of large supercapacitor banks in electric vehicle applications. *IET Electr Syst Transp* 7: 318–326. https://doi.org/10.1049/iet-est.2017.0013.

11 Institute of Electrical & Electronics Engineers (2019). IEEE Recommended Practice for Monitoring Electric Power Quality. IEEE Std 1159-2019 (Revision IEEE Std 1159-2009), pp. 1–98. https://doi.org/10.1109/IEEESTD.2019.8796486.

12 Monkowski, J.R., Bloem, J., Giling, L.J., and Graef, M.W.M. (1979). Comparison of dopant incorporation into polycrystalline and monocrystalline silicon. *Appl Phys Lett* 35: 410–412. https://doi.org/10.1063/1.91143.

13 Zekry, A., Shaker, A., and Salem, M. (2018). Solar cells and arrays. In: *Advances in Renewable Energies and Power Technolgies*, 3–56. Elsevier https://doi.org/10.1016/B978-0-12-812959-3.00001-0.

14 Kurtz, S.R., Myers, D., Townsend, T. et al. (2000). Outdoor rating conditions for photovoltaic modules and systems. *Sol Energy Mater Sol Cells* 62: 379–391. https://doi.org/10.1016/S0927-0248(99)00160-9.

15 Castañer, L., Bermejo, S., Markvart, T., and Fragaki, K. (2012). Energy Production by a PV Array. In: *Practical Handbook of Photovoltaics*, 645–658. Elsevier https://doi.org/10.1016/B978-0-12-385934-1.00018-0.

16 Nayan, M.F., Ullah, S.M.S., and Saif, S.N. (2016). Comparative analysis of PV module efficiency for different types of silicon materials considering the effects of environmental parameters. In: *2016 3rd International Conference on Electrical Engineering and Information Communication Technology*, 1–6. IEEE https://doi.org/10.1109/CEEICT.2016.7873089.

17 El Bassam, N. (2021). Solar energy. In: *Distributed Renewable Energies for Off-Grid Communities*, 123–147. Elsevier https://doi.org/10.1016/B978-0-12-821605-7.00015-5.

18 Krishnan, U., Kaur, M., Kumar, M., and Kumar, A. (2019). Factors affecting the stability of perovskite solar cells: a comprehensive review. *J Photonics Energy* 9: 1. https://doi.org/10.1117/1.JPE.9.021001.

19 Idoko, L., Anaya-Lara, O., and McDonald, A. (2018). Enhancing PV modules efficiency and power output using multi-concept cooling technique. *Energy Reports* 4: 357–369. https://doi.org/10.1016/j.egyr.2018.05.004.

20 Said, S.A.M., Hassan, G., Walwil, H.M., and Al-Aqeeli, N. (2018). The effect of environmental factors and dust accumulation on photovoltaic modules and dust-accumulation mitigation strategies. *Renew Sustain Energy Rev* 82: 743–760. https://doi.org/10.1016/j.rser.2017.09.042.

21 Gupta, V., Sharma, M., Pachauri, R.K., and Dinesh Babu, K.N. (2019). Comprehensive review on effect of dust on solar photovoltaic system and mitigation techniques. *Sol Energy* 191: 596–622. https://doi.org/10.1016/j .solener.2019.08.079.

22 Bhattacharjee, S. and Saharia, B.J. (2014). A comparative study on converter topologies for maximum power point tracking application in photovoltaic generation. *J Renew Sustain Energy* 6: 053140. https://doi.org/10.1063/1.4900579.

23 Ba, A., Ehssein, C.O., Mahmoud, M.E.M.O.M. et al. (2018). Comparative study of different DC/DC power converter for optimal PV system using MPPT (P&O) method. *Appl Sol Energy* 54: 235–245. https://doi.org/10.3103/ S0003701X18040047.

24 Boxwell, M. (2013). *Solar Electricity Handbook*, vol. 197. Greenstream Publishing Limited. Http://SolarelectricityhandbookCom/Solar-IrradianceHtml.

25 Dogga, R. and Pathak, M.K. (2019). Recent trends in solar PV inverter topologies. *Sol Energy* 183: 57–73. https://doi.org/10.1016/j.solener.2019.02.065.

26 Zeb, K., Khan, I., Uddin, W. et al. (2018). A review on recent advances and future trends of transformerless inverter structures for single-phase grid-connected photovoltaic systems. *Energies* 11: 1968. https://doi.org/10 .3390/en11081968.

27 Hassaine, L., OLias, E., Quintero, J., and Salas, V. (2014). Overview of power inverter topologies and control structures for grid connected photovoltaic systems. *Renew Sustain Energy Rev* 30: 796–807. https://doi.org/10.1016/j.rser .2013.11.005.

28 Spataru, S., Sera, D., Kerekes, T., and Teodorescu, R. (2015). Diagnostic method for photovoltaic systems based on light I-V measurements. *Sol Energy* 119: 29–44. https://doi.org/10.1016/j.solener.2015.06.020.

29 Solórzano, J. and Egido, M.A. (2013). Automatic fault diagnosis in PV systems with distributed MPPT. *Energy Convers Manag* 76: 925–934. https://doi.org/10 .1016/j.enconman.2013.08.055.

30 We SG, One M, Pid PIDE, Degradation PI (2017). The five most common problems with solar panels. *Greensolver Blog* (25 April 2017): 1–2. https://blog.greensolver.net/en/the-five-most-common-problems-with-solar-panels/ (accessed 14 January 2022).

31 Degraaff, D., Lacerda, R., Campeau, Z., and Corp, S. (2011). Degradation mechanisms in Si module technologies observed in the field; their analysis and statistics. *NREL 2011 Photovolt Modul Reliab Work* 1–25. http://www.irishellas.com/files/Degradation-Mechanisms-in-Si-Module.pdf.

32 Saadsaoud, M., Ahmed, A.H., Er, Z., and Rouabah, Z. (2017). Experimental study of degradation modes and their effects on reliability of photovoltaic modules after 12 years of field operation in the steppe region. *Acta Phys Pol A* 132: 930–935. https://doi.org/10.12693/APhysPolA.132.930.

33 International Energy Agency, Photovoltaic Power System Programme (2013). *Review on Failures of Photovoltaic Modules*. Task 13, Subtask 3.2. International Energy Agency, Photovoltaic Power System Programme.

34 Westerlund, P., Hilber, P., Lindquist, T., and Kraftnat, S. (2014). A review of methods for condition monitoring, surveys and statistical analyses of disconnectors and circuit breakers. In: *2014 International Conference on Probabilistic Methods Applied to Power Systems PMAPS 2014 – Conference Proceedings*, 7–10. https://doi.org/10.1109/PMAPS.2014.6960621.

35 Kurukuru, V.S.B., Haque, A., Khan, M.A. et al. (2021). A review on artificial intelligence applications for grid-connected solar photovoltaic systems. *Energies* 14: 4690. https://doi.org/10.3390/en14154690.

36 Gautam, S. and Brahma, S.M. (2013). Detection of high impedance fault in power distribution systems using mathematical morphology. *IEEE Trans Power Syst* 28: 1226–1234. https://doi.org/10.1109/TPWRS.2012.2215630.

37 Ancuta, F. and Cepisca, C. (2011). Failure analysis capabilities for PV systems. *Recent Res Energy, Environ Entrep Innov* 1: 109–115.

38 Kim, D.E. and Lee, D.C. (2009). Fault diagnosis of three-phase PWM inverters using wavelet and SVM. *J Power Electron* 9: 377–385.

39 Tsanakas, J.A., Ha, L., and Buerhop, C. (2016). Faults and infrared thermographic diagnosis in operating c-Si photovoltaic modules: a review of research and future challenges. *Renew Sustain Energy Rev* 62: 695–709. https://doi.org/10.1016/j.rser.2016.04.079.

40 Dhoke, A., Sharma, R., and Saha, T.K. (2018). PV module degradation analysis and impact on settings of overcurrent protection devices. *Sol Energy* 160: 360–367. https://doi.org/10.1016/j.solener.2017.12.013.

41 Dhimish, M., Holmes, V., Mehrdadi, B., and Dales, M. (2018). Comparing Mamdani Sugeno fuzzy logic and RBF ANN network for PV fault detection. *Renew Energy* 117: 257–274. https://doi.org/10.1016/j.renene.2017.10.066.

42 Tadj, M., Benmouiza, K., Cheknane, A., and Silvestre, S. (2014). Improving the performance of PV systems by faults detection using GISTEL approach. *Energy Convers Manag* 80: 298–304. https://doi.org/10.1016/j.enconman.2014.01.030.

43 Takashima, T., Yamaguchi, J., Otani, K. et al. (2009). Experimental studies of fault location in PV module strings. *Sol Energy Mater Sol Cells* 93: 1079–1082. https://doi.org/10.1016/j.solmat.2008.11.060.

44 Takashima, T., Yamaguchi, J., Otani, K. et al. (2006). Experimental studies of failure detection methods in PV module strings. In: *2006 IEEE 4th World*

Conference on Photovoltaic Energy Conversion, 2227–2230. IEEE https://doi .org/10.1109/WCPEC.2006.279952.

45 Sheuly, S.S., Barua, S., Begum, S. et al. (2021). Data analytics using statistical methods and machine learning: a case study of power transfer units. *Int J Adv Manuf Technol* 114: 1859–1870. https://doi.org/10.1007/s00170-021-06979-7.

46 Johnson, J., Montoya, M., McCalmont, S. et al. (2012). Differentiating series and parallel photovoltaic arc-faults. In: *2012 38th IEEE Photovoltaic Specialists Conference*, 000720–000726. IEEE https://doi.org/10.1109/PVSC.2012.6317708.

47 Dini, D.A., Brazis, P.W., and Yen, K.-H. (2011). Development of arc-fault circuit-interrupter requirements for photovoltaic systems. In: *2011 37th IEEE Photovoltaic Specialists Conference*, 001790–001794. IEEE https://doi.org/10 .1109/PVSC.2011.6186301.

48 Suryanarayana R. Causes for Insulation faults in PV systems and detection methods 2018:1–3. https://www.researchgate.net/publication/328871157_ Causes_for_Insulation_faults_in_PV_systems_and_detection_methods

49 Falvo, M.C. and Capparella, S. (2015). Safety issues in PV systems: design choices for a secure fault detection and for preventing fire risk. *Case Stud Fire Saf* 3: 1–16. https://doi.org/10.1016/j.csfs.2014.11.002.

50 Behrends, H., Millinger, D., Weihs-Sedivy, W. et al. (2022). Analysis of resid-ual current flows in inverter based energy systems using machine learning approaches. *Energies* 15: 582. https://doi.org/10.3390/en15020582.

51 Flicker, J. and Johnson, J. (2016). Photovoltaic ground fault detection recom-mendations for array safety and operation. *Sol Energy* 140: 34–50. https://doi .org/10.1016/j.solener.2016.10.017.

52 Alam, M.K., Khan, F., Johnson, J., and Flicker, J. (2013). PV ground-fault detection using spread spectrum time domain reflectometry (SSTDR). In: *2013 IEEE Energy Conversion Congress and Exposition*, 1015–1102. IEEE https://doi .org/10.1109/ECCE.2013.6646814.

53 Saleh, M.U., Deline, C., Benoit, E. et al. (2020). An overview of spread spec-trum time domain reflectometry responses to photovoltaic faults. *IEEE J Photovoltaics* 10: 844–851. https://doi.org/10.1109/JPHOTOV.2020.2972356.

54 Chine, W., Mellit, A., Lughi, V. et al. (2016). A novel fault diagnosis technique for photovoltaic systems based on artificial neural networks. *Renew Energy* 90: 501–512. https://doi.org/10.1016/j.renene.2016.01.036.

55 Platon, R., Martel, J., Woodruff, N., and Chau, T.Y. (2015). Online fault detec-tion in PV systems. *IEEE Trans Sustain Energy* 6: 1200–1207. https://doi.org/ 10.1109/TSTE.2015.2421447.

56 Silvestre, S., Da, S.M.A., Chouder, A. et al. (2014). New procedure for fault detection in grid connected PV systems based on the evaluation of current and voltage indicators. *Energy Convers Manag* 86: 241–249. https://doi.org/10 .1016/j.enconman.2014.05.008.

57 Silvestre, S., Kichou, S., Chouder, A. et al. (2015). Analysis of current and voltage indicators in grid connected PV (photovoltaic) systems working in faulty and partial shading conditions. *Energy* 86: 42–50. https://doi.org/10 .1016/j.energy.2015.03.123.

58 Jenitha, P. and Immanuel, S.A. (2017). Fault detection in PV systems. *Appl Sol Energy* 53: 229–237. https://doi.org/10.3103/S0003701X17030069.

59 Kumar, B.P., Ilango, G.S., Reddy, M.J.B., and Chilakapati, N. (2018). Online fault detection and diagnosis in photovoltaic systems using wavelet packets. *IEEE J Photovoltaics* 8: 257–265. https://doi.org/10.1109/JPHOTOV.2017 .2770159.

60 Ali, M.H., Rabhi, A., El, H.A., and Tina, G.M. (2017). Real time fault detection in photovoltaic systems. *Energy Procedia* 111: 914–923. https://doi.org/10.1016/ j.egypro.2017.03.254.

61 Dhimish, B.M., Holmes, V., Mehrdadi, B. et al. (2017). Detecting defective bypass diodes in photovoltaic modules using mamdani fuzzy logic system. *Global Journal of Researches in Engineering: F Electrical and Electronics Engineering* 17 (5): 33–44. http://eprints.hud.ac.uk/id/eprint/33779/.

5

Failure Mode Effect Analysis of Power Semiconductors in a Grid-Connected Converter

V S Bharath Kurukuru[1] and Irfan Khan[2]

[1] *Advance Power Electronics Research Lab, Department of Electrical Engineering, Jamia Millia Islamia (A Central University), New Delhi, India*
[2] *Clean and Resilient Energy Systems (CARES) Lab, Texas A&M University, Galveston, TX, USA*

5.1 Introduction

Grid-connected operation of distributed generation systems (DGs) provide dynamic operations, low power losses, better reliability, and does not allow any negative or unwanted effects such as poor voltage quality to harm the load side. Under such precise operating conditions, the failure of power electronic converters (PECs) associated with the DGs lead to major power losses and operational failures. This chapter aims at identifying the failure modes of PECs and their criticality by establishing the failure mode effect analysis (FMEA) for power semiconductor devices (PSDs) in PECs.

Generally, the failure mechanisms are considered as the critical aspects of determining the reliability of PECs and deal with the physical, chemical, and electrical processes that occur in the system [1]. Based on the type of failure process, these failure mechanisms can be modeled when appropriate material and environmental information are available. Moreover, with the advancements in monitoring technologies, the failure data along with the modeled mechanisms can be used to identify the operating state of the PEC. This helps monitor the operation of the system, performing risk analysis, estimating the reliability of the product, and reducing the probability that a customer is exposed to a potential product failure and/or process problem [2]. Further, following the development of FMEA for a PEC, the criticality analysis is also performed to prioritize failure modes and mechanisms, respectively. Such prioritization allows for efficient allocation of resources for enabling and improving reliability of the system. One difficulty for prioritizing failure mechanisms for component-level failure classification is that the

Fault Analysis and its Impact on Grid-connected Photovoltaic Systems Performance, First Edition.
Edited by Ahteshamul Haque and Saad Mekhilef.

information necessary to make the decision is highly application-dependent. Hence, methods for defining and estimating criticality and establishing component-level information-based guidance for ranking failure mechanisms are also discussed.

5.2 Power Electronics Converters

There are many different types of PECs that operate in a grid-connected system and they range from DC–DC buck/boost conversion process to AC–AC rectifier-inverter operation with AC–DC (rectifier) and DC–AC (inverter) conversion [3–5]. A brief representation of these converters in a grid-connected system is shown in Figure 5.1.

Most of these PECs are achieved by assembling a variety of active and passive components. These components are switches, diodes, capacitors, inductors, and resistors. The use of switches allows for the converter to operate in discrete states depending on whether any given switch is in the open or closed state. Electrical analysis of the various states of the switches shows that the input voltage is then converted to an output voltage of differing voltage level and/or frequency. As a result of the switching between multiple operating states, ripples in the output voltage and current profiles develop. Passive components such as capacitors and inductors act to reduce this ripple across the output of the converter [6]. However, the use of passive components increases the size of the converter. There is general trend for reduction in size weight power, and cost throughout the market. One

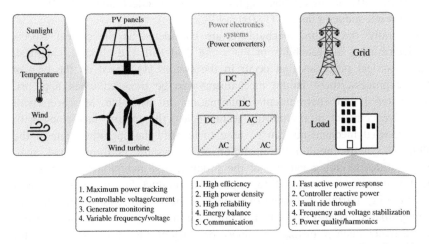

Figure 5.1 Block diagram representation of power electronics converters interfaced in a grid-connected system.

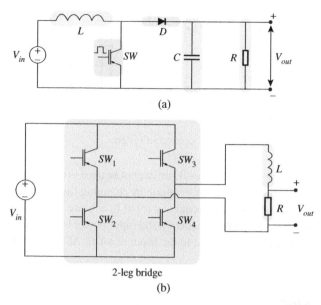

Figure 5.2 Power electronic converters used in a single-phase grid-connected system. (a) DC–DC boost converter. (b) DC–AC single phase converter.

method for reducing the size and weight of the components is by reducing the ripple caused by the switching and therefore allowing for the use of smaller and lighter passive components. To reduce the ripple, PECs are being manufactured with switches that operate at higher frequencies which reduces the time required between operating states.

An example of the most widely used topologies in a single-phase grid connected systems are shown in Figure 5.2. The first is a DC–DC boost converter which converts one form of DC power to DC power at a higher voltage. During a switching period, the switch is in the on-state for some portion of the time and the off-state otherwise. The portion of time that the switch is in the on-state divided by the total switching period is defined as the duty cycle (D).

The second converter shown in Figure 5.2b is a DC–AC single phase converter. This converter operates on a more complex switching principle than the DC–DC boost converter shown in Figure 5.2a, called pulse-width modulation (PWM). The name refers to the fact that to achieve the desired AC output power, the converter is continually modulating the duty cycle of each of the switches. The use of PWM techniques allows for the AC output voltage to take on any frequency up to practical limits of the switching devices.

There are a variety of AC–AC converters, the most well-known of which are transformers. However, transformers are limited to converting AC power of one

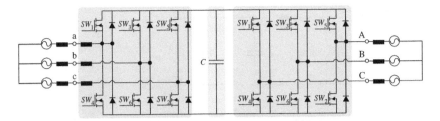

Figure 5.3 AC–AC converter interfaced in a grid-connected system.

voltage level to AC power of another voltage level and cannot be used to change the frequency of the AC power. Many applications operate by changing the input AC frequency, such as an AC motor changing frequency input to achieve a different output torque. To achieve variable output frequency, the input power is sent to an AC–DC converter whose output is connected to the input of a DC–AC converter, see Figure 5.3.

5.2.1 Components in PECs

From this discussion, it is clear that all the PECs include similar components in different topologies. One of the components is the power switch. To achieve the high switching frequencies that are desired in PECs, solid state switches, or transistors, are used. There are many types of transistors including power metal-oxide semiconducting field effect transistors (MOSFET) and insulated gate bipolar transistors (IGBT). These transistors are vertically structured and designed to handle the significant currents that power electronic systems must provide. The work presented herein focuses on these solid-state switches. The second most widely used solid-state component used in PECs are diodes. In some instances, as in the DC–DC boost converter, the diode is essential to the working principle of the converter. In other instances, the function of the diodes is to prevent voltage spikes across the solid-state switches due to their parasitic inductance. Diodes typically used in power electronics are vertically structured like the transistors that are used. Specifically, Schottky and PIN diodes are used in these applications. In addition to the power switches, and diodes, the capacitors are also used in the design of PECs. These components serve a variety of purposes including the suppression of the ripple at the output of DC converters. For this purpose, electrolytic capacitors are typically used as they are able to achieve the large capacitance values necessary for this function. Another function that capacitors serve is electromagnetic interference (EMI) suppression at the output of AC converters. Electrolytic capacitors cannot be used in these applications as they are polar devices therefore thin film capacitors, which are nonpolar, are often selected to serve this function.

5.2.2 Integrated Application Environment

Considering the operating and conversion efficiency of the PECs, and their day-by-day reduction in size and weight, these devices have been called upon to integrate in a wide variety of environments. Some of these integrated applications involve harsh high temperature and high humidity environments, while some applications require their operations in outdoor environments [7–9]. Due to the high currents causing joule heating and switching losses at high frequencies, power electronic systems can regularly experience temperatures in excess of 100°C. Further, 80% of respondents to an industry survey of PECs said that their power converters are rated for over 1 kW, and 12% of respondents use power converters rated for over 1 MW [10]. Even with aggressive cooling regimes such as liquid cooling maintaining the power components used in these converters at a low temperature is a challenge.

Additionally, due to varying operational loads, including planned downtimes, power electronics may experience significant swings in temperature. Over 85% of respondents to the earlier mentioned industry survey said that their power electronics can experience temperature swings in excess of 30°C and 47% of respondents said the temperature swings for their application were in excess of 80°C [10]. Due to the large thermal cycles observed in their life cycle profiles, thermomechanical stresses caused by mismatches in coefficients of thermal expansion (CTE) of the different materials which make up power electronics packages are of significant concern in PECs. In particular, the power switches which dissipate the most heat are of highest concern. Besides, many power electronics applications are operated in outdoor environments. Here, in addition to uncontrolled environmental temperature, the humidity in outdoor environments is uncontrolled and can swing between high and low relative humidity (RH) values or stay at high average values.

5.2.3 Power Semiconductor Devices

To provide a better understanding of the FMEA of the PSDs, the design and operation of the power MOSFETs and IGBTs are discussed in the further sections. From the literature, it is identified that these two components make up over two-thirds of PSDs which are used in PECs [10].

5.2.3.1 Metal-Oxide-Semiconductor Field-Effect Transistor

The design technologies of the Power MOSFETs which are widely used in the development of the present-day PSDs are shown in Figure 5.4. Generally, the operating principle of the power MOSFETs is similar to that of the traditional MOSFET devices; where the off-state corresponds to the cut-off region, and the on-state corresponds to the maximum characteristics. Further, the vertical structure of the

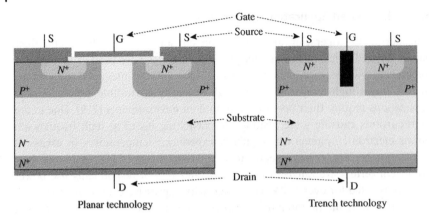

Planar technology   Trench technology

Figure 5.4 Design technologies of power metal-oxide-semiconductor field-effect transistor.

power MOSFETs allows them to block higher voltages than traditional MOSFETs by giving the devices a large drift region at the expense of increased on-state resistance. Figure 5.4 shows a single elementary cell; in a full device, thousands of these cells are connected in parallel, allowing the device to conduct currents up to 100 A [11].

5.2.3.2 Insulated-Gate Bipolar Transistor
The cell structure of the Power IGBTs which are widely used in the development of the present-day PSDs are shown in Figure 5.5. These high-power switching devices have their application in many medium- and high-power systems. The power IGBTs are a vertically structured semiconductors which consist of an NPN-MOSFET driving the gate of a PNP bipolar junction transistor. Similar to a power MOSFET, an IGBT is made up of many of these elementary cells connected in parallel [11]. By combining the switching characteristics of a MOSFET and the current-handling capabilities of a BJT, these devices reach fast switching speeds of 1 – 150 kHz and high collector/emitter current handling of up to 1500 A. They also are able to operate with high collector-emitter voltages of 600 – 6500 V [12, 13]. When compared with power MOSFETS, IGBTs have higher current and voltage ratings; however, they have lower switching speeds in the range of 10–100 kHz.

5.2.4 Power Semiconductor Packaging

The packaging of PECs serves four fundamental purposes: to provide electrical interconnection, to provide a path for thermal dissipation, to protect the circuit from damage due to environmental exposure, and to provide mechanical support

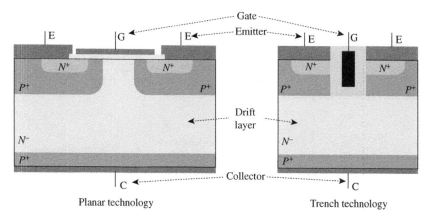

Figure 5.5 Design technologies of power insulated-gate bipolar transistor.

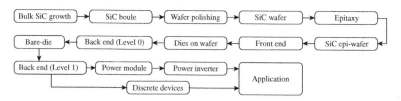

Figure 5.6 Steps and materials required for packaging of power semiconductor device.

for the circuit that facilitates handling and assembly. A brief overview of the steps for packaging of PSDs is shown in Figure 5.6.

For PSDs, these steps are accomplished through three configurations: electronic packaging, module packaging, and press-packs. In this section, the major focus is on the discrete component, and module packaging of the PSDs.

5.2.4.1 Electronic Packaging

In the electronic packaging, the discrete components are packaged individually and typically house as a single semiconductor die. Functionally, this packaging has a single switch with three leads: a controlling gate lead, and two leads which are connected in series with the circuit in which they switch. Generally, the transistors housed in discrete packages are typically rated for lower operating conditions than those used in module packages. One of the main advantages of these types of packages is that they come in a few well-defined packages such as TO-220, and TO-247 as shown in Figure 5.7. They come in leaded or surface mount packages for connection to the printed circuit board.

A cross section of a TO-220 part is shown in Figure 5.8 which will be necessary for understanding the stresses within the package in subsequent sections. Internal

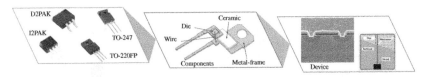

Figure 5.7   Discrete component packaging of a transistor.

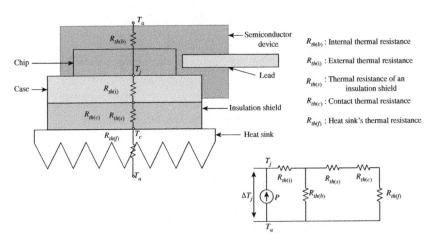

Figure 5.8   Detailed discrete component packaging cross-section of a transistor.

to the package, a semiconductor-die, typically power MOSFET or IGBT, is soldered to a collector baseplate made of copper. Wire bonds, which connect the die to the leads, are typically made of aluminum for power packages; however, there is potential to change to copper. Finally, the die and wire bonds are encapsulated in an epoxy molding compound (EMC) to enable mechanical rigidity and protect the die from the environment.

5.2.4.2 Module Packaging

The second common form of power semiconductor packaging is module packaging. These packages as shown in Figure 5.9, typically house many power semiconductors in a half or full bridge power converter configuration. These packages are typically designed to handle significantly higher electrical and thermal stresses than discrete packages.

A cross section of a typical power semiconductor module is shown in Figure 5.10. The semiconductor dies are soldered to a substrate metallization which is typically aluminum or copper. This metallization is on a ceramic substrate, typically alumina or aluminum nitride, which electrically insulates the semiconductor-die from the heat sink. Together the metallization and the

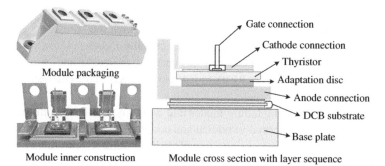

Figure 5.9 Module packaging of a power semiconductor device. Source: Semikron/Distrelec Schweiz AG.

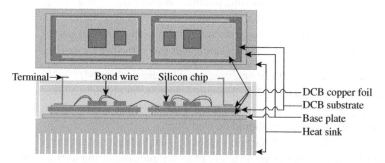

Figure 5.10 Module packaging cross-section of a power semiconductor device.

substrate are referred to as direct copper bonding (DCB) or direct aluminum bonding (DAB) depending on the choice of metallization. This substrate is able to withstand significantly more thermal and electrical stress than would be expected for discrete components. The DCB/DAB is then attached to a larger baseplate or heat sink which is used to cool the semiconductors on the substrate. A silicone gel is used to encapsulate the package after it has been assembled. This has a higher breakdown strength than air and thus increases the insulated between wire-bonds and the metallization.

5.3 Failure Mode Effect Analysis of Power Semiconductors

5.3.1 Failure Mode Effect Analysis

The FMEA is a method for developing comprehensive lists of failure modes for components in the system and analyzing the effects of the failure mode to the larger system. The failure mode effect and criticality analysis (FMECA) expands upon FMEA by introducing a criticality metric through which the failure modes are ranked based on the severity, occurrence and detectability of each failure mode. The criticality analysis allows engineers to focus on these critical failure modes, identified by a risk priority number (RPN), to reduce the effects to the end user or system manufacturer.

These methods for identifying failure modes and were first established in the mid-1900s [14, 15], but were formally codified and adopted as the FMEA process when Joint Electron Device Engineering Council (JEDEC) solid state technology association released JEDEC publication (JEP)131 in 1998 [16] and the latest revision is done in 2018 [17]. This document defines FMEA as an anticipatory thought process designed to utilize as much knowledge and experience of an organization as possible toward the end of addressing potential issues defined in a new project. The objective is to reduce the probability that a customer is exposed to a potential product and or process problem by performing a thorough risk analysis [17]. For any given system, the FMEA is developed through the following steps:

Step 1: The system which is to be analyzed must be clearly defined.

Step 2: The system should be broken down into subsystems either in a functional, geographical/architectural, or combination of the two. Here, all the functions of the subsystems should be identified.

Step 3: Identify all the possible failure modes that the subsystem can experience which may be done using a variety of techniques including testing, engineering judgement, and simulation.

Step 4: For each failure mode, identify the possible causes of failure. Here, a life cycle profile for the product should be developed to help understand the various stresses the component may see not only during operation but also during manufacturing, storage, and transportation.

Step 5: In this final step, identify how the failure effects the end user. Like the life cycle profile, this information is application-specific.

Based on the analysis from all these steps, attempts were made to prioritize the maintenance of the system to allow for effective usage of resources and to address reliability concerns [18]. Here again, the analysis is highly application-specific.

5.3.2 Failure Modes, Mechanisms, and Effects Analysis

As opposed to FMEA which identifies the high-risk failure modes to update the design and reduce risks to acceptable levels, failure modes, mechanisms, and effects analysis (FMMEA) takes the FMEA an additional step and identifies the failure mechanisms associated with failure causes and modes. For failure mechanisms, relevant failure model(s) can be identified which can illustrate how

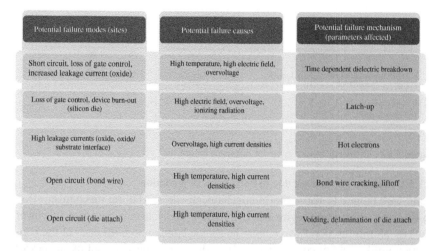

Potential failure modes (sites)	Potential failure causes	Potential failure mechanism (parameters affected)
Short circuit, loss of gate control, increased leakage current (oxide)	High temperature, high electric field, overvoltage	Time dependent dielectric breakdown
Loss of gate control, device burn-out (silicon die)	High electric field, overvoltage, ionizing radiation	Latch-up
High leakage currents (oxide, oxide/ substrate interface)	Overvoltage, high current densities	Hot electrons
Open circuit (bond wire)	High temperature, high current densities	Bond wire cracking, liftoff
Open circuit (die attach)	High temperature, high current densities	Voiding, delamination of die attach

Figure 5.11 Failure modes, mechanisms, and effects analysis. Source: Adapted from Mathew et al. [19].

the stress leads to the failure of a system. Failure mechanisms are highly dependent on the materials, geometries, and stresses within a system. The FMMEA for silicon power devices discussed in the literature are shown in Figure 5.11. However, this FMMEA is limited to only discrete IGBT parts, power cycled at high mean temperatures with large junction temperature swings [19]. Besides, the FMMEA was developed for the purposes of identifying failure precursor parameters for prognostics applications and it served that function well. Such an FMMEA is limited in several respects, first as will be discussed in failure mechanism criticality analysis, FMMEA requires application knowledge, therefore this is not truly an FMMEA and it makes no effort to account for failure mechanism criticality. Second, the list does not include all relevant failure mechanisms for silicon power devices in the reasonably expected operating conditions.

5.3.3 Failure Mechanisms of Power Semiconductor Devices

The lifecycle profile of a component or device provides a list of failure modes and mechanisms that may be precipitated from the stresses for establishment in an FMMEA. While there are many potential sources of thermal, mechanical, and electrical stresses on the PSD components, the mechanisms which may be precipitated are dependent on the geometry and material characteristics. Hence, an overview of the failure mechanisms which have been reported for PSDs are reported in this section. A brief overview of different types of failure modes at different locations is shown in Figure 5.12.

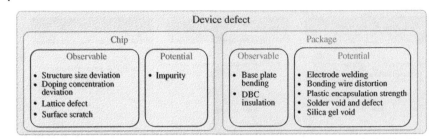

Figure 5.12 Different failure types at different locations.

5.3.3.1 Failure Mechanism due to Bond-Wire Degradation

Bond-wire fatigue is also a thermomechanically driven failure mechanism. Due to the large diameter of wire used to handle high current densities in power devices, aluminum bond-wires must be wedge-bonded on both ends of the connection unlike gold or copper bond-wires which are typically used in low power applications with smaller wires that can be ball bonded. Fatigue of these wedge bonds occurs due to joule heating, switching losses, changes in the temperature of the environment, and the mismatch of thermal coefficients of expansion of the aluminum wire and the silicon die [20–25]. Bond-wire fatigue manifests itself as either lift off or heel cracking. The heel of the wedge bond acts to as a stress concentrator and a crack propagates through the heel. In bond-wire lift off, the wedge delaminates from the surface of the bond pad at the heel and toe sides of the wedge bond and propagates inward. Bond-wire fatigue results in an increase in the on-state resistance of the power device and can lead to an open circuit. In addition to the Bond-wire fatigue, the melting of bond-wire is also a substantial failure mechanism. While the die is the most significant source of joule heating within the power package, the parasitic resistance of the bond-wires also acts as a heat source. The bond-wires are encapsulated in either silicone gel or an EMC depending on whether or not the package is a discrete or a module. Both silicone gel and EMC are poor conductors of heat. This means heat generated within the bond-wires is difficult to dissipate and the wire can see elevated temperatures. If the power package is used near or at its current rating and effective measures are not taken to cool the package, the bond-wire can melt [26]. The result of this failure is an open circuit of the transistor.

5.3.3.2 Failure Mechanism Causing Aluminum Degradation

The die metallization for power semiconductors is typically aluminum. Additionally, if the substrate is DAB, then there is aluminum metallized on the ceramic substrate. Due to thermal cycling of the component from joule heating, switching losses, changes in the temperature of the environment, and the mismatch of CTE between the aluminum and silicon and ceramic substrate, thermomechanical stresses are generated within the package. These stresses can be significant enough to cause yielding of the aluminum metallization, causing it to buckle and form hillocks. This mechanism is referred to as aluminum reconstruction [20–24, 27–29]. The reconstruction of aluminum can increase in the resistance of the metallization layer. Aluminum reconstruction can be exacerbated by electromigration which can happen if significant current densities are present in the metallization. Aluminum metallization that is coated with a passivation layer, typically silicon nitride, has been shown to resist reconstruction; however, bond pads, which make up a significant portion of the total metallized area, are not passivated and therefore remain unprotected. Further, when moisture is present within the package, corrosion of the aluminum bond-wires and bond pads can be of concern [30, 31]. In the presence of moisture, aluminum reacts to form $Al(OH)_3$ which passivates its surface and passivates the aluminum. This passivation layer can become soluble in the presence of contaminants such as halogens. During power cycling of a component, the encapsulant layer can delaminate from the base plate thus developing a path for moisture and contaminants to ingress into a component. Similar to the thermomechanical fatigue mechanisms, corrosion would cause an increase in the on-state resistance of the power semiconductor and is a wear-out failure mechanism.

5.3.3.3 Failure Mechanisms Degrading Die-Attach

The die-attach which connects the die to the substrate is another location of thermomechanical stress due to CTE mismatch. As the die is vertically conductive, the die-attach must be conductive as it is part of the electrical path of the power semiconductor component. Similar to the aluminum metallization and bond-wires, the die-attach is in intimate contact with the die and undergoes thermomechanical fatigue and possible delamination [22, 24, 25, 32, 33]. Delamination occurs when the separation is between the die itself and the die-attach material; however, fatigue can occur and propagate through the die-attach. Additionally, the delamination of the die-attach increases the thermal resistance of the die-attach, decreasing the ability of the die-attach to dissipate heat generated at the die. Increased thermal resistance results in higher die temperature impacting the electrical characteristics of the device. In addition to the die-attach delamination and fatigue, the voiding of the die-attach [34–36] also relates to a potential failure mechanism. Small amounts of voiding are residual in the die-attach from the manufacturing process. Power cycles grow and coalesce the smaller distributed voids which are initially present in the solder. Like die-attach fatigue, die-attach voiding results in an increase in the on-state resistance of the power package and increases the thermal resistance of the packaging, thus increases the junction temperature of the package.

Every so often, these die-attaches contain silver, as a sintered silver paste or a tin-silver alloy solder, which has a strong propensity to migrate under a variety of conditions [37–40]. In the presence of moisture, silver and other metals show some slight solubility. If an electric field is also present, the silver and other metals will migrate from the anode to the cathode through electrochemical migration (ECM). At the cathode, the silver will deposit and form dendritic structures back toward the anode. Given time, these dendrites can grow long enough to short the cathode and the anode. Mass transport of silver can also happen through corrosion, particularly in the presence of sulfur. Unlike ECM, in silver migration due to corrosion, there is no expectation of growth direction. Silver migration and ECM manifest themselves in an increased leakage current and are wear-out failure mechanisms.

5.3.3.4 Failure Mechanisms due to Breakdown

Three different types of breakdown mechanisms are seen in PSDs. These relate to the avalanche breakdown, dielectric breakdown, and time-dependent dielectric breakdown. The avalanche breakdown [41–44] mechanism can precipitate, often during switching, when the drain-source or collector-emitter voltage exceeds the breakdown voltage of the power device. Electrons within the device gain sufficient energy to impact atoms within the device and ionize the atoms and releasing additional electrons. If these impacts continue, the device can "avalanche" as an increasing number of electrons are freed and able to impact atoms to free addition electrons. Avalanche breakdown often occurs during switching of a device when the inductance of the power semiconductor or the system within which it is operating, creates a voltage spike on the system. Avalanche breakdown manifests itself as a short circuit of the device and is considered an overstress mechanism.

Further, it is essential that the insulated dielectric that forms the gate of voltage-controlled devices maintains dielectric integrity for the device to operate. If the electric field through the gate exceeds the dielectric strength of the insulating material, then the terminals will short and permanent damage will be done to the gate [45]. For silicon devices, the gate is made of silicon dioxide. Dielectric breakdown can occur due to an overvoltage event on the die for a short period of time. One possible cause of this overvoltage is an electrostatic discharge (ESD). Such cases would be overstress events and likely result in a shorting of the gate to one of the conduction terminals. Besides, dielectric breakdown can also occur over time through a process called time-dependent dielectric breakdown [46, 47]. One leading explanation for this mechanism is that Si–Si bonds within the dielectric are weak and over time, the application of an electric field breaks down these bonds, creating locations within the dielectric through which electrons can jump to and travel through the insulating gate. Time dependent dielectric breakdown manifests itself as high gate leakage current and is a wear-out mechanism.

Generally, silicone gel is used to encapsulate the metallization and bond-wires within power modules for increasing the breakdown strength of the PSDs. However, due to the high voltage and geometries of the conductors the electric field within the package can still be enhanced and cause a partial discharge within the silicone gel [48, 49]. Over time, these discharge events can develop a carbonized, conductive path within the gel leading to increased leakage. Locally, partial discharge can cause bubbles in the gel to form due to the local heating events. Partial discharge within the gel is observed as increased leakage, developing toward a short circuit and is a wear-out mechanism.

5.3.3.5 Failure Mechanisms with the Packaging

Generally, the substrate acts to insulate the conductive paths of the power package from the heat sink or other cooling mechanisms. But the literature identifies it as a possible location of failure within a power module. The ceramics in the DCB and DAB substrates can crack when subjected to thermomechanical cycling due to operational and environmental loading [50, 51]. Therefore, when cracking occurs, the insulation properties of the ceramic break down and a reduced insulation strength is observed. Depending on the electrical connection of the heat sink, this can create a significant leakage path within the power package.

In addition to the substrate cracking, there are parasitic circuit elements associated with the power package as semiconductors and metals have non-ideal material properties associated with them. One particularly harmful parasitic element within power components are parasitic thyristors. The general behavior of thyristors indicates a continuous conduction after the removal of the gate voltage or current pulse until the electric potential between the anode and cathode is zero. But switches like MOSFETs and IGBTs are expected to conduct only when a gate voltage is applied, therefore the activation of the thyristor may cause the loss of gate control of the device. Such an event is referred to as latch-up and is observed as a short circuit between the conduction terminals [41, 52]. The thyristor is activated when the current exceeds the so-called latching current of the device. In an IGBT, this overcurrent forces current into the base of the parasitic NPN transistor, see Figure 5.13, as the local high-current density in the P region at the base increases the resistance locally. Latch-up of a device does not inherently cause a destructive failure of the component. In the unlikely event that the latch-up event is detected and measures are taken to remove the current from the device, the event can be stopped and the device will function normally. However, if the latch-up event is not identified, it can lead to a thermal runaway causing the device to burn-out. Latch-up is an overstress failure mechanism and is observed as a short circuit of the device.

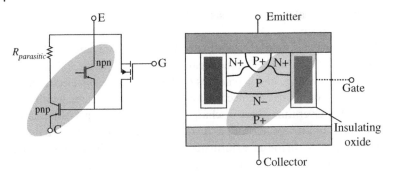

Figure 5.13   Parasitic thyristor within an IGBT.

5.3.3.6   Other Failure Mechanisms

In PSDs, some electrons while travelling through the MOS channel may gain sufficient energy to be able to tunnel through the gate oxide layer [53]. These electrons become hot, referring to their individual speed, and as an energy consequence which is opposed to the bulk temperature of the device itself, the electron travels along through the gate channel when the device is conducting. This can cause impact ionization near the end of the channel which can produce electrons which can inject themselves into the gate dielectric. These failure mechanisms are reported to be more common at low temperatures, unlike most other mechanisms which are thermally accelerated. At lower temperatures, lattice scattering is reduced allowing longer free paths for electrons to accelerate, gaining energy to create hot carriers. Besides, all these steps can cause damage at the interface of the silicon and silicon dioxide or allow the carriers to become trapped within the dielectric itself. Further, such a phenomenon also causes parameters associated with the gate such as the gate threshold voltage to shift. Under this damage, gate threshold voltage would drift higher, requiring higher gate voltage to be applied to achieve the same level of conduction in an otherwise healthy device.

All the failure mechanisms discussed above are not all independent of each other. In many cases one failure mechanism may lead the device to failure through another mechanism. For example, the degradation of the die-attach can cause a latch-up event on the die. Power cycling through delamination and voiding causes both an increase in electrical and thermal resistance. This leads to an increase in the temperature of the die during operation as more power is dissipated due to the resistance increase and heat cannot leave the package as easily. Additionally, as portions of the die-attach have "disconnected" from the die, current crowding occurs, leading portions of the device susceptible to latch up failure event though the device itself is still conducting the same amount of current. In such an instance, it is evident that one failure mechanism drove the device to failure through another

Table 5.1 Potential failure modes and mechanisms of silicon power devices.

Possible failure mode	Possible failure location	Possible failure causes	Possible failure mechanisms
Short circuit	Collector-emitter	High temperature, cosmic rays, collector-emitter current above latching trigger current	Latch-up
		High-frequency switching, unclamped inductive switching, collector-emitter voltage surpasses breakdown voltage	Avalanche breakdown and Secondary breakdown
	Gate oxide	The gate voltage is higher than the gate's breakdown voltage.	Electrostatic Discharge and Electrical overstress
	Encapsulant	The electric field between bond wires is greater than the encapsulant's dielectric strength.	Partial Discharge
Increased leakage current (collector-emitter)	The periphery of die	Humidity, extreme heat, mobile ions, and a strong electric field	Electrochemical migration
		Silver in the packaging, wetness, high temperature, and a strong electric field	Silver migration
dielectric strength reduction	Substrate Insulation	CTE mismatch, temperature, and power cycling	Substrate cracking
Gate threshold voltage and gate leakage current increment	Gate oxide	Implementation of a high gate voltage for a long time and at a high temperature	Time-based dielectric breakdown
		Low temperature, high MOS-channel currents	Hot carrier injection

(continued)

Table 5.1 (Continued)

Possible failure mode	Possible failure location	Possible failure causes	Possible failure mechanisms
Raise in On-state resistance	Bond wire	CTE mismatch, temperature, and power cycling	Cracking and lift-off in bond wire
		Humidity and pollutants such as halogens are present.	Al corrosion
	Surface Metallization	CTE mismatch, temperature, and power cycling	Al reconstruction
	Die-attach	CTE mismatch, temperature, and power cycling	Voiding, delamination of die
Open circuit	Bond wire	Due to power dissipation, there is a high temperature.	Melting of bond wire

Wear out Overstress Package Die

mechanism. Due to the potential of failure mechanisms to convolute each other, it is important for engineers to understand such mutually accelerating factors when designing systems. Considering the list of relevant failure mechanisms identified for the PSDs the corresponding failure mode, cause, and location are identified as shown in Table 5.1. Systems integrators will find this information useful for identifying the failure causes and mechanisms that should be considered in the design of the system. For example, if they are aware that moisture will be present in the application, they should consider relevant measures to prevent ECM and silver migration. The corollary to this is that if they are confident that no significant moisture will be present then such measure may not be necessary in the design of the system. Another possible application of this table is for failure analysis engineers. Based on the information that they establish during the failure analysis; this table can help lead the failure analysis team to identify the cause and mechanism associated with the failure. Once the cause has been identified, the proper steps can be taken to reduce the likelihood of future failures or identify risks for fielded systems.

5.4 Failure Analysis

Failure analysis yields insights into what caused the devices to fail, whether there was a manufacturing defect that caused a latent failure or a design issue that caused the failure in nominal use conditions, and how to reduce and remove these defects to prevent future problems. Having a comprehensive FMMA information is useful in the process of developing failure analysis steps and drawing conclusions from the results. It also aids in compiling information regarding the failure of the component within the power semiconductor package. Life cycle and operating conditions such as storage temperature, operating temperatures, humidity levels, and electrical characteristics, should be noted. This information should be compared with the information in the FMMEA to determine critical failure mechanisms that may have caused the failure. The identified potential failure mechanisms will be the point of focus for the ensuing inspection of the device.

Further with the advancements in intelligent learning approaches, the use of machine learning has widely improved the failure analysis with the FMEA, and FMMEA data. A brief application of one such method is discussed in this section along with the requirements for criticality and severity analysis.

5.4.1 Failure Mechanism Detection Approach

The occurrence of failure mode depends on the working and operating conditions of the component in the system. Failure mechanisms are only expected in the

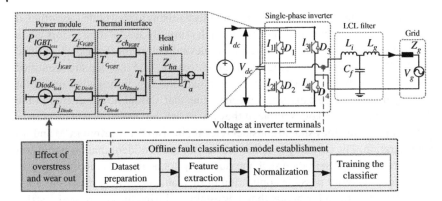

Figure 5.14 Block diagram of proposed failure detection approach.

presence of certain stressors. There are two approaches to calculating the value of an event. One is for the wear mechanism and the other is for the overuse mechanism. For wear mechanisms, it is necessary to recognize a failure model that associates the stressor with the material and shape of the system. This can result in a failure or equivalent amount of time that indicates the occurrence of a mechanism in your application. An example of this is to use the Norris Landsberg model to calculate fatigue. The failure model represents the time to failure or equivalent time as a function of the impact of the load on the system. Overuse failures are given a superior priority in terms of occurrence because they need to be interpreted for a voltage that is reasonably predictable in the lifecycle profile. Assuming that the correct design precautions are taken, the overuse mechanism is not likely to occur in the field and can only be quantified by determining the likelihood of an overuse condition. To simulate the occurrence of these failure mechanisms, the circuit in Figure 5.14 is simulated in MATLAB/Simulink and PLECS integration environment with the component details provided in Table 5.2.

The data corresponding to healthy and failure mechanism in PSDs is simulated using the simulation environment. All these conditions correspond to nine different modes with one mode as normal operating mode, and the remain eight modes as failure mechanism effect modes. The failure mechanism effect modes are considered with all the power modules and diodes in the converter circuit. The scenario for bond-wire failure in the four IGBTs are labelled with Fault 1 to Fault 4, and Fault 5 to Fault 8 indicate the failure in the diodes. For all the modes, the inverter terminal voltage is measured as shown in Figure 5.15 to aggregate the data for developing the failure detection approach using machine learning algorithm.

The inverter terminal voltage in Figure 5.15a corresponds to the normal operating condition of the power modules in the inverter. Further, the measured voltage

Table 5.2 Simulation parameters for generating the normal and failure modes in single phase inverter of grid connected system.

Parameter		Value
DC link voltage		400 V
Infineon (IGBT SGP15N60)	Dissipation of Maximum Power	139 W
	Maximum Collector – Emitter Voltage	600 V
	Collector – Emitter saturation Voltage	2.3 V
	Maximum Collector Current	31 A
	Thermal resistance, junction to case	$0.9°C/W$
	Thermal resistance, junction to ambient	$62 - 40°$ C/W
Microsemi (Diode APT15D60BG)	Maximum D. C. Reverse Voltage, Peak Repetitive Reverse Voltage, and Working Peak Reverse Voltage	600 V
	Maximum Average Forward Current	15 A
	RMS Forward Current	32 A
	Non – Repetitive Forward Surge Current	110 A
	Junction – to – Case Thermal Resistance	$1.35°C/W$
	Junction – to – Ambient Thermal Resistance	$40°C/W$

in Figure 5.15b and c identify the effect of bond-wire fatigue through switches I_1, and I_2, I_3, respectively. The failure mechanism is achieved by changing the resistance of the simulated IGBT in switches I_1 and I_3 while operating in an inverter circuit, whereas, the failure through I_2 is achieved by disrupting the gating pulse to the switch. Similarly, the effect of diode failure is achieved through D_1, and D_4 and the corresponding inverter terminal voltage measurements are shown in Figure 5.15d and e.

Further, the failure mechanism in the component of a system can be classified by training a machine learning framework. Generally, the conventional idea of detection is focussed at determining the capability of an algorithm to identify errors, and defects prior to the end of manufacturing. This conventional description does not apply to discussions at the component level. Therefore, detection is regard as a feature of detecting a working failure before it occurs. With the expected load profile, you can avoid the overstress mechanism by selecting the right part and correspondingly degrading performance. Still, random and unpredictable load fluctuations such as lightning strikes and collisions can cause overload failures. Not

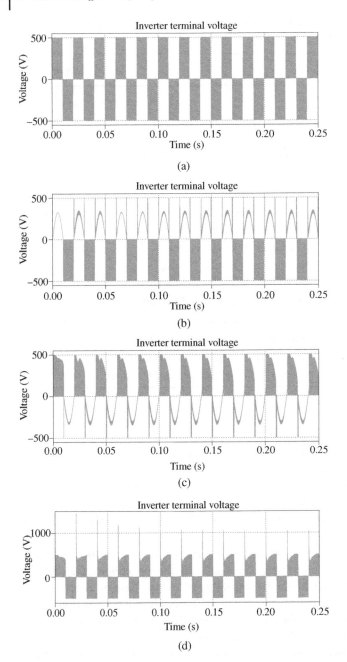

Figure 5.15 Inverter terminal voltage for normal operation and power module failure conditions. (a) Normal operation. (b) Inverter I_1 failure. (c) Inverter I_2 and I_3 failure. (d) Inverter D_4 failure. (e) Inverter D_1 and D_4 failure.

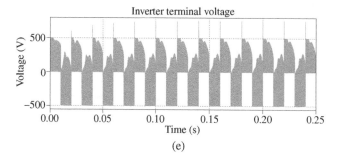

Figure 5.15 (*Continued*)

all methods are unpredictable, as damage accumulation varies some observable and detectable factors. This capability of monitoring and forecast failure is also known as prognosis and health care (PHM). PHM monitors and analyses field data to identify anomalies, classify defects, and enable calculations of remaining useful lives. The wear mechanism is detectable, and various groups have successfully executed PHM in silicon power devices [54–56].

Wear mechanisms are more noticeable than other wear and abuse mechanisms, depending on their probability and relationship with damage to the electrical constraints related with the mechanism. PHM methods and methodologies are evolving rapidly, reducing implementation costs and simplicity. However, developing and implementing the PHM framework is not yet easy, so it can only be cost-effective for certain critical components and applications. In addition, competing error mechanisms have the same or similar error modes, which can tamper with the measured signal. This makes it difficult to differentiate between error modes and take the necessary corrective actions. Hence, to ease the process of failure detection, the machine learning framework is developed using the artificial neural networks. The details of neural network for implementing pattern recognition and classification are discussed in the authors previous works. A brief overview of the various steps involved in designing and training the framework is shown in Figure 5.16. To proceed with the development of the failure detection approach, the measured inverter terminal voltage in a sample window of two cycles is logged for all the modes of operation. Further, the wavelet transform is applied for the measured data set at each operation modes to extract 13 different features as discussed in [57–59].

The final feature matrix obtained for training with the machine learning algorithm is of size $14\,400 \times 13$. Here, each class holds a feature matrix of size 1600×13. This dataset is further normalized from 0 to 1 using the minmax approach to avoid the misfitting of the data during the training process. This normalized data is trained with the ANN classifier, where the data is initially divided randomly in

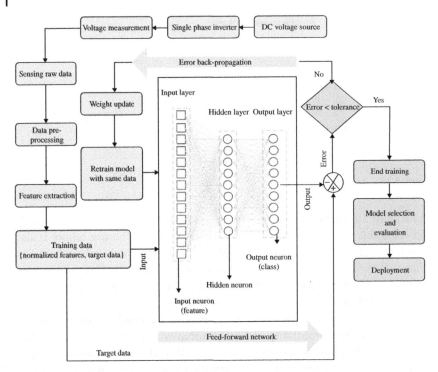

Figure 5.16 Framework for classifier training.

the size of 60 : 20 : 20%, These three divisions correspond to training, testing, and validation respectively. Further, the network structure forms three layers with 13 neurons in input layer (features), the hidden layer has 9 nodes, and the output layer has 9 neurons (classes). The training of the dataset with ANN is achieved with the scaled conjugate gradient back-propagation algorithm. The results of the training process are estimated using the truly and falsely classified samples which are identified through the confusion matrix shown in Figure 5.17.

Figure 5.17a identifies the truly and falsely classified samples of normal and failure mode operating IGBT scenarios. The combined falsely classified samples of the normal operating condition and the 8 different failure classes is 27 samples. This estimates the accuracy of the trained classifier to be approximately 99.8%. Further, under the testing conditions, the trained classifier has a prediction speed of approximately 3300 observations per second. This provides a cumulative testing time of 0.02 seconds with a testing accuracy of 100%. Similarly, the results in Figure 5.17b illustrate the positive prediction value and false detection rate with the trained samples. The false detection rate identifies the misclassification of 0.3% at fault 5 with reference to the failure conditions 1 and 2. Further, the true positive rate and

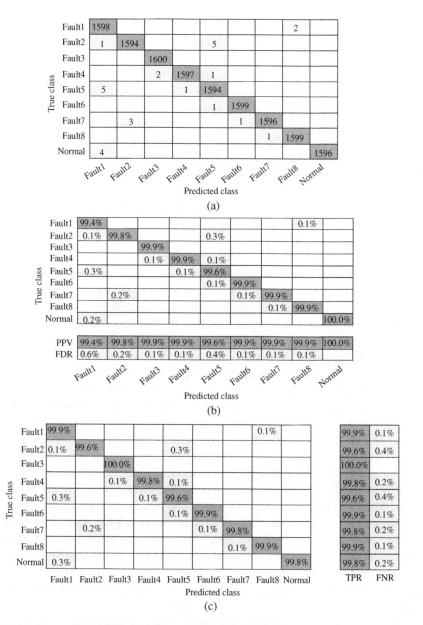

Figure 5.17 Classifier training results for IGBT failure detection. (a) Truly and falsely classified samples. (b) Positive prediction value and false detection rate. (c) True positive rate and false negative rate.

false negative rate of the classifier training process are show in Figure 5.17c. The true positive rate identifies the percentage of acceptably classified samples for a true class and the false negative rate indicates the inaccurately classified samples for true class.

Further, the degradation analysis is carried out by measuring the data corresponding to the inverter operation with both healthy and degraded PSDs. To simulate these conditions, the accelerated aging of IGBT and Diode is achieved such that 50% degradation rate is identified in the module operation. The degradation scenarios during the classifier training are identified as Degraded I_1 = Fault 1, Degraded I_2 = Fault 2, Degraded I_3 = Fault 3, Degraded I_4 = Fault 4, Degraded D_1 = Fault 5, Degraded D_2 = Fault 6, Degraded D_3 = Fault 7, Degraded D_4 = Fault 8. This is realized by subjecting the power modules to high-temperature stresses which changes the internal resistances of bond wire and solder joints. Further, the effect of degradation for different components is identified at the terminal voltage of the inverter as shown in Figure 5.18.

The measured inverter terminal voltage in Figure 5.18a and b identify the effect of bond wire resistance degradation in switches I_1, and I_3, respectively. The change in resistance is achieved by varying the $Z_{ch_{IGBT}}$ of the power module operating in the inverter circuit. Further, the inverter terminal voltage for the degraded operation of the diodes is shown in Figure 5.18c and d. The effect of diode degradation is achieved by varying the $Z_{ch_{Diode}}$ of the power module operating in the inverter circuit. To proceed with the development of the failure detection approach, the measured inverter terminal voltage in a sample window of two cycles is logged for all the modes of operation. Further, the wavelet transform is applied for the measured data set at each operation modes to extract 13 different features. The final feature matrix obtained for training with the machine learning algorithm is of size $14\,400 \times 13$. Here, each class holds a feature matrix of size 1600×13. This dataset is further normalized from 0 to 1 using the minmax approach to avoid the misfitting of the data during the training process. This normalized data is trained with the ANN classifier, where the data is initially divided randomly in the size of 60 : 20 : 20%, These three divisions correspond to training, testing, and validation respectively. Further, the network structure forms 3 layers with 13 neurons in input layer (features), the hidden layer has 9 nodes, and the output layer has 9 neurons (classes). The training of the dataset with ANN is achieved with the scaled conjugate gradient back-propagation algorithm. The results of the training process are estimated using the truly and falsely classified samples which are identified through the confusion matrix shown in Figure 5.19.

Figure 5.19a identifies the truly and falsely classified samples of normal and degraded operation of power modules in the inverter. The combined falsely classified samples of the normal operating condition and the eight different failure classes is 117 samples. This estimates the accuracy of the trained classifier to be

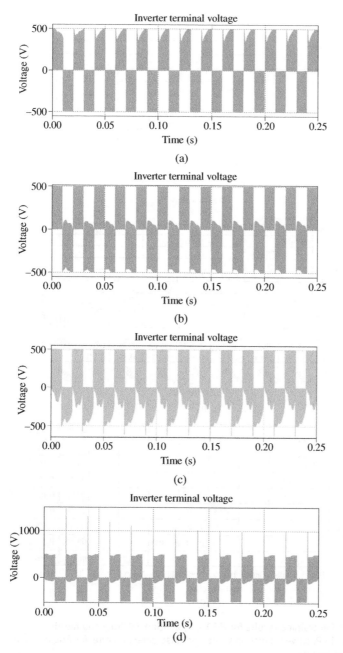

Figure 5.18 Inverter terminal voltage for normal operation and degraded power module operation. (a) Inverter I_1 degraded. (b) Inverter I_3 degraded. (c) Inverter I_2 and D_2 degraded. (d) Inverter D_4 degraded.

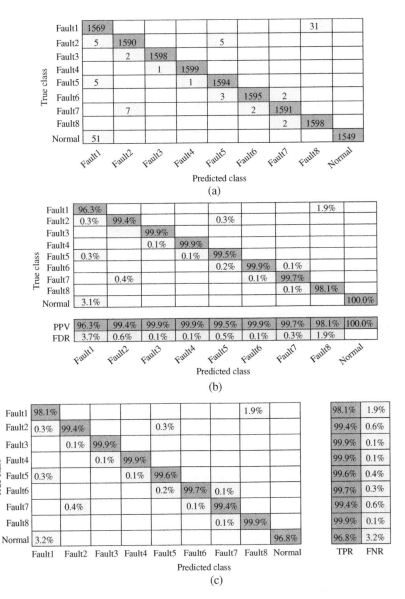

Figure 5.19 Classifier training results for IGBT degradation. (a) Truly and falsely classified samples. (b) Positive prediction value and false detection rate. (c) True positive rate and false negative rate.

approximately 99.2%. Further, under the testing conditions, the trained classifier has a prediction speed of approximately 2100 observations per second. This provides a cumulative testing time of 0.03 seconds with a testing accuracy of 99.6%. Similarly, the results in Figure 5.19b illustrate the positive prediction value and false detection rate with the trained samples. The false detection rate identifies the misclassification of 3.2% at normal operation with reference to the fault 1 condition. Further, the true positive rate and false negative rate of the classifier training process are show in Figure 5.19c. Similar to the results in Figure 5.17c, the true positive rate identifies the percentage of acceptably classified samples for a true class and the false negative rate indicates the inaccurately classified samples for true class.

5.4.2 Failure Mechanism Criticality Analysis

Both FMEA and FMMEA require importance assessment to prioritize failure mechanisms. Such prioritization enables efficient distribution of resources and improves system reliability. The difficulty in prioritizing FMMEA failure mechanisms at the component level is that the data needed to make decisions is highly application dependent. This section describes traditional methods of defining and estimating importance and provides component-level information-based guidance for classifying failure mechanisms based on importance.

With JEP131B, JEDEC outlines three components for critical classification of failure modes: severity, occurrence, and detection [60]. The three classifications are individually ranked from 1 to 10 established on the evaluation of the team that completed the analysis, with 10 being the extremely serious, the maximum incidence, and the most challenging to detect. Then multiply these three metrics to establish a RPN. It turns out that a failure mode with a high RPN is more annoying than a failure mode with a low RPN. Priority corrective action to decrease the RPN in the highest failure mode. After taking corrective action, the RPN should be revised, and the design reassessed. Do not compare RPN rankings to RPN rankings from different groups on the same system or different systems, as rankings are based on the judgment of the team.

The seriousness of the failure mode depends on the impact on the end user. Initially, seriousness must consider the possibility for harm to users of the system. If the effects of a particular mechanism can be harmful, you should assign a higher severity to that mechanism. The next consideration is the cost of users and system manufacturers. Costs come in many forms, including legal costs, warranty and returns, related maintenance, and brand reputation. Based on your team's assessment, you need to create a severity assessment that takes these factors into account. Obviously, the severity of the error depends on the application. The following describes component-level severity.

The existence of a failure mode indicates the possible of its occurrence. Occurrence considerations should include load conditions, as well as system materials, shapes, and component types. You can use this information to determine the likelihood of a mechanism. It can be categorized according to the evaluation of the FMEA development team.

Lastly, failure mode detection metrics have conventionally been defined as the capability to distinguish failure modes prior to the product is shipped to the customer. Conventionally, in the electronics industry, detection has been associated with specific test or screen escape rates. JEDEC suggests using the reverse of the escape rate as a way to quantify mode detection. Since the FMEA scope was developed at the system level, it is not expected that individual power semiconductors will be tested at the system integration level. Therefore, in the discussion below, we will choose a different approach to detection.

5.4.3 Failure Mechanism Severity Analysis

Severity depends on the impact on the end user. If this information is missing, the severity needs to be considered in another context. Nevertheless, all Silicon Power devices are used in bigger circuits, so there are other electrical components in proximity. For component-level prioritization, seriousness is ascertained by the potential for a failure to be catastrophic and affect adjacent components.

For Silicon Power devices, overload mechanisms that lead to short circuit failures are most likely to harm the surrounding system. Short circuits can generate considerable Joule heat and, if left uncontrolled, can damage adjacent components, and cause a fire, rising the associated failure costs. The wear mechanism can be deemed less serious as only silicon power devices fail, other nearby components are not damaged, and the system can be refurbished.

5.5 Summary

In this chapter a detailed overview of the failure modes, mechanisms, and effect analysis for PSDs operating in a grid connected system are discussed. Initially, the information related to design and operation of PSDs along with their properties and packaging are analyzed. Further, different failure mechanisms in the PSDs are studied and a brief analysis is performed by developing a detection approach. The results of the failure approach proved to be effective in early detection of the occurrence of the failure with high accuracy. The failure detection approach can be further extended to detect more and multiple failure mechanisms in a PSD.

References

1 Falck, J., Felgemacher, C., Rojko, A. et al. (2018). Reliability of power electronic systems: an industry perspective. *IEEE Ind. Electron. Mag.* 12 (2): 24–35. https://doi.org/10.1109/MIE.2018.2825481.

2 Poon, J., Jain, P., Konstantakopoulos, I.C. et al. (2017). Model-based fault detection and identification for switching power converters. *IEEE Trans. Power Electron.* 32 (2): 1419–1430. https://doi.org/10.1109/TPEL.2016.2541342.

3 Khan, I., Xu, Y., Sun, H., and Bhattacharjee, V. (2018). Distributed optimal reactive power control of power systems. *IEEE Access* 6: 7100–7111. https://doi.org/10.1109/ACCESS.2017.2779806.

4 Khan, I., Li, Z., Xu, Y., and Gu, W. (2016). Distributed control algorithm for optimal reactive power control in power grids. *Int. J. Electron. Power Energy Syst.* 83: 505–513. https://doi.org/10.1016/j.ijepes.2016.04.004.

5 Chang, X., Xu, Y., Sun, H., and Khan, I. (2021). A distributed robust optimization approach for the economic dispatch of flexible resources. *Int. J. Electron. Power Energy Syst.* 124: 106360. https://doi.org/10.1016/j.ijepes.2020.106360.

6 Nasir, M., Anees, M., Khan, H.A. et al. (2019). Integration and decentralized control of standalone solar home systems for off-grid community applications. *IEEE Trans. Ind. Appl.* 55 (6): 7240–7250. https://doi.org/10.1109/TIA.2019.2911605.

7 Kurukuru, V.S.B., Haque, A., Khan, M.A. et al. (2021). A review on artificial intelligence applications for grid-connected solar photovoltaic systems. *Energies* 14 (15): 4690. https://doi.org/10.3390/en14154690.

8 Shahzad, M., Bharath, K.V.S., Khan, M.A., and Haque, A. (2019). Review on reliability of power electronic components in photovoltaic inverters. In: 2019 *International Conference on Power Electronics, Control and Automation (ICPECA)*, 1–6. IEEE https://doi.org/10.1109/ICPECA47973.2019.8975585.

9 Kurukuru, V.S.B., Haque, A., Tripathi, A.K., and Khan, M.A. (2021). Condition monitoring of IGBT modules using online TSEPs and data-driven approach. *Int. Trans. Electr. Energy Syst.* 31 (8): 1–24.

10 Yang, S., Bryant, A., Mawby, P. et al. (2011). An industry-based survey of reliability in power electronic converters. *IEEE Trans. Ind. Appl.* 47 (3): 1441–1451. https://doi.org/10.1109/TIA.2011.2124436.

11 Hoffmann, K.F. and Karst, J.P. (2005). High frequency power switch - improved performance by MOSFETs and IGBTs connected in parallel. In: 2005 *European Conference on Power Electronics and Applications*, 6–11. IEEE https://doi.org/10.1109/EPE.2005.219594.

12 Li, D., Qi, F.; Packwood, M. et al. (2018). New Developed 3.3 kV/1500 A IGBT Module. *PCIM Europe 2018; International Exhibition and Conference for Power*

Electronics, Intelligent Motion, Renewable Energy and Energy Management, Nuremberg, Germany (5–7 June 2018).

13 Liu, G., Ding, R., and Luo, H. (2015). Development of 8-inch key processes for insulated-gate bipolar transistor. *Engineering* 1 (3): 361–366. https://doi.org/10.15302/J-ENG-2015043.

14 Department of Defense (1940). Military Standard, Procedures for performing a failure mode, effects and criticality analysis (24 November 1980). MIL-STD-1629A. United States of America: Department of Defense. https://doi.org/10.1093/nq/s7-X.252.326-b.

15 Society of Automotive Engineers (2018). Fault/Failure Analysis Procedure, 70. ARP926C. S-18 Aircraft and Sys Dev and Safety Assessment Committee. SAE International https://doi.org/10.4271/ARP926C.

16 JEP (1998). Potential Failure Mode and Effects Analysis (FMEA). JEP 131. Joint Electron Device Engineering Council Solid State Technology Association.

17 JEP (2018). Potential Failure Mode and Effects Analysis (FMEA). JEP 131. Joint Electron Device Engineering Council Solid State Technology Association.

18 Rastayesh, S., Bahrebar, S., Blaabjerg, F. et al. (2019). A System Engineering Approach Using FMEA and Bayesian Network for Risk Analysis—A Case Study. *Sustainability* 12 (1): 77. https://doi.org/10.3390/su12010077.

19 Mathew, S., Alam, M., and Pecht, M. (2012). Identification of Failure Mechanisms to Enhance Prognostic Outcomes. *Journal of Failure Analysis and Prevention* 66–73. https://doi.org/10.1007/s11668-011-9508-2.

20 Nguyen, T.A., Lefebvre, S., Joubert, P.-Y. et al. (2015). Estimating current distributions in power semiconductor dies under aging conditions: bond wire liftoff and aluminum reconstruction. *IEEE Trans. Components, Packag. Manuf. Technol.* 5 (4): 483–495. https://doi.org/10.1109/TCPMT.2015.2406576.

21 Martineau, D., Mazeaud, T., Legros, M. et al. (2009). Characterization of ageing failures on power MOSFET devices by electron and ion microscopies. *Microelectron. Reliab.* 49 (9–11): 1330–1333. https://doi.org/10.1016/j.microrel.2009.07.011.

22 Smet, V., Forest, F., Huselstein, J.-J. et al. (2011). Ageing and failure modes of IGBT modules in high-temperature power cycling. *IEEE Trans. Ind. Electron.* 58 (10): 4931–4941. https://doi.org/10.1109/TIE.2011.2114313.

23 Bouarroudj, M., Khatir, Z., Ousten, J.P. et al. (2007). Degradation behavior of 600V–200A IGBT modules under power cycling and high temperature environment conditions. *Microelectron. Reliab.* 47 (9–11): 1719–1724. https://doi.org/10.1016/j.microrel.2007.07.027.

24 Held, M., Jacob, P., Nicoletti, G. et al. Fast power cycling test of IGBT modules in traction application. In: *Proceedings of Second International Conference on Power Electronics and Drive Systems*, vol. 1, 425–430. IEEE https://doi.org/10.1109/PEDS.1997.618742.

25 Lai, W., Chen, M., Ran, L. et al. (2016). Low stress cycle effect in IGBT power module die-attach lifetime modeling. *IEEE Trans. Power Electron.* 31 (9): 6575–6585. https://doi.org/10.1109/TPEL.2015.2501540.

26 Teasdale, K. (2009). Continuous dc Current Ratings of International Rectifier's Large Semiconductor Packages, 1–17. AN-1140. International Rectifier Corporation.

27 Eni, E.-P., Bęczkowski, S., Munk-Nielsen, S. et al. (2017). Short-circuit degradation of 10-kV 10-A SiC MOSFET. *IEEE Trans. Power Electron.* 32 (12): 9342–9354. https://doi.org/10.1109/TPEL.2017.2657754.

28 Black, J.R. (1970). RF power transistor metallization failure. *IEEE Trans. Electron Devices* 17 (9): 800–803. https://doi.org/10.1109/T-ED.1970.17077.

29 Santoro, C.J. (1969). Thermal cycling and surface reconstruction in aluminum thin films. *J. Electrochem. Soc.* 116 (3): 361. https://doi.org/10.1149/1.2411847.

30 Ciappa, M. (2002). Selected failure mechanisms of modern power modules. *Microelectron. Reliab.* 42 (4–5): 653–667. https://doi.org/10.1016/S0026-2714(02)00042-2.

31 Huang, J., Hu, Z., Gao, C., and Cui, C. (2014). Analysis of water vapor control and passive layer process effecting on transistor performance and aluminum corrosion. *Proceedings of the. 2014 Prognostics and. System Health Management Conference. PHM 2014*, Zhangjiajie, China (24–27 August 2014), no. 61201028, pp. 26–30. 10.1109/PHM.2014.6988126.

32 Navarro, L.A., Perpiñà, X., Vellvehi, M. et al. (2012). Thermal cycling analysis of high temperature die-attach materials. *Microelectron. Reliab.* 52 (9–10): 2314–2320. https://doi.org/10.1016/j.microrel.2012.07.022.

33 Quintero, P.O. and McCluskey, F.P. (2011). Temperature cycling reliability of high-temperature lead-free die-attach technologies. *IEEE Trans. Device Mater. Reliab.* 11 (4): 531–539. https://doi.org/10.1109/TDMR.2011.2140114.

34 Katsis, D.C. and van Wyk, J.D. (2003). Void-induced thermal impedance in power semiconductor modules: Some transient temperature effects. *IEEE Trans. Ind. Appl.* 39 (5): 1239–1246. https://doi.org/10.1109/TIA.2003.816527.

35 Zhu, N. (1999). Thermal impact of solder voids in the electronic packaging of power devices. *Fifteenth Annual IEEE Semiconductor Thermal Measurement and Management Symposium (Cat. No.99CH36306)*, San Diego, CA, USA (9–11 March 1999), pp. 22–29. doi: 10.1109/STHERM.1999.762424.

36 Fleischer, A.S., Chang, L., and Johnson, B.C. (2006). The effect of die attach voiding on the thermal resistance of chip level packages. *Microelectron. Reliab.* 46 (5–6): 794–804. https://doi.org/10.1016/j.microrel.2005.01.019.

37 Kohman, G.T., Hermance, H.W., and Downes, G.H. (1955). Silver migration in electrical insulation. *Bell Syst. Tech. J.* 34 (6): 1115–1147. https://doi.org/10.1002/j.1538-7305.1955.tb03793.x.

38 Cosiansi, F., Mattiuzzo, E., Turnaturi, M., and Candido, P.F. (2016). Evaluation of electrochemical migration into an isolated sintered power module. *CIPS 2016; 9th International Conference on Integrated Power Electronics Systems*, Nuremberg, Germany (8–10 March 2016), pp. 1–5.

39 Lu, G.Q., Yang, W., Mei, Y.H. et al. (2014). Migration of sintered nanosilver on alumina and aluminum nitride substrates at high temperatures in dry air for electronic packaging. *IEEE Trans. Device Mater. Reliab.* 14 (2): 600–606. https://doi.org/10.1109/TDMR.2014.2304737.

40 Medgyes, B., Szivos, D., Adam, S. et al. (2016). Electrochemical migration of Sn and Ag in NaCl environment. *2016 IEEE 22nd International Symposium for Design and Technology in Electronic Packaging (SIITME)*, Oradea, Romania (20–23 October2016), pp. 274–278. https://doi.org/10.1109/SIITME .2016.7777294.

41 Benbahouche, L., Merabet, A., and Zegadi, A. (2012). A comprehensive analysis of failure mechanisms: latch up and second breakdown in IGBT(IXYS) and improvement. *2012 19th International Conference on Microwaves, Radar & Wireless Communications*, May 2012, pp. 190–192. doi: https://doi.org/10.1109/ MIKON.2012.6233539.

42 Knipper, U., Pfirsch, F., Raker, T., Niedermeyr, J., and Wachutka, G. (2008). Destruction in the Active Part of an IGBT Chip Caused by Avalanche-Breakdown at the Edge Termination Structure. *2008 International Conference on Advanced Semiconductor Devices and Microsystems*, Smolenice, Slovakia (12–16 October 2008), pp. 159–162. https://doi.org/10.1109/ASDAM .2008.4743305.

43 Spirito, P., Maresca, L., Riccio, M. et al. (2015). Effect of the collector design on the IGBT Avalanche Ruggedness: a comparative analysis between punch-through and field-stop devices. *IEEE Trans. Electron Devices* 62 (8): 2535–2541. https://doi.org/10.1109/TED.2015.2442334.

44 Jahdi, S., Alatise, O., Bonyadi, R. et al. (2015). An analysis of the switching performance and robustness of power MOSFETs body diodes: a technology evaluation. *IEEE Trans. Power Electron.* 30 (5): 2383–2394. https://doi.org/10 .1109/TPEL.2014.2338792.

45 Duvvury, C., Rodriguez, J., Jones, C., and Smayling, M. (1994). Device integration for ESD robustness of high voltage power MOSFETs. *Proceedings of 1994 IEEE International Electron Devices Meeting*, pp. 407–410. doi: 10.1109/IEDM.1994.383381.

46 McPherson, J.W. (2012). Time dependent dielectric breakdown physics – models revisited. *Microelectron. Reliab.* 52 (9–10): 1753–1760. https:// doi.org/10.1016/j.microrel.2012.06.007.

47 Liu, T., Zhu, S., White, M.H. et al. (2021). Time-dependent dielectric break-down of commercial 1.2 kV 4H-SiC power MOSFETs. *IEEE J. Electron Devices Soc.* 9: 633–639. https://doi.org/10.1109/JEDS.2021.3091898.

48 Sato, M., Kumada, A., Hidaka, K. et al. (2016). Surface discharges in silicone gel on AlN substrate. *IEEE Trans. Dielectr. Electr. Insul.* 23 (1): 494–500. https://doi.org/10.1109/TDEI.2015.005412.

49 Fabian, J.-H., Hartmann, S., and Hamidi, A. (2005). Analysis of insulation failure modes in high power IGBT modules. *Fourtieth IAS Annual Meeting. Conference Record of the 2005 Industry Applications Conference,* vol. 2, pp. 799–805. doi: https://doi.org/10.1109/IAS.2005.1518425.

50 McCluskey, P. (2012). Reliability of power electronics under thermal loading. *7th International Conference on Integrated Power Electronics Systems CIPS 2012,* vol. 9.

51 Park, J., Kim, M., and Roth, A. (2014). Improved thermal cycling reliability of ZTA (Zirconia Toughened Alumina) DBC substrates by manipulating metallization properties. *CIPS 2014; 8th International Conference on Integrated Power Electronics Systems,* pp. 1–9.

52 Heumann, K. and Quenum, M. (1993). Second breakdown and latch-up behavior of IGBTs. *1993 Fifth European Conference on Power Electronics and Applications,* vol. 2, pp. 301–305.

53 Su, P., Goto, K.I., Sugii, T., and Hu, C. (2002). A thermal activation view of low voltage impact ionization in MOSFETs. *IEEE Electron Device Lett.* 23 (9): 550–552. https://doi.org/10.1109/LED.2002.802653.

54 Oh, H., Han, B., McCluskey, P. et al. (2015). Physics-of-failure, condition monitoring, and prognostics of insulated gate bipolar transistor modules: a review. *IEEE Trans. Power Electron.* 30 (5): 2413–2426. https://doi.org/10.1109/TPEL .2014.2346485.

55 Ji, B., Song, X., Cao, W. et al. (2015). In-situ diagnostics and prognostics of solder fatigue in IGBT modules for electric vehicle rrives. *IEEE Trans. Power Electron.* 30 (3): 1535–1543. https://doi.org/10.1109/TPEL.2014.2318991.

56 Patil, N., Das, D., Yin, C. et al. A fusion approach to IGBT power module prognostics. *EuroSimE 2009 – 10th International Conference on Thermal, Mechanical and Multi-Physics Simulation and Experiments in Microelectronics and Microsystems,* April 2009, pp. 1–5. doi: https://doi.org/10.1109/ESIME.2009 .4938491.

57 Dadhich, K., Kurukuru, V. S. B., Khan, M.A., and Haque, A. (2019). Fault identification algorithm for grid connected photovoltaic systems using machine learning techniques. *2019 International Conference on Power Electronics, Control and Automation (ICPECA),* New Delhi, India (16–17 November 2019), pp. 1–6. doi: 10.1109/ICPECA47973.2019.8975397.

58 Bharath, K.V.S., Blaabjerg, F., Haque, A., and Khan, M.A. (2020). Model-based data driven approach for fault identification in proton exchange membrane fuel cell. *Energies* 13 (12): 3144. https://doi.org/10.3390/en13123144.

59 Khan, M.A., Haque, A., and Kurukuru, V.S.B. (2019). An efficient islanding classification technique for single phase grid connected photovoltaic system. *2019 International Conference on Computer and Information Sciences (ICCIS)*, Sakaka, Saudi Arabia (3–4 April 2019), pp. 1–6. https://doi.org/10.1109/ICCISci.2019.8716438.

60 JEDEC Solid State Technology Association (2005). JEP131A-Potential Failure Mode and Effects Analysis (FMEA). Arlington [Online]. http://www.jedec.org/Catalog/catalog.cfm.

6

Fault Classification Approach for Grid-Tied Photovoltaic Plant

V S Bharath Kurukuru and Ahteshamul Haque

Advance Power Electronics Research Lab, Department of Electrical Engineering, Jamia Millia Islamia (A Central University), New Delhi, India

6.1 Introduction

This chapter aims at providing a detailed information about the functioning of grid-tied solar photovoltaic (PV) plants, identify the faults which can be potential hurdle in achieving high yield and develop a fault classification (FC) mechanism for timely indication of the fault alarm. The alarm indication enables the PV power plant operator to act upon the fault indication and take necessary measures as per the nature of the fault. From the literature it is identified that the several presently available FC algorithms are formulated while considering very small sized PV plants [1–4]. Besides, these algorithms are not tested on PV plants which are beyond 100 kW size. Today, PV power plants are being installed which are in capacities of megawatts and giga watts. So, there is need for developing FC systems which is designed by focusing on large scale PV plants. This will facilitate the power plant operators to have better picture of how the power plants are performing. Further this chapter models a 125 kW on-grid PV power plant as the reference power plant to design the FC approach. Two identical PV plants are modeled and named as Actual PV plant and Theoretical PV plant. The modeling and simulation aids in understanding the performance deviations in presence of a certain fault. Efforts were made to keep the design of FC mechanism simple, easy to implement, and robust by testing its implementation under several conditions. This algorithm compares the power outputs of actual PV plant and theoretical PV plant using mathematical equations. These mathematical equations are developed with basic statistical tools of mean and standard deviation. The algorithm performs its computations and gives indications whenever a fault is occurred.

The objective of this chapter is to develop a simple and practically viable approach to a FC mechanism which is able to trace a fault and diagnose the fault

Fault Analysis and its Impact on Grid-connected Photovoltaic Systems Performance, First Edition.
Edited by Ahteshamul Haque and Saad Mekhilef.

type. The aim is to develop and test an algorithm that would be applicable to variety of solar power plants, especially large-scale power plants ranging from 100 kW up to a capacity in megawatts. The size of power plants should not affect the algorithm's operation and it should be simple enough to be deployed without any special requirements. The key objectives of this work are:

- Understanding the operation of a grid-connected PV power plant and tracing out the most commonly occurring faults in the DC side of power plant.
- Modeling a grid-connected solar PV plant based on the information of a physically installed PV power plant.
- Analyze the performance of the modeled grid-connected PV plant under different weather conditions with varying irradiance and temperature.
- Design the FC mechanism that is capable of detecting and identifying faults in a large-scale grid-tied solar PV power plant.
- Test the developed FC system under different scenarios to observe the performance and reliability of developed system.

6.2 Solar Power Plants

A brief overview of solar power plants is presented in this chapter in order to discuss the major components and their significance in a power plant. The main focus is the PV power plants. Further the components of solar PV plants and topologies of PV plants are described here [5–7].

6.2.1 Types of Solar Power Plants

Solar power plants are generally categorized under two broad categories, solar PV power plants and concentrated solar thermal power plants. The difference is based on the fact that the sunlight carries two types of energies that can be utilized in energy production. One of these is the photon energy carried by the photons in the light, second is the heat energy that comes along.

6.2.1.1 Concentrated Solar Thermal Power Plants

The concentrated solar thermal power plants utilize the heat energy of the sunlight [8]. The sunlight is converged on a single point, usually on top of a high tower. At the top of that tower, water tank is constructed. The heat causes the vessel to heat up to a level that the water starts reaching its boiling point. The steam produced from the water heating is used to run the turbines and electricity is produced by the alternator coupled with the turbine. Also, the steam is used for heating purposes. Concentrated solar thermal plants are usually installed in GW capacity. These power plants have a problem that the ecosystem gets affected as the flying birds get struck with concentrated rays of light from heliostats die immediately because of excessive heat and this causes damage to the surrounding flora as well [9, 10].

6.2.1.2 Solar Photovoltaic Power Plants

In case of solar PV plants, the heat energy is not the driver of energy generation but the photons [11]. The photons fall on to the surface of solar cells and penetrate into the atomic level, where electrons are hit with these photons and holes are created. The phenomenon of the separation of charge carriers causes ow of current, in other words production of electricity. Solar cells are combined in to a series-connected set which is encapsulated in a vacuumed enclosure, called a PV module or PV panel. Now grown and a mature technology, the PV modules are normally created with wafers based crystalline silicon cells or thin films [12].

6.2.2 Major Components of PV Plants

6.2.2.1 Solar PV Modules

The solar PV modules produce DC electricity, which is then converted to AC or consumed directly as DC as per requirement. The PV modules are of many types today, based on the composition, sunlight to electricity conversion efficiency and power output rating, etc. [13, 14]. The most common and commercially available technologies of PV modules are monocrystalline silicon, polycrystalline silicon and thin film-based PV modules. These types have their own positives and negatives and naturally the selection depends on the application. The PV modules are connected in combinations of series and parallel connections based on the voltage and current requirements. Generally, the complete set of PV modules installed in a PV plant are referred to as PV array [15, 16].

Crystalline Silicon PV Modules: The monocrystalline PV modules are made from silicon wafers formed with a single crystal. The unbroken crystal makes the appearance uniform and continuous structure with no grain boundaries [17, 18]. The polycrystalline PV modules are made with multiple small crystals called crystallites. The surface of a polycrystalline cell has visible grain boundaries that give the polycrystalline PV modules their different appearance [19].

Thin-Film PV Modules: The thin-film PV modules are made by depositing thin films of semiconductor material on glass. There are several further types of thin-film PV modules, including cadmium telluride (CdTe), copper indium gallium selenide (CIGS) and amorphous silicon (a-Si). The PV modules vary in efficiency and performance under different conditions. Monocrystalline PV modules are considered better than other types in terms of performance. Thin-film PV modules are cheaper but larger in size when compared with other technology PV modules at same power rating [20–23].

Performance of PV Modules: The performance of PV modules depends on two very important factors, the sunlight intensity/irradiance and the module temperature. The irradiance is the measure of power that can be produced by the sunlight inclining on unit surface area. Irradiance is measured in Watts per square meter. The PV module temperature is important because as the temperature rises the collisions at the atomic level increase which results in resisting the flow of charge. Higher the irradiance, higher is the PV module output, whereas, if the temperature increases beyond 25 °C, then the output starts degrading. The PV modules are normally rated in Watts. This rating is based on the output of PV module that it could give under certain values of irradiance and module temperature. These values are referred to as "standard test conditions" (STCs) and are used for quality testing in laboratories as well. 1000 W/m^2 of irradiance and 25 °C are the STC and a PV module rated at 270 W capacity would yield a power of 270 W under these conditions [24].

Aging is also a factor which reduces the PV module outputs [25]. Generally, the PV modules are considered to live up to 25 years. However, these still are able to produce power but their efficiency is reduced so much that they are considered for replacement with new ones. PV modules drop to about 2–25% of their nameplate capacity after 1 year of operation and then a linear decrease is seen through rest of 24 years [26, 27].

The effects of irradiance, module temperature, and aging are provided in the datasheet of PV modules by the manufacturers. This helps in better prediction of output of PV power plants upon different weather conditions and gradual effect of aging can also be incorporated in the PV output forecasting [28]. The output characteristics of PV modules are represented with help of current to voltage (IV) and power to voltage (PV) curves. Also, the tolerance against the harsh environment conditions severe wind storms, hail storms and pressure, etc. are also mentioned in the datasheet of PV modules in order to have knowledge and understanding for the PV plants designers to select proper type of PV module for their specific conditions.

6.2.2.2 Inverters

Inverter is the power electronics device that converts the DC electricity coming from the PV modules in to AC electricity for use. Inverters are built as per the mode the operation [29]. The mode of operation can either be grid-connected [30–32], standalone or bi-modal [33–35]. These modes of operation are discussed in detail in further sections of the chapter. Inverters can have single or multiple input ports for DC power input and the PV array is installed in accordance with the specification of inverters in order to conform with the input voltage and current threshold limitations. The inverters play critical role in a grid-connected PV plant because there is no DC load and if inverter is not functioning, then the PV plant stops working completely.

6.2.2.3 Batteries and Charge Controllers

Batteries are used in the PV power plants to store energy which could be used later. The batteries provide a backup to PV plant. So, when sunlight is not available, the charge stored in the batteries may power up the connected load. The charge controllers are the components that connect the batteries with PV modules. Energy passes through the charge controller in order to be stored in the batteries. The charge controllers, as the name suggests, control the charge and prevent batteries from over charging. Charge controllers may also provide other features like battery health and temperature monitoring, etc.

6.2.2.4 System and Environment Monitoring Units

The system monitoring units provide the energy yield reports of the power plants. The power plant operator may be able to observe any abnormalities with the help of system monitoring units [36]. The environment monitoring systems provide data of the environment conditions especially the sunlight intensity/irradiance and the PV module temperature. The environment conditions are usually helpful in order to understand the performance of PV plant under different weather conditions [37–39].

6.2.2.5 Maximum Power Point Tracking

Maximum power point tracking (MPPT), is normally a built-in functionality of the inverter. The function of MPPT is to track the maximum power point of the PV array in real-time [40, 41]. The MPPT unit continuously monitors the power output of PV array and performs certain actions to obtain maximum yield under prevailing weather conditions.

6.2.3 Topologies of Solar PV Plants

There are three common topologies which are used while installing a PV power plant. The topologies form standalone PV plants, bi-modal PV plants, and grid-connected PV plants.

6.2.3.1 Standalone PV Plants

The standalone PV plants are those which are isolated from the grid supply [35, 42]. There is no external power connection and only the PV modules are power producing/supplying source. Further the energy produced from the PV array is stored in the batteries which may be used for later purpose [7]. Standalone PV plants are also called "off-grid" or "grid-isolated" PV plants as shown in Figure 6.1.

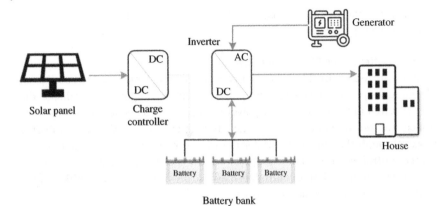

Figure 6.1 Design of standalone PV power plant. Source: Helman [10]/with permission of Elsevier.

6.2.3.2 Bi-Modal PV Plants
A bi-modal PV plant is a combination of standalone and the grid-connected PV system. The plant design is shown in Figure 6.2 [43]. The PV plant is connected with the grid and can provide its surplus energy to the grid and at the same time it can store energy to provide for later usage [7]. Bi-modal PV plants are also called "hybrid" PV plants.

6.2.3.3 Grid-Connected PV Plants
The grid-connected PV plants are those which have their AC output coupled with the grid power supply. The inverters which are designed to operate in

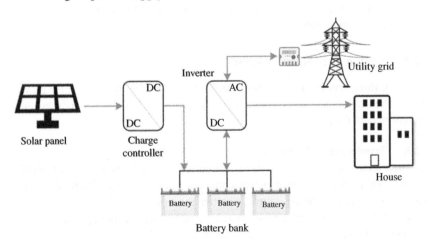

Figure 6.2 Design of bi-modal PV power plant. Source: Helman [10]/with permission of Elsevier.

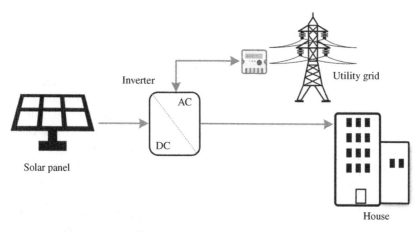

Figure 6.3 Design of grid-connected PV power plant. Source: Helman [10]/with permission of Elsevier.

this mode sense the voltage and frequency from the grid and give their output accordingly [44, 45]. The grid-connected PV plants do not have any battery backup for charge storage, all the energy produced is consumed on spot or fed in to the grid. Grid-connected PV plants are also called "on-grid," "grid-tied" or "grid-integrated" PV plants. Basic design of grid-connected PV plant is shown in Figure 6.3 [46, 47]. Grid-connected PV plants are the most obvious and preferred choice when designing a large-scale PV power plant. The simple design, easy installation and lesser limitations make the financial feasibility of grid-connected PV systems far better than rest of topologies. The feed-in energy to grid concept helps industries to deploy grid-connected PV systems and supply their surplus energy to grid and consume from grid when their requirement exceeds. Further, the net-metering concept helps achieve better pay-back times [48].

6.3 Modeling of PV Power Plant and FC System

This section discusses the modeling and simulation of PV power plant based on an actually operational solar power plant which acts as a reference for testing of the proposed FC algorithm.

6.3.1 Reference Solar PV Power Plant

The 125 kW system installed at the engineering and architecture campuses of Jamia Millia Islamia Central University, New Delhi, India is considered the reference system for developing the FC mechanism. The system considered is a part of the 2.25 MW roof top solar power plant installed at different faculties of the university. The said power plant was installed in the year 2019 and is integrated with the local grid network. The energy incoming from grid and outgoing from the solar plant is measured with a bidirectional energy meter that is used for billing purpose by the distribution company. The subject power plant is taken as reference as its installed capacity is higher than 100 kW satisfying our definition of large-scale PV plants. The details of installed equipment are given as follows:

PV Modules: The PV modules used in the power plant are 60 Cell, 270 W multi-crystalline modules. A total of 462 number of PV modules are installed in this plant section. The PV modules are connected in series forming up a string. One string has 22 number of PV modules and three strings are connected in parallel in a combiner box (CB). There are total of 21 strings and 7 CBs. Each CB has equal three number of PV strings. The seven DC positive and negative pairs of connections terminate at the DC busbar of inverter as parallel connections. The connections of PV strings and inverter are illustrated in the single line diagram shown in Figure 6.4. The specifications under the STCs of PV module are given in Table 6.1.

Inverter: The grid-tied inverter has DC input capacity of 141 kW and it yields a maximum of 125 kW of 3-phase AC power with 415 V and 50 Hz output. The inverter specifications are given in Table 6.2.

Monitoring and Measurement Units: In addition to these components, the pyranometer and temperature sensor are used in the environment monitoring unit for acquiring the real-time environment conditions, i.e. solar irradiance and PV module temperature [49, 50]. It also provides ambient temperature, wind speed, and direction. Further, a data logger is also installed to log the power output data from the connected inverter. The CBs which connect the PV strings have separate voltmeters and current transformers to measure the voltages and currents of

Table 6.1 PV module specifications.

Parameter	Value
Module type	Multi-crystalline module
Power output at STC	270 W
Module efficiency	16.4%
Voltage at Pmax (V_{mpp})	31.71 V
Current at Pmax (I_{mpp})	8.53 A
Open-circuit voltage (V_{oc})	38.81 V
Short-circuit voltage (V_{sc})	9.1 A

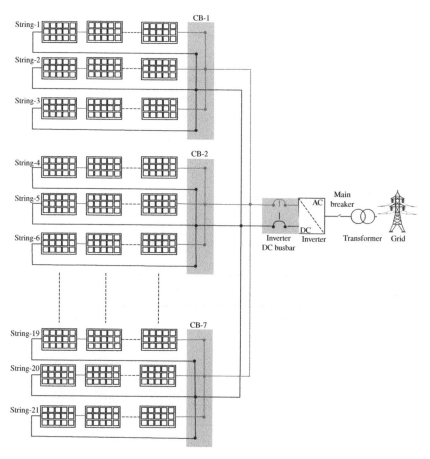

Figure 6.4 Single line diagram of reference PV power plant.

all strings [51]. The output of these measurement devices is connected with the communication module of inverter. Inverter sends the information of CBs and its output to the data logger. The data logger is connected to a work station for real-time display of DC power output and AC energy [52–54]. It keeps log of daily data and keeps record of several parameters. If there are any abrupt changes in the average values of the power, then the system raises an alarm of deviation from the average outputs. This helps the power plant operator know that there is an abnormality in the operation of the system. However, the type of fault is not indicated through this device.

The single line diagram of the reference PV power plant is shown in Figure 6.4. It is seen that 22 number of series-connected PV modules form up a string. Three number of strings are connected in parallel at a CB. The seven CBs output the DC

Table 6.2 Inverter specifications.

Parameter	Value
Maximum DC input power	141 kW
Maximum DC voltage	1000 V
C-rated power	125 kW
Maximum AC apparent power	137.5 kVA
Grid voltage	415 i
System frequency	50 Hz
Maximum output current	275 A

Table 6.3 Ranges of Z_{score}, T_{score} and PR_{str} for the identification of faults.

Fault type	Z_{score}	T_{score}	PR_{str}
F1	$-2600 \leq Z \leq 0$	$-25 < T < -9$ –	–
F2	$Z < -2900$	–	–
F3	$Z < -2900$	–	–
F4	$-1150 < Z < -850$	–	$1.22 \leq P \leq 2.4$
	$-2600 < Z < -1150$ –	–	
F5	$-2600 < Z < -850$	–	–

power and the inverter DC busbar combines the outputs of all CBs. The inverter converts the DC power to AC power as per the voltage and frequency of the grid. The main breaker acts as the isolation circuit breaker between the inverter output and the grid supply. The transformer is a 500 kVA rated 230/415 V *AC* 3 phase transformer which steps up the inverter output to 11 kV for power feed-in to the grid.

6.3.2 Faults in DC Side of PV Power Plant

DC side faults in the grid-tied PV plant are important because these faults do not always cause a system shutdown and no indication may be there for the plant operator to know if the output is normal or it has any fault [55–61]. AC side faults are critical in terms of safety but normally the inverters and isolation components have protection from AC side faults and the system shuts down to prevent any damage to equipment, also power plant shutting down itself is a good enough indication to let the operator know of an AC side failure [57, 59]. AC side faults like grid voltage fluctuations [62], frequency fluctuations [63–65] and anti-islanding [66–70] protection usually cause inverter shut down, hence the system remains safe and the operator is well aware of the fault. The types of DC side fault that are addressed in this study are module faults, faults in PV string, loss of MPPT, partial shading, and module soiling. The module faults refer to a short-circuited PV module, or PV module removed from the string. This means that whenever a PV module has short-circuited, only the number of PV modules is reduced in the respective string, rest of the modules in string keep on producing DC electricity [23, 71–74]. Consider a case of a large-scale PV plant where modules are in hundreds. Then a single PV module failure is a negligible effect on the power output. As the series-connected PV modules contribute their voltage to the string and current remains same for any number of modules in that string. So, if a single module is short-circuited, the bypass diode will come into play passing the same amount of current and the MPPT module, which is regulating the DC voltage will slightly increase its output voltage to remain at the MPP level. Therefore, a single module fault has not been considered significant in this study. It is observed that a measurable effect occurs on the PV string when at least 5 PV modules are gone off in the string. In addition, if size of PV string is reduced to a certain level that the operational modules of the string are unable to compensate the voltage and meet the minimum required DC voltage by the inverter, then the complete string stops working. In this situation, the current of string will increase so much that the protection circuit breakers in CB will trip off and string will get disconnected.

The proposed algorithm considers and detects module fault of 5 PV modules in a string. This fault is referred to as fault type 1 and abbreviated as F1 for further classification in this chapter. Further, a faulty PV string means the string that has been disconnected from the system, either due to cable breakage, tripping or any other malfunctioning. A single PV string failure or multiple PV string failure are considered separate faults to highlight that the algorithm is able to diagnose simultaneous failures occurring in the DC side. A single string failure means a significant size of PV array not yielding any power. In case of the reference power plant, where a single PV module is 270 W in size and there are 22 number of PV modules in a string, ideally a 5940 W of power has gone from the PV plant. Single PV String fault is referred to as fault type 2 and abbreviated as F2 and multiple string's fault is referred to as fault type 3 and abbreviated as F3 for further classification in this chapter. Moreover, the MPPT unit is a very important part in the DC side of the PV plant. The MPPT unit regulates the DC voltage output from the PV array to optimize the performance of the PV plant. If the MPPT does not work, then the DC voltage may not be operating at the optimum value and a significant drop in power generation will occur. This fault is referred to as fault type 4 and abbreviated as F4 for further classification in this chapter. Finally, the partial shading on a PV module means that some portion of a module is under shading effect.

The PV modules are manufactured such that certain number of cells is connected in series and a bypass diode is connected in parallel to these cells. If a set of cells has lower output current than the rest of cells, then the bypass diode will pass the current bypassing or disconnecting the faulty cells. It is observed in a large sized PV power plant that partial shading on a single module or a part of module in a total of hundreds of PV modules will not put any effect on power output, because such small power deviations are not only insignificant but quite difficult to observe as well. A faulty bypass diode has not been considered a fault here. The bypass diode comes into action when there is partial shading on a single module. Soiling loss is also critical when large-scale PV plants are under focus. PV modules usually require cleaning or washing of modules surface with water to remove the dust deposited on it. The dust particles settled on the PV modules prevent sunlight's photon particles to reach to the solar cell and perform the sunlight-to-electricity conversion efficiently. Therefore, there may be soil and dust deposited on the surface of PV modules resulting in reduction in DC power output, which is called as soiling loss. The soiling loss has similar effect on PV output as that of partial shading. The partial shading or soiling loss having similar impact on power are taken as one fault. This fault is referred to as fault type 5 and abbreviated as F5 for further classification in this chapter.

6.3.3 System Modelling

The modelling and simulation of the system is carried out in the MAT-LAB/Simulink environment [75]. The PV components, DC–DC boost converter configuration, voltage-source converter (VSC) (inverter), Step-up Transformer, and Grid, available in the Simulink library are used to model the system. The PV component receives input of irradiance and module temperature and gives an output of DC power and measurements, which give the values of PV Voltage, PV Current, Diode Current, Irradiance and Temperature on the PV array. The DC power from the PV strings is combined by connecting in parallel and then the combined DC power is fed in to the inverter for conversion to AC power. The inverter unit contains the DC–DC converter, which is controlled with output from MPPT unit, inverter control unit to provide controlling function for switches to produce AC electricity.

6.3.3.1 PV Modelling
Further, two PV power plants are modeled in the simulation. One of the PV power plants is tagged as an actual PV plant which is subjected to operate with a

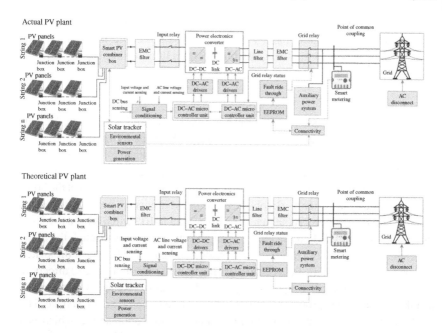

Figure 6.5 Schematic of the actual and theoretical plant models.

fault in the system. The second is the theoretical PV plant, which remains same throughout the simulation. Both these power plants have their separate step-up transformers and grids modeled so no connection of either PV plant is established, which may affect the results. For simulation purpose, the irradiance and module temperature are obtained from the reference power plant's monitoring system with intervals of 15 minutes for a complete day. The schematic of both the theoretical and actual power plants as shown in Figure 6.5.

The actual PV plant corresponds to the physically installed solar power. It contains complete PV array including the 21 number of PV strings and Inverter unit. The PV strings are combined in parallel at seven points and then finally merged in one DC connection which is terminated from the PV array. Since the reference power plant is operating with the real-time conditions, the degradation factor of 4% is applied on the PV output to account for the aging effect of PV modules; therefore, a gain of 0.96 is applied on the current of PV array. The factor 4% degradation was calculated based on the degradation pattern as defined in the PV modules manufacturer datasheet. Inside each PV sub-array, there are 3 PV strings, which are designed such that it is able to yield output as close to the real-life PV module as per the values of irradiance and module temperature. In order to create a realistic approach in the PV system to incorporate the fact that all PV modules do not yield a uniform and identical power output and there is always some variation in output of 1 PV module from the other, random numbers are generated and multiplied with the output of all PV string current values. This helped in obtaining a varying yield of all the strings within a certain limit to simulate the factor of nonlinearity.

6.3.3.2 Inverter Unit

The inverter unit consists of a DC–DC boost converter and a VSC. The DC–DC boost converter is used to step-up the DC voltage from the PV array. The step-up voltage is input in to the VSC which inverts the DC voltage to AC voltage. The VSC is based on the three-phase bridge inverter. The DC–DC boost converter is essential to maintain a value of DC voltage. Figure 6.5 shows the circuit of the boost converter. Boost converter works on the principle of inductor's property to resist a sudden change of current. As shown in Figure 6.5, an IGBT switch is used to trigger the on and off states. When the switch is on, the current passes through the diode and energy is stored in the inverter. This creates a magnetic field. When the switch is turned off, the magnetic field decays causing flow of current in reverse direction. This results in increase in output voltage than the input DC voltage. The switching signal to the switch is provided by the MPPT module to optimize the voltage. The MPPT technique used in the modeling is incremental conductance, which is a quite common and efficient technique. This technique is used because it gives the output power very similar to the reference power plant. The boosted DC voltage is given in to the VSC circuit where the three-level IGBT-switched bridge outputs an alternating three-phase voltage supply. The three-level bridge takes switching signals from the inverter control to output voltage matching with the grid power supply. The inverter unit output of 415 V is given in to the step-up transformer and an 11 kV AC is then fed in to the grid network.

6.3.3.3 Fault Injection

To generate the failure or short-circuiting of PV modules, number of PV modules in series of Simulink PV component series-connected modules per string is changed during running of simulation to reduce the string size. This value gets normal once fault is removed. Similarly, in order to enable or disable single or multiple strings, constant value of one or zero is multiplied with the input irradiance and temperature to enable or disable the string. The MPPT unit active or inactive is done in the same way, a constant one or zero is given input to a selector block, which enables or disables the output of MPPT unit to control boost converter. Further, the partial shading or soiling loss means less irradiance is being incident on the surface of PV module; therefore, a gain is multiplied with the irradiance on the string, which reduces the irradiance representing partial shading or soiling effect on that particular string.

6.3.4 Fault Classification Mechanism

The proposed FC system uses a statistical approach which detects an abnormality in the PV plant and the ratios of power and voltage give the range to indicate what fault has occurred. The principle of proposed algorithm is based on the student's T-test to detect the outlier in a given set of data [76–79]. The strings in an actual PV plant, which have different behavior compared to their respective counterparts in theoretical PV system are pointed out. Among the numerous applications of the T-test method, outlier detection is one of them. Outlier detection enables the detection of an odd sample among as many samples as possible.

6.3.4.1 Statistical Tools

T-test: The T-test statistical tool [80] is based on time average of several samples to find an outlier in the given samples. The outlier is a sample that differs from the rest of the samples. The T-test is popular in statistical applications where it is desired to test if two sets of data are different from each other or not. The T-test approach is easy for implementation and is perfectly applicable on any size of PV string and overall PV plant. The most common form of the formula to compute T-test is:

$$T = \frac{\sqrt{n} \times (x - \mu)}{\sigma} \tag{6.1}$$

where x is the current sample, μ is the mean of samples, n is the number of samples and σ is the standard deviation [77, 78]. The use of T-test based approach to detect and identify the fault has made the system robust that faults are highlighted under any kind of weather conditions, irrespective of size of strings or complete PV array.

Requirements: The key difference of T-test [80] and the statistical tools proposed in this study is that the T-test in (2.1) uses certain samples distributed at a known time interval. However, the proposed system uses somewhat a spatial approach. Samples of power of 21 strings is taken at a given instant of time to compute the T_{score} and highlight the outlier. Another tweaking applied to the test method is that the mean and standard deviation to calculate the score is taken from the theoretical plant. This approach detects the fault whenever a string in Actual PV plant has some abnormality as compared to the corresponding string in the theoretical PV plant model. The resultant of this test is called as Z_{score} and is estimated as:

$$Z_{score} = \frac{P_{str}(k) - \mu_{P_{str(Th)}}}{\sigma_{Th}} \tag{6.2}$$

where Z_{score} is the resultant score for kth string in actual PV plant, which has power of P_{str} at the given instant of time. The mean value $\mu_{P_{str(Th)}}$ and σ_{Th} are the mean and standard deviation of the 21 strings of theoretical PV plant at the same instant.

The T_{score} value with mean and standard deviation of actual plant is computed, which is used to locate if there are faulty PV modules in a string. This is estimated as:

$$T_{score}(k) = \frac{4 \times (P_{str}(k) - \mu_{P_{str}})}{\sigma} \tag{6.3}$$

where $T_{score}(k)$ is the resultant score for kth string in actual plant, which has power of $P_{str}(k)$ at the given instant of time. The mean value $\mu_{P_{str}}$ and σ are the mean and standard deviation of the 21 strings of actual plant at the same instant. The power ratio is calculated on each string for identification of MPPT unit fault, i.e. ratio of each string of theoretical plant to the corresponding string of actual plant. The string power ratio (PR_{str}) is estimated as:

$$PR_{str}(k) = \frac{P_{str_{Th}}(k)}{P_{str}(k)} \tag{6.4}$$

6.3.4.2 Boundary Conditions

The defined statistical approaches are used to identify the type of fault. All the faults are distinguished using different values of Z_{score}, T_{score}, and PR_{str}, some faults only require Z_{score} to be diagnosed, whereas, some faults require second level of detection using T_{score} and PR_{str} for the accurate detection. But there are some limitations which are discussed below while adapting these aspects.

Module Failure: For a condition where the reference PV plant has 22 number of series-connected 270 W PV modules in one string, if some PV modules stop working, the string voltage will reduce eventually reducing the PV string power. When the string is working in normal conditions and without any faulty PV modules in the string, ideally the string power should be 270 multiplied with 22 resulting in about 5.94 kW DC power and if one PV module is not working in the string, then the DC power will reduce to 5.67 kW. Further, if the string gets affected with 10 of the faulty PV modules, then the string DC power reduces to 3.24 kW. These values of DC powers in case on 1 PV module and 10 PV modules is put in to the (2.2) and Equation (2.3), which results in a range of values of Z_{score} and T_{score} to be in limits of 0 to −2600 and −9 to −25, respectively. If a complete string stops working, then the DC power is zero. Computing the value of Z score under such condition yields the values below −2900. Further, a counter in the algorithm keeps a record of error on number of strings. The value of counter determines the number of strings under faulty condition.

Control Unit Failure: The MPPT unit failure affects the complete actual PV plant's DC voltage. To compute the values of mathematical operations, the value of DC voltage of 250 V because of fault is taken to calculate the DC power Actual PV plant. This results in DC power of all strings to be 2.28 kW. The value of Z_{score} and PR_{str} of each string in such case comes out to be between range of −1150 to −850 and 1.22 to 2.4, respectively. This fault type count is compulsory to be equal to number of strings for the identification of MPPT fault, otherwise other faults may occur.

Shading/Soiling Faults: The partial shading or soiling loss is generated in the actual PV plant by reducing the input irradiance by factory or 0.6 for a PV string. As the irradiance has a direct effect on the DC power output of PV module, so in order to compute the values of mathematical operations, the DC power of actual PV plant is reduced by the same factor and then the values of Z_{score} are found to be between range of -2600 and -850. If the fault count value is greater than 0 and less than number of strings, then it is the partial shading or soiling fault. However, if the fault count value is equal to number of strings, then it is taken as MPPT unit fault. The threshold values of Z_{score}, T_{score}, and PR_{str} for the studied faults are given in Table 6.3.

6.3.4.3 Algorithm and Flowchart of FC System

The FC mechanism calculates the Z_{score} at every 15 minutes time when both the actual and theoretical PV plants generate power. The Z_{score} is located in certain ranges of values which define the type of fault. T_{score} and PR_{str} are also required to identify the faults of PV modules failure and MPPT unit failure, respectively. The rules of FC are: If $-2600 \leq Z_{score}(k) \leq 0$ and $-25 < T_{score}(k) < -9$, then increment F1 count by one. If the value of F1 count is greater than zero, then at least one string has faulty PV modules in it.

When one sample has below -2900 of Z_{score} value, F2 count is incremented by one. If the F2 count value is one, then one string is faulty and if the value of F2 count is more than one that means multiple strings are faulty. Further, if $-1150 < Z_{score}(k) < -850$ and $1.22 \leq PR_{str}(k) \leq 2.4$, then F4 count is incremented by one. If the condition is fulfilled for all k, i.e. F4 count is 21, then the MPPT unit has turned off or not working. Finally, if $-2600 < Z_{score}(k) < -850$, this range overlaps with the range used for MPPT FC. If $0 < F5$ count < 21, then there is partial shading on number of strings equal to the value of F5 count. If F5 count is 21, then this also correspond to MPPT unit fault. The flowchart of simulation and FC system is shown in Figure 6.6.

This FC system is implemented in the Simulink using MATLAB function block. The voltages and currents of both actual and theoretical PV Plants are passed to the function block where script coding implements the algorithm of FC and identification. Z_{score}, T_{score}, and PR_{str} for all strings are calculated at time instants at 15 minutes' intervals when irradiance and temperature information are given from the environment-monitoring unit.

6.4 Result Evaluation and Discussion

This section discusses the simulation results of the modeled PV power plant with faults induced at different time intervals and different weather conditions.

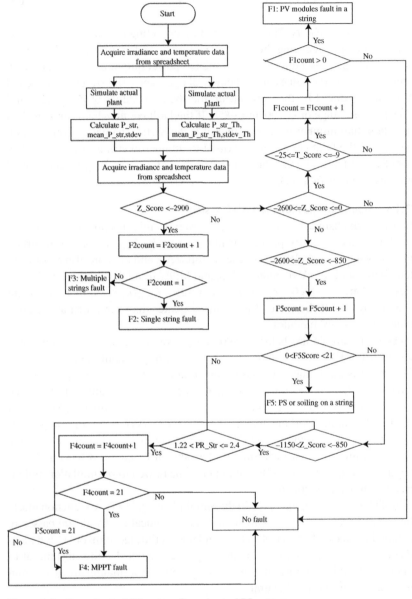

Figure 6.6 Flow chart of FC system flow chart of FC system.

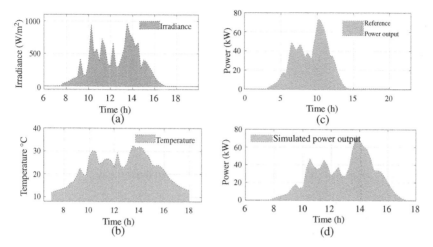

Figure 6.7 Comparison of power outputs of reference and simulated PV plants. (a) Irradiance on a general sunny day. (b) Temperature on a general sunny day. (c) Power output of the reference PV plant. (d) Power output of simulated PV power plant.

6.4.1 Simulation of Modeled Grid-tied PV Plant

In order to analyze the test results of the proposed FC system, it is important to examine the DC power outputs of the reference PV plant to confirm that the modeled PV plant has very similar behavior at the DC side. This is required because if the theoretical PV plant output does not match closely with reference PV plant, then algorithm may indicate faults in normal conditions as well. To compare the DC power of reference PV plant and the simulated PV plant, the results of PV power from real-time monitoring system of reference PV plant are obtained. Figure 6.7a–d presents irradiance and temperature experienced by the PV module on a general sunny day and the DC power graph of reference PV plant and simulated PV plant.

6.4.2 Results Evaluation

The FC mechanism is evaluated on a sunny day with a peak irradiance of 833 W/m^2, and highest temperature of 40 °C. The irradiance and temperature of the test day are shown in Figures 6.8 and 6.9, respectively.

Three faults were introduced in the actual PV plant. These faults, their location in the PV system and timings are mentioned in Table 6.4.

At intervals of 15 minutes, the simulation receives irradiance and temperature value and corresponding power of all strings are passed to the FC system. The FC system runs its algorithm and finds the faults correctly as set before the simulation.

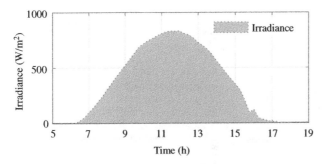

Figure 6.8 Irradiance obtained at 15 minutes intervals on the test day.

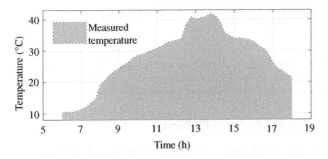

Figure 6.9 Temperature obtained at 15 minutes intervals on the test day.

Table 6.4 Details of faults injected in the actual PV plant.

S. no.	Fault type	Fault location	Time of fault
1.	PV modules fault in a string	String 2	$09:00-10:30$
2.	Single PV string fault	String 3	$12:00-13:00$
3.	MPPT fault	Complete PV array	$14:00-16:00$

The variations between the voltage and power of actual PV plant and theoretical PV plant show the effects of these faults and how the power degrades when such faults occur in the PV power plants. The comparison of voltage and power output of Actual and Theoretical PV plants is pictured in Figures 6.10 and 6.11, respectively. It is clearly visible there are reasonable deviations in the power output when the 1 string fault occurs at 12 hours and when the MPPT fault arises at 14 hours. It can be seen in Figure 6.11 that the PV module fault is not visible in the power outputs; however, this fault is clearly detected with the help of the proposed algorithm.

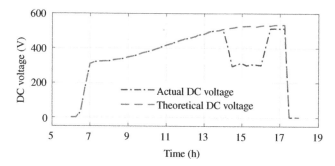

Figure 6.10 DC voltage of theoretical and actual PV plants under the influence of different faults.

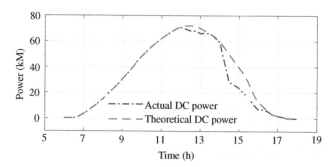

Figure 6.11 DC power of theoretical and actual PV plants under the influence of different faults.

Based on the algorithm, we can analyze the results of the FC system and the performance of the algorithm under the applied conditions. The calculation of Z_{score}, T_{score}, and PR_{str} are done by the system and their values for all strings are collected. The Z_{score}, T_{score}, and PR_{str} during the fault times and for the affected strings are shown in Figures 6.12, 6.13, and 6.14, respectively.

When the type 1 fault occurs in string 2 of actual PV plant from 9 hours to 10:30 hours, the Z_{score} of all string's ranges between the values of −63 and 56. This condition takes to the inspection of T_{score}. T_{score} of all strings are checked and the value for string 2 comes out to be ranging around −16 during this time. It can be seen in Figure 6.12 that the Z_{score} during this time is within the range of −2600 to 0. Then it is required to look for T_{score} at the corresponding time shown in Figure 6.13. The T_{score} is less than −9 and higher than −25. This indicates the PV module failure. The Z_{score} deviation was observed on the String number 2. So, it can be concluded that there is F1 fault on the string 2. Similarly, at 12 hours, the system encounters value of Z_{score} for string 3 to be less than −3200. This drop in value of Z_{score} remains till 13 hours' time, whereas the rest of strings are not showing such drop

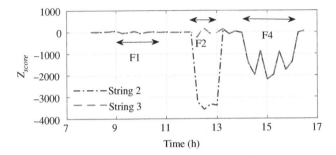

Figure 6.12 Z_{score} for string 2 and string 3 affected by the faults.

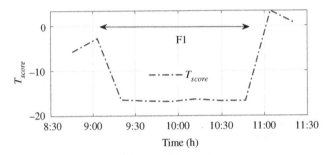

Figure 6.13 T_{score} for string 2 indicating faulty PV modules.

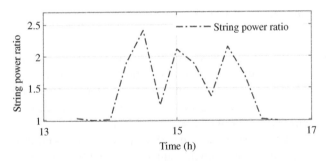

Figure 6.14 P_{str} indicating MPPT unit fault.

in the values of their respective Z_{scores}. Z_{score} of −3200 is far less than −2900. Then it means the fault is either a single string or multiple strings failure. The system internally processes the final value of F2 count which is equal to 1. This highlights the single string fault. Z_{score} of all strings is examined and only one string was the outlier so that string is pointed out to be faulty. Further, at 14 hours, the Z_{score} value for all strings has dropped to −1400 and reaches up to −2180. It is to be noted that the Z_{score} is in the range of −2600 and −850. Then the F5 count is checked to be 21. This corresponds to the range for MPPT unit failure fault. Here ranges of Z_{score} are not directly falling under the MPPT range but when the F5 count is equal to 21, then such fault is not the partial shading but MPPT fault as per the algorithm. From these results, the advantages of using the proposed FC system are: it is simple and easy to implement. The only requirement for its computation is the power and voltages of all strings which can be easily obtained from PV plants. The FC is able to detect the DC side faults, which are critical for proper functioning of a solar PV plant. The statistical tools that are used in the FC algorithm do not require complex calculations. The FC system can be used with small- and large-scale power plants. The simple equations of Z_{score}, T_{score}, and PR_{str} can be implemented for any size of PV power plants.

6.5 Summary

This chapter developed an FC mechanism incorporated for real-time monitoring of PV power plants. The proposed FC system is a different approach to detect the faults in a DC side of the grid-connected PV power plant, and has gone through several tests under various conditions. The performance and accuracy of the detection system is good as no false alarms were raised during the testing. With the help of this study, it can be concluded that a grid-connected PV power plant can have possible faults, which may include, but are not limited to, some faulty PV modules in a string, complete PV strings not yielding any power, partial shading on PV array, soiling on PV array and a faulty MPPT unit. The power output reductions during occurrence of these faults are significant enough to put an impact on energy yields and immediate remedial actions are essential to achieve maximum output. The proposed FC system is a suitable addition to the available options for FC systems. The advantage of the proposed FC system over the rest is that the proposed system has been tested on a power plant with capacity higher than 100 kW size, whereas, maximum of the FC algorithms discussed in the literature are implemented based on power plants of small size. This makes the proposed FC system to have an edge of being tested on a larger system. Further, this system has not given wrong indications and false alarms under the tested scenarios. This can be taken as evidence to the accuracy and precise results of this system. The implementation is simple as no other than irradiance and temperatures are required to simulate the theoretical PV plant, and, second, the current and voltages of each string are required.

References

1 Mayr, S., Grabmair, G., and Reger, J. (2019). Fast model-based fault detection in single-phase photovoltaic systems. In: IECON 2019 – 45th Annual Conference of the IEEE Industrial Electronics Society, 4615–4622. IEEE https://doi .org/10.1109/IECON.2019.8927172.

2 Pei, T. and Hao, X. (2019). A fault detection method for photovoltaic systems based on voltage and current observation and evaluation. *Energies* 12: 1712. https://doi.org/10.3390/en12091712.

3 Kurukuru, V.S.B., Haque, A., Khan, M.A. et al. (2021). A review on artificial intelligence applications for grid-connected solar photovoltaic systems. *Energies* 14: 4690. https://doi.org/10.3390/en14154690.

4 Kurukuru, V.S.B., Khan, M.A., and Malik, A. (2021). Failure mode classification for grid-connected photovoltaic converters. In: Reliability of Power Electronics Converters for Solar Photovoltic Applications, 205–249. Institution of Engineering and Technology https://doi.org/10.1049/pbpo170e_ch8.

5 Dogga, R. and Pathak, M.K. (2019). Recent trends in solar PV inverter topologies. *Sol. Energy* 183: 57–73. https://doi.org/10.1016/j.solener.2019.02.065.

6 El Bassam, N. (2021). Solar energy. In: Distributed Renewable Energies for Off-Grid Communities, 123–147. Elsevier https://doi.org/10.1016/B978-0-12-821605-7.00015-5.

7 Cabrera-Tobar, A., Bullich-Massagué, E., Aragüés-Peñalba, M., and Gomis-Bellmunt, O. (2016). Topologies for large scale photovoltaic power plants. *Renew. Sustain. Energy Rev.* 59: 309–319. https://doi.org/10.1016/j.rser .2015.12.362.

8 Wang, Z. (2019). Introduction. In: Design of Solar Thermal Power Plants, 1–46. Elsevier https://doi.org/10.1016/B978-0-12-815613-1.00001-8.

9 Pitz-Paal, R. (2020). Concentrating solar power. In: Future Energy, 413–430. Elsevier https://doi.org/10.1016/B978-0-08-102886-5.00019-0.

10 Helman, U. (2014). Economic and reliability benefits of large-scale solar plants. In: Renewable Energy Integration, 327–345. Elsevier https://doi.org/10.1016/ B978-0-12-407910-6.00026-0.

11 Ulrich, T. and Gerken, K. (1986). Photovoltaic power systems – A user's guide to reliability in sizing and design. In: 36th IEEE Vehicular Technology Conference, 398–404. IEEE https://doi.org/10.1109/VTC.1986.1623466.

12 Letcher, T. and Fthenakis, V.M. (2018). A Comprehensive Guide to Solar Energy Systems. Academic Press doi: https://doi.org/10.1016/C2016-0-01527-9.

13 Zekry, A., Shaker, A., and Salem, M. (2018). Solar cells and arrays. In: Advances in Renewable Energies Power and Technologies, 3–56. Elsevier https://doi.org/10.1016/B978-0-12-812959-3.00001-0.

14 Soga, T. (2006). Fundamentals of solar cell. In: Nanostructured Materials for Solar Energy Conversion, 3–43. Elsevier https://doi.org/10.1016/B978-044452844-5/50002-0.

15 Mishra, S. and Sharma, D. (2016). Control of photovoltaic technology. In: Electric Renewable Energy Systems, 457–486. Elsevier https://doi.org/10.1016/B978-0-12-804448-3.00019-0.

16 Castañer, L., Bermejo, S., Markvart, T., and Fragaki, K. (2012). Energy production by a PV array. In: Practical Handbook of Photovoltaics, 645–658. Elsevier https://doi.org/10.1016/B978-0-12-385934-1.00018-0.

17 Kalogirou, S.A. (2018). McEvoy's Handbook of Photovoltaics. Elsevier https://doi.org/10.1016/C2015-0-01840-8.

18 Franco, F. and Técnico, I.S. (2018). Performance Assessment of Photovoltaic Systems: Monitoring their Abnormal Operating Conditions, 1–91. Universidade de Lisboa.

19 Wang, W., Liu, A.C.-F., Chung, H.S.-H. et al. (2016). Fault diagnosis of photovoltaic panels using dynamic current–voltage characteristics. *IEEE Trans. Power Electron.* 31: 1588–1599. https://doi.org/10.1109/TPEL.2015.2424079.

20 Kharseh, M. and Wallbaum, H. (2020). Comparing different PV module types and brands under working conditions in the United Kingdom. In: Reliability and Ecological Aspects Photovoltaic Modules. IntechOpen https://doi.org/10.5772/intechopen.86949.

21 Benda, V. and Černá, L. (2020). PV cells and modules – state of the art, limits and trends. *Heliyon* 6: e05666. https://doi.org/10.1016/j.heliyon.2020.e05666.

22 Powalla, M., Paetel, S., Hariskos, D. et al. (2017). Advances in cost-efficient thin-film photovoltaics based on Cu(In,Ga)Se 2. *Engineering* 3: 445–451. https://doi.org/10.1016/J.ENG.2017.04.015.

23 Haque, A., Bharath, K.V.S., Khan, M.A. et al. (2019). Fault diagnosis of photovoltaic modules. *Energy Sci. Eng.* 7: https://doi.org/10.1002/ese3.255.

24 Kurtz, S.R., Myers, D., Townsend, T. et al. (2000). Outdoor rating conditions for photovoltaic modules and systems. *Sol. Energy Mater. Sol Cells* 62: 379–391. https://doi.org/10.1016/S0927-0248(99)00160-9.

25 Alves dos Santos, S.A., JPN, T., Fernandes, C.A.F., and Marques Lameirinhas, R.A. (2021). The impact of aging of solar cells on the performance of photovoltaic panels. *Energy Convers. Manag. X* 10: 100082. https://doi.org/10.1016/j.ecmx.2021.100082.

26 Pang, W., Cui, Y., Zhang, Q. et al. (2019). Comparative investigation of performances for HIT-PV and PVT systems. *Sol Energy* 179: 37–47. https://doi.org/10.1016/j.solener.2018.12.056.

27 Kim, J., Rabelo, M., Padi, S.P. et al. (2021). A review of the degradation of photovoltaic modules for life expectancy. *Energies* 14: 4278. https://doi.org/10.3390/en14144278.

28 Rahman, M.M., Hasanuzzaman, M., and Rahim, N.A. (2015). Effects of various parameters on PV-module power and efficiency. *Energ. Conver. Manage.* 103: 348–358. https://doi.org/10.1016/j.enconman.2015.06.067.

29 Ho, B.M.T. and Chung, H.S.-H. (2005). An integrated inverter with maximum power tracking for grid-connected PV systems. *IEEE Trans. Power Electron.* 20: 953–962. https://doi.org/10.1109/TPEL.2005.850906.

30 Sreedevi, J., Ashwin, N., and Naini, R.M. (2017). A study on grid connected PV system. In: 2016 National Power System Conference (NPSC), 1–6. IEEE https://doi.org/10.1109/NPSC.2016.7858870.

31 Li, W., Gu, Y., Luo, H. et al. (2015). Topology review and derivation methodology of single-phase transformerless photovoltaic inverters for leakage current suppression. *IEEE Trans. Ind. Electron.* 62: 4537–4551. https://doi.org/10.1109/TIE.2015.2399278.

32 Zhou, Q., Xun, C., Dan, Q., and Lin, S. (2015). Grid-connected inverter reliability considerations: a review. In: 16th International Conference on Electronic Packaging Technology, 266–274. IEEE.

33 Khan, M.A., Haque, A., and Kurukuru, V.S.B. (2019). Performance assessment of stand alone transformerless inverters. *Int. Trans. Electr. Energy Syst.* https://doi.org/10.1002/2050-7038.12156.

34 Khan, M.A., Haque, A., and Bharath, K.V. (2018). Voltage balancing control for stand-alone H5 transformerless inverters. In: Lecture Notes in Electrical Engineering. Germany: Springer.

35 Khan, M.A., Haque, A., Kurukuru, V.S.B. et al. (2021). Stand-alone operation of distributed generation systems with improved harmonic elimination scheme. *IEEE J. Emerg Sel. Top. Power Electron.* 9: 6924–6934. https://doi.org/10.1109/JESTPE.2021.3084737.

36 Tyagi, A., Dubey, M., and Gawre, S. (2018). Advance monitoring of electrical and environmental parameters of PV system: a review. In: 2018 International Conference on Sustainable Energy, Electronics and Computing Systems, 1–5. IEEE https://doi.org/10.1109/SEEMS.2018.8687366.

37 Paredes-Parra, J., Mateo-Aroca, A., Silvente-Niñirola, G. et al. (2018). PV module monitoring system based on low-cost solutions: wireless raspberry application and assessment. *Energies* 11: 3051. https://doi.org/10.3390/en11113051.

38 Chouder, A., Silvestre, S., Taghezouit, B., and Karatepe, E. (2013). Monitoring, modelling and simulation of PV systems using LabVIEW. *Sol. Energy* 91: 337–349. https://doi.org/10.1016/j.solener.2012.09.016.

39 Lindig, S., Louwen, A., Moser, D., and Topic, M. (2020). Outdoor PV system monitoring—input data quality, data imputation and filtering approaches. *Energies* 13: 5099. https://doi.org/10.3390/en13195099.

40 Haque, A. and Zaheeruddin. (2013). Research on solar photovoltaic (PV) energy conversion system: an overview. *IET Conf. Publ.* 2013: 605–611. https://doi.org/10.1049/cp.2013.2653.

41 Haque, A. (2014). Maximum Power Point Tracking (MPPT) scheme for solar photovoltaic system. *Energy Technol. Policy* 1: 115–122. https://doi.org/10.1080/23317000.2014.979379.

42 Vairavasundaram, I., Varadarajan, V., Pavankumar, P.J. et al. (2021). A review on small power rating PV inverter topologies and smart PV inverters. *Electronics* 10: 1296. https://doi.org/10.3390/electronics10111296.

43 Ghenai, C., Merabet, A., Salameh, T., and Pigem, E.C. (2018). Grid-tied and stand-alone hybrid solar power system for desalination plant. *Desalination* 435: 172–180. https://doi.org/10.1016/j.desal.2017.10.044.

44 Carbone, R. (2009). Grid-connected photovoltaic systems with energy storage. In: 2009 International Conference on Clean Electrical Power, ICCEP 2009, 760–767. IEEE https://doi.org/10.1109/ICCEP.2009.5211967.

45 Shiva Kumar, B. and Sudhakar, K. (2015). Performance evaluation of 10 MW grid connected solar photovoltaic power plant in India. *Energy Rep.* 1: 184–192. https://doi.org/10.1016/j.egyr.2015.10.001.

46 Buticchi, G., Barater, D., Lorenzani, E., and Franceschini, G. (2012). Digital control of actual grid-connected converters for ground leakage current reduction in PV transformerless systems. *IEEE Trans. Ind. Informatics* 8: 563–572. https://doi.org/10.1109/TII.2012.2192284.

47 Marion, B., Adelstein, J., Hadyen, H. et al. (2005). Performance parameters for grid-connected PV systems. In: 31st IEEE Photovoltaics Specialists Conference and Exhibition, 1601–1606. NREL https://doi.org/10.1109/PVSC.2005.1488451.

48 Shayestegan, M. (2018). Overview of grid-connected two-stage transformer-less inverter design. *J. Mod. Power Syst. Clean Energy* 6: 642–655. https://doi.org/10.1007/s40565-017-0367-z.

49 Yagi, Y., Kishi, H., Hagihara, R. et al. (2003). Diagnostic technology and an expert system for photovoltaic systems using the learning method. *Sol Energy Mater. Sol Cells* 75: 655–663. https://doi.org/10.1016/S0927-0248(02)00149-6.

50 Usamentiaga, R., Venegas, P., Guerediaga, J. et al. (2014). Infrared thermography for temperature measurement and non-destructive testing. *Sensors* 14: 12305–12348. https://doi.org/10.3390/s140712305.

51 Littwin, M.P., Baumgartner, F., Green, M., and van Sark, W. (2021). Performance of New Photovoltaic System Designs 2021, Task 13, subtask 1.3. EA-PVPS T13-15:2021. International Energy Agency (IEA), Photovoltaic Power Systems Programme.

52 Singh, T. and Thakur, R. (2019). Design and development of PV solar panel data logger. *Int. J. Comput. Sci. Eng.* 7: 364–369. https://doi.org/10.26438/ijcse/v7i4.364369.

53 de GCG, M., Torres, I.C., de ÍBQ, A. et al. (2021). A low-cost IoT system for real-time monitoring of climatic variables and photovoltaic generation for smart grid application. *Sensors* 21: 3293. https://doi.org/10.3390/s21093293.

54 Wibawa, A.S., Kumara, I.N.S., and Sukerayasa, I.W. (2020). Instruments and data logger for measuring electrical parameters: Indonesian market review and research direction. *J. Electr. Electron. Informatics* 4: 20. https://doi.org/10.24843/JEEI.2020.v04.i01.p04.

55 Harrou, F., Sun, Y., Taghezouit, B. et al. (2018). Reliable fault detection and diagnosis of photovoltaic systems based on statistical monitoring approaches. *Renew. Energy* 116: 22–37. https://doi.org/10.1016/j.renene.2017.09.048.

56 Azizi, K., Farsadi, M., and Farhadi Kangarlu, M. (2017). Efficient approach to LVRT capability of DFIG-based wind turbines under symmetrical and asymmetrical voltage dips using dynamic voltage restorer. *Int. J. Power Electron. Drive Syst.* 8: 945. https://doi.org/10.11591/ijpeds.v8.i2.pp945-956.

57 Dhimish, M., Holmes, V., and Dales, M. (2017). Parallel fault detection algorithm for grid-connected photovoltaic plants. *Renew. Energy* 113: 94–111. https://doi.org/10.1016/j.renene.2017.05.084.

58 Subramaniam, U., Vavilapalli, S., Padmanaban, S. et al. (2020). A hybrid PV-battery system for ON-grid and OFF-grid applications—controller-in-loop simulation validation. *Energies* 13: 755. https://doi.org/10.3390/en13030755.

59 Madeti, S.R. and Singh, S.N. (2017). A comprehensive study on different types of faults and detection techniques for solar photovoltaic system. *Sol Energy* 158: 161–185. https://doi.org/10.1016/j.solener.2017.08.069.

60 Davarifar, M., Rabhi, A., and El, H.A. (2013). Comprehensive modulation and classification of faults and analysis their effect in DC side of photovoltaic system. *Energy Power Eng.* 05: 230–236. https://doi.org/10.4236/epe.2013.54B045.

61 Mohamed, S., Jeyanthy, P., Devaraj, D. et al. (2019). DC-link voltage control of a grid-connected solar photovoltaic system for fault ride-through capability enhancement. *Appl. Sci.* 9: 952. https://doi.org/10.3390/app9050952.

62 Yang, F., Yang, L., and Ma, X. (2014). An advanced control strategy of PV system for low-voltage ride-through capability enhancement. *Sol. Energy* 109: 24–35. https://doi.org/10.1016/j.solener.2014.08.018.

63 Wadhwa, K., Bharath, K.V.S., Pandey, K., and Sehrawat, S. (2017). Controlling of frequency deviations in interconnected power systems using smart techniques. In: 1st IEEE International Conference on Power Electronics, Intelligent Control and Energy Systems. ICPEICES 2016. IEEE https://doi.org/10.1109/ICPEICES.2016.7853222.

64 Yi, Z., Dong, W., and Etemadi, A.H. (2018). A unified control and power management scheme for PV-Battery-based hybrid microgrids for both grid-connected and islanded modes. *IEEE Trans. Smart Grid* 9: 5975–5985. https://doi.org/10.1109/TSG.2017.2700332.

65 Silvestre, S., Kichou, S., Chouder, A. et al. (2015). Analysis of current and voltage indicators in grid connected PV (photovoltaic) systems working in faulty and partial shading conditions. *Energy* 86: 42–50. https://doi.org/10.1016/j.energy.2015.03.123.

66 Haque, A., Alshareef, A., Khan, A.I. et al. (2020). Data description technique-based islanding classification for single-phase grid-connected photovoltaic system. *Sensors* 20: 3320.

67 Khan, M.A., Haque, A., Blaabjerg, F. et al. (2021). Intelligent transition control between grid-connected and standalone modes of three-phase grid-integrated distributed generation systems. *Energies* 14: 3979. https://doi.org/10.3390/en14133979.

68 Khan, M.A., Haque, A., and Kurukuru, V.S.B. (2019). Machine learning based islanding detection for grid connected photovoltaic system. In: 2019 International Conference on Power Electronics, Control and Automation, 1–6. IEEE https://doi.org/10.1109/ICPECA47973.2019.8975614.

69 Khan, M.A., Haque, A., and Kurukuru, V.S.B. (2022). Islanding detection techniques for grid-connected photovoltaic systems – a review. *Renewable and Sustainable Energy Reviews* 154. https://doi.org/10.1016/j.rser.2021.111854.

70 Khan, M.A., Kurukuru, V.S.B., Haque, A., and Mekhilef, S. (2020). Islanding classification mechanism for grid-connected photovoltaic systems. *IEEE J. Emerg Sel. Top. Power Electron.* 1–1.

71 Jaffery, Z.A., Dubey, A.K., and Irshad, H.A. (2017). Scheme for predictive fault diagnosis in photo-voltaic modules using thermal imaging. *Infrared Phys. Technol.* 83: 182–187. https://doi.org/10.1016/j.infrared.2017.04.015.

72 Irshad, J.Z.A. and Haque, A. (2018). Temperature measurement of solar module in outdoor operating conditions using thermal imaging. *Infrared Phys. Technol.* 92: 134–138. https://doi.org/10.1016/j.infrared.2018.05.017.

73 Kurukuru, V.S.B., Haque, A., Tripathy, A.K., and Khan, M.A. (2022). Machine learning framework for photovoltaic module defect detection with infrared images. *Int. J. Syst. Assur. Eng. Manag.* https://doi.org/10.1007/s13198-021-01544-7.

74 Kurukuru, V.S.B., Haque, A., Khan, M.A., and Tripathy, A.K. (2019). Fault classification for photovoltaic modules using thermography and machine learning techniques. In: 2019 International Conference on Computer and Information Science, 1–6. IEEE https://doi.org/10.1109/ICCISci.2019.8716442.

75 Savita Nema, R.K. and Nema, G.A. (2010). Matlab/simulink based study of photovoltaic cells/modules/array and their experimental verification. *Int. J. Energy Environ.* 1: 487–500.

76 Livingston, E.H. (2004). Who was student and why do we care so much about his t-test?1. *J. Surg. Res.* 118: 58–65. https://doi.org/10.1016/j.jss.2004.02.003.

77 Owen, D.B. (1965). The power of student's t-test. *J. Am. Stat. Assoc.* 60: 320–333. https://doi.org/10.1080/01621459.1965.10480794.

78 Efron, B. (1969). Student's t-test under symmetry conditions. *J. Am. Stat. Assoc.* 64: 1278–1302. https://doi.org/10.1080/01621459.1969.10501056.

79 Mishra, P., Singh, U., Pandey, C. et al. (2019). Application of student's t-test, analysis of variance, and covariance. *Ann. Card. Anaesth.* 22: 407. https://doi.org/https://doi.org/10.4103/aca.ACA_94_19.

80 Jankowski, K.R.B., Flannelly, K.J., and Flannelly, L.T. (2018). The t-test: an influential inferential tool in chaplaincy and other healthcare research. *J. Health Care Chaplain.* 24: 30–39. https://doi.org/https://doi.org/10.1080/08854726.2017.1335050.

7

System-Level Condition Monitoring Approach for Fault Detection in Photovoltaic Systems

Younes Zahraoui[1], Ibrahim Alhamrouni[2], Barry P. Hayes[3], Saad Mekhilef[4], and Tarmo Korõtko[1]

[1] Department of Electrical Power Engineering and Mechatronics Tallinn University of Technology Smart City Center of Excellence (Finest Twins), Tallinn, Estonia
[2] British Malaysian Institute, Universiti Kuala Lumpur, Selangor, Malaysia
[3] School of Engineering and Architecture, University College Cork, Cork, Ireland
[4] School of Science, Computing and Engineering Technologies, Swinburne University of Technology, Hawthorn, Australia

7.1   Introduction

Currently, energy is the primary driver of global economic development at the moment due to its importance in enabling economic and social growth. The majority of businesses that contribute to a country's economic development are entirely reliant on electrical energy. The global trends related to the electricity production systems mentioned a considerable growth in renewable energy sources as a low-cost and environmentally friendly source of electricity generation supply [1]. Figure 7.1 shows the development of renewable energy production from 2017 to 2020. The fundamental goal of research into renewable energy technology is to transform renewable resources into electrical power that may be used to generate utility systems or loads. Solar energy generation is considered the best technique to extract energy from nature among all renewable energy sources [2].

Worldwide solar photovoltaic (PV) deployment is expanding dramatically due to the cost reduction of PV panels and the governmental policies that encourage the utilization of green energy [3]. Solar PV contributes 2.8% to the total global energy. PV generation increased by 139 GW (22.38%) in 2020 from 621 to 760 GW [4]. Much of the anticipated increase in PV generation is attributed to large-scale PV plants with ever-increasing capacity. However, due to the situation of the PV plants in an outdoor environment, constant exposure to harsh environmental conditions (sunbeam, precipitation, etc.) may reduce the optimal performance of the system and its reliability. The challenge is not

Fault Analysis and its Impact on Grid-connected Photovoltaic Systems Performance, First Edition.
Edited by Ahteshamul Haque and Saad Mekhilef.

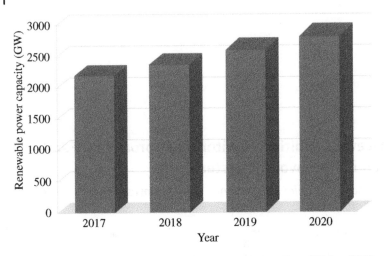

Figure 7.1 Global renewable energy generation capacity from 2017 to 2020.

solely environmental conditions that impact the reliability of the PV but also the faults that occur in the PV power plant. Moreover, in the large-scale solar farms that comprise thousands of solar panels spread, there will be hundreds of PV performance computational streams to monitor in real-time or periodically [5]. This is typically caused by external interferences or faults caused by dust deposition on PV modules, PV module aging, shading, MPPT inaccuracy, and inverter failures. Faults on the DC side of PV systems are frequently difficult to avoid, resulting in energy loss, system shutdown, or major safety problems. It is straightforward to detect anomalies in energy production. However, it is very complex to find and diagnose the source of the anomalies accurately [6]. Thus, condition monitoring and fault detection in large-scale solar farms is essential to ensure equipment longevity and maximized power yield. Condition monitoring is key for determining system state to find its effect on energy production and avoid the high cost of maintenance, for example, panel cleaning and replacements, and diode checks [7].

In grid-connected with PV systems, keeping a PV system running smoothly and safely while generating the power demand is considered a major challenge. Therefore, monitoring of the PV plants has recently attracted considerable attention from safety engineering academics and practitioners. Increased focus on fault detection and safety has resulted in numerous fault detection approaches, which can be divided into model-based fault detection techniques and data-based fault detection techniques. Data-based approaches rely on computational intelligence and machine learning (ML) methodologies, and implicit empirical models built from data analysis. The data-based approaches are largely based on different techniques and methods. for example, Artificial Neural Networks (ANNs) [8, 9], Monte Carlo [10, 11], and ML [12, 13]. However, accurate and early detection of faults in a PV system using those techniques are required a huge dataset collection, including condition data with attached labels. On other hand, the model-based fault detection methods include a high-voltage and a low-voltage fault diagnosis [14], thermal cameras [15, 16], earth capacitance measurements [17, 18], and diode-based fault detection technique [19]. These techniques and their limitations are discussed in Table 7.1.

Comprehensive solutions from these methods are commonly referred to as a monitoring system along with fault diagnosis techniques, whose functions are to maximize the operational reliability of the PV system with minimum system costs and to detect the causes affecting the performance of the PV system.

Generally, faults that occur in the DC part of a PV system have a negative impact on the power dispatch, which is highly irregular and sometimes damages the power system utilities, even if the system is equipped with protection devices. Meanwhile, the faults that occur by irradiation levels may not be determined in the PV system and lead to significant energy losses and degradation in the PV station [24]. According to the studies of the National Renewable Energy Laboratory, ultraviolet (UV) light in the natural sunlight can cause the degradation of polymeric materials that are used for PV. Thus, the module degradation rate can be as high as 4%/year, but the median and average degradation rates are calculated to be 0.5 and 0.8%/year, respectively. Therefore, it is important to confirm whether the performance of polymeric materials changes characteristics after long-term UV irradiation [25]. In this context, standard qualification tests such as the International Electrotechnical Commission (IEC) 61215 may apply to verify the general application of the new materials for the initial phase of testing the product cycle life and reduce the O&M costs. These reliability tests are intended to evaluate failures, quantify them, and help in understanding the failure factors to enhance the PV module's reliability [26].

This chapter provides a comprehensive overview of the system-level condition monitoring approach for fault detection in the PV system. The structure of this chapter is described as follows, Section 7.2 describes the aging and degradation effects of components on the PV system. Section 7.3 presents the effect of the temperature on PV system operation. Section 7.4 details the irradiance impact on the PV system. Section 7.5 describes the capacitor ESR impact on PV. Sections 7.6 and 7.7 present the data acquisition for failure modes in PV and the fault classifier development and monitoring, respectively. Section 7.8 summarizes this chapter.

Table 7.1 Summary of fault detection techniques reported in the literature.

Approach	Technique	Ref.	Description	Limitation
Data-based	Machine learning	[9, 10]	These algorithms can learn the system and predict after it is trained, depending on a large amount of labeled data.	• Requires very large datasets. • High complexity and implementation cost. • Cannot localize faulty modules. • Accuracy depends on the quality of the PV model.
	Artificial neural networks	[11, 12]		
	Monte Carlo	[13, 14]		
	Virtual imaging	[20]	The thermal images of PV modules are captured using infrared cameras. This approach is highly accurate.	• Consistent monitoring of PV modules is a difficult method. • Only applicable to PV array without blocking diodes. • Can apply in offline mode.
	Signal analysis	[21]	This approach is based on the analysis of the signal.	• Requires extra hardware to perform computation and features extraction. • Cannot classify faults and locate faulty modules. • Sensitive to partial shading.
Model-based	Earth capacitance measurements	[17]	Applied between the anode (or cathode) terminal of the PV string and the grounding terminal, and the cathode (or anode) terminal of the string is kept open.	• Consistent monitoring of PV modules is a difficult task. • High complexity and implementation cost.
	High voltage and a low voltage	[8]	These methods measure the voltage of PV that evolve patterns of the first and the last module voltages in each string.	• These methods need additional hardware and lots of wiring, which increase the cost of investment.
	Current sensors	[22]	These methods use the current sensor to obtain the faults of PV.	• Current sensors are costlier than voltage sensors. • These methods require a lot of wiring.
	Capture losses	[23]	Mainly used at the DC side of the PV conversion chain.	• Cannot detect high-impedance faults. • Cannot define the locations of the faults.

7.2 Aging and Degradation Effects of Components on PV System Operation

7.2.1 PV Cells Modeling

The single diode is the most widely used solar cell model in PV panels. The PV solar cell is modeled as a light-generated current source connected in parallel with a diode with series and parallel resistance, as shown in Figure 7.2.

The power dispatch from a PV cell is predicted using the well-known "five parameters" of the model in the system, where the relationship between output current and voltage is described by the nonlinear equation as follows:

$$I = I_{ph} - I_0 \left[\exp\left(\frac{q(V + R_s I)}{n K_B T} \right) - 1 \right] - \frac{V + R_s I}{R_{sh}} \tag{7.1}$$

where V and I are the voltage and the generated current of the solar PV cell, respectively. I_{ph} is the photogenerated current which is the value of the current generated by the incoming photons, which is related to the irradiance surface of the solar PV cell. I_0 is the diode saturation current. R_s and R_{sh} are series and parallel resistances, respectively. n is the diode ideality factor. K_B is Boltzmann constant where ($K_B = 1.3806503 \times 10^{-23}$ j/k). T is the solar cell/PV module temperature. q is the electronic charge where ($q = 1.60217646 \times 10^{-19}$ c).

The impacts of meteorological conditions such as temperature and solar irradiation on the current and the voltage dispatch from the PV cell module by the one-diode model. Eq. (7.1) shows the exponential term of $\exp\left(\frac{q(V + R_s I)}{n K_B T} \right)$. Furthermore, the photocurrent is given by Eq. (7.2), which is related to the solar irradiation and the PV module temperature.

$$I_{ph} = \frac{G}{G_{ref}} \left[I_{ph,ref} + \mu_I \left(T - T_{ref} \right) \right] \tag{7.2}$$

where $G_{ref} = 1000$ (W/m^2) is irradiance at Standard Test Conditions (STCs), G is the irradiance (W/m^2). $T_{ref} = 298.15$ K, which is the PV cell temperature at STC.

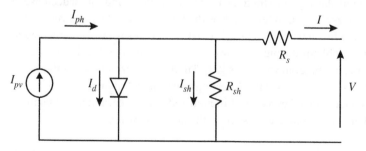

Figure 7.2 Equivalent circuit of the single-diode PV model.

T is the temperature of the PV cell. $I_{ph,\ ref}$ is the photocurrent at STC. μ_I is the temperature coefficient of short-circuit current (A/K), which is given by the PV manufacturer, which is proportional to the materials used for manufacturing the PV cell.

7.2.2 Impact of PV Materials on the Degradation of PV Cells

The PV modules consist of multilayer-structured devices that have solar cells and electronics embedded in transparent polymeric encapsulants. For advanced protection, these internal materials are covered by glass or polymeric backsheets. Precise materials are used to assure a service lifetime of PV cells is 25–30 years. To achieve long-term PV module durability, each layer must preserve its attributes and function over the module's lifetime. PV modules are subjected to a variety of environmental stressors throughout their outside service, including UV, irradiation, temperature and humidity cycles, rain, snow, and wind loads, hail, sand, and/or salt. Furthermore, internal stressors like the materials selection and design (additives, morphology, and material compatibility) might contribute to PV module degradation [27]. Nowadays, revolutionary technologies enable the development of novel PV cells based on nanomaterials that are easier and less costly in production, resulting in ultrathin, lightweight, and flexible solar cells that are much easier to carry and install and have outstanding transparency. However, crystalline silicon (Si) technologies are now widely used around the world, posing waste management and recycling issues. This type of PV panel is made up of various solar cells, wires, glass, and frames, with ethylene-vinyl acetate (EVA) which is most often used as encapsulation material [28].

Accelerated testing such as IEC-61215, IEC-61646 is regarded as the major testing tool for detecting modules prone to early-life failures or severe deterioration. The researchers concluded from the standard testing that environmental tests such as thermal cycling (TC50 and TC200) cause thermomechanical failure in the interconnections, contributing to an increase in series resistance and thus a decrease in efficiency, whereas irradiation tests such as the UV test affect the properties of the PV cell, resulting in a lower I_{sh}. While the first type of loss is rather connected to quality control problems during the power dispatch from the PV cell, the second type is attributed to material weakness [28]. Till date, UV irradiance in the range of 295–385 nm is generally used in PV testing to predict UV damage.

The aging efficiency μ takes a curve function and not a linear function as shown in Figure 7.3, due to several effects such as the glass and wires.

$$\mu = \mu z_{25°C} \cdot \frac{AE}{100} \qquad (7.3)$$

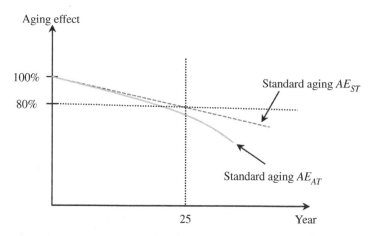

Figure 7.3 PV panels aging evolution.

The standard AE_{ST} takes *EVA*, glass and wires aging evolution as independent parameters:

$$AE_{ST} = 100 - \sqrt{AE_{ST}^2 + AE_{glass}^2 + AE_{wire}^2}\, n_{year} \tag{7.4}$$

where n_{year} is the total operating years.

Typically, the *EVA* aging evolution is related to the glass aging, and the wire aging evolution (AE_{wire}) is related to the *EVA* aging evolution. However, the glass aging evolution (AE_{glass}) is independent.

Many research papers have been worked in the field of PV technologies by using accelerated testing to improve the efficiency of the PV operation, considering the impact of the materials selection and design such as in [29] proposed the copper-based corrole as a hole transporting material (HTM) to improve the stability of the device in the prolonged 85 °C stress conditions. Corrole-based devices have been compared to the initial devices for several periods. The experiment work by the authors shows the proposed device structure can represent a possible method to pass thermal stress tests proposed which have been investigated using the IEC-61646 standard. To address the issues posed by soiling, authors in [30] present a novel anti-soiling coating with excellent weather stability. The proposed anti-soiling coating characteristics provide stability for the PV plants to encounter weather conditions due to the bonding of functionalized PV characteristics with the glass. The coating's weather stability has been analyzed using the IEC-61646 standards in conditions of 1500 hours with a temperature range of 85 and 40 °C and maximum relative humidity of 85%. In [31], the authors have investigated the expected impact of certain defects in Si modules using accelerated testing. Specifically, the behavior of boron-oxygen (BO) light-induced degradation (LID) and light and elevated temperature-induced degradation (LeTID) was simulated during some of the stress tests in IEC-61215.

In the following sections, we will provide a brief overview of the most commonly used materials for PV components, their composition, and the impact of their degradation on the overall performance of PV modules.

7.2.2.1 Encapsulant

To achieve maximum efficiency, sensitive PV components, such as brittle solar cells and metallization, must be shielded from direct outdoor factors such as dust, humidity, and UV irradiation. Polymeric materials are considered the best option for that purpose from the start due to their low weight, low price, and specific characteristics. Nevertheless, the polymers materials remain for aging and degradation under extreme environmental and climatic conditions. Therefore, the combination of the polymer materials with a different set of additives like the UV absorbers may prevent the main materials in the PV module from degradation and meet the expected 25 years lifetime. To date, the experiments have shown that the performance of PV modules may be enhanced by adding encapsulation materials [32], backsheet [27], or transparent materials [33].

The most important component in the PV module is the solar cell, which must be covered using encapsulation materials. The encapsulant materials can offer the optimal optical coupling, electrical isolation, structural support, environmental protection, and interconnecting of all components. The encapsulant's composition also influences heat dissipation among the module's layers, which can be critical for PV modules working at high temperatures. Only a few polymers, such as EVA copolymer, poly (vinyl butyral) (PVB), pol (dimethyl siloxane) (PDMS), silicones, ionomers, and thermoplastic elastomers (TPEs). However, thermoplastic polyolefins (TPOs), were found to be suitable for application in PV modules due to the obvious meeting requirements. The polymer film based on EVA is considered the most widely used as encapsulant material because it is low-cost, low coefficient of water absorption, wet steam resistance, and has reasonable durability. Therefore, the current state of the art shows that it is important to understand the mechanisms of encapsulant materials so that more researchers may concentrate their efforts on developing novel encapsulant materials for solar cells with promoted reliability [34]. Table 7.2 shows the characteristics of encapsulating the EVA material for PV.

To investigate any developed encapsulant for the PV module, various key and basic parameters must be analyzed and met. For instance, the optical transmission should be more than 90%, and adhesion strength to glass more than 80 N/cm are typical parameters often cited by the encapsulant manufacturers and described in the product datasheet. Similarly, to passaging testing like damp heat, humidity freeze, and thermal cycling test, an encapsulant must have an

Table 7.2 The characteristics of encapsulating the EVA material for PV.

Characteristics	Requirements
Manufacturing temperature	$\leq 171\,°C$
Total hemispherical emissivity	$> 91\%$
Water absorption	$< 0.5\,wt\%$ to 100% RH
Glass transition temperature	$< -40\,°C$
Degradation by UV absorption	No wavelength $> 350\,nm$
Pressure lamination manufacturing	$\leq 1\,atm$
Shear modulus	$<20.7\,MPa$ (3000 psi) at 25 °C
Clouding	None at 80 °C and 100% RH

adequately high melting point greater than 85 °C. Because of module components' varying thermal expansion coefficients, thermal stresses are continuously generated in the PV module. The encapsulant should have a low modulus of elasticity ($<50\,MPa$ industrial suggested) in order to disperse these heat stresses. Volume resistivity is another key encapsulant feature for preventing current leakage-/potential-induced deterioration in PV modules. Most PV producers may lose out on greater knowledge of the material behavior of encapsulant film due to the suppliers of the resin and recipe frequently operating as just raw material suppliers across the whole value chain [35].

7.2.2.2 Backsheet
Generally, the PV backsheets consist of a multilayer (mostly three layers), with each layer performing a distinct function. The backsheets can provide electrical insulation, prevent moisture ingress, and protect against mechanical damage. The geometry and thermophysical properties of the PV materials, and especially the backsheet, have a direct impact on the thermal behavior of PV modules. It is well known that elevated operating PV cell temperatures negatively affect the energy yield of systems by decreasing efficiency. Backsheets in PV are often composed of multiple polymer layers: the main layer that offers electrical insulation, mechanical strength, and other protective features, an inner layer that facilitates module adhesion, and an outer layer that protects against weathering. These layers are often combined using adhesives. Only a limited number of material combinations may be co-extruded into multilayer backsheets. Because each of the backsheet layers is subjected to a unique set of pressures during outdoor exposure, each individual performance has an impact on the overall performance of the backsheet and, ultimately, the complete PV module [36].

For several years, the fluoropolymers such as polyvinyl fluoride (PVF) and polyvinylidene fluoride (PVDF) have been the dominating material of choice for the backsheet outer layer, due to their chemical stability. However, a number of factors, including cost reduction, have led to the adoption of non-fluoropolymer-based materials for the outer layer, such as polyamide (PA) and polyethylene terephthalate (PET). Nevertheless, the latest reports of fielded modules using these polymers have demonstrated signs of decreased weatherability in the form of cracking, formation of bubbles and discoloration, yellowing, and loss of mechanical features [37].

The delamination between individual backsheet layers may result in deterioration and lack of adhesion of the adhesives used to link backsheet layers. Specifically, the adhesives have a strong affinity for water, and hydrolytic breakdown of the adhesive layer can occur, which results in its depletion. In addition, the adhesive layer also impacts the dimensional stability of the backsheet owing to increased internal residual stresses. Thus, cracking of the outer layer may occur, which reduces the effectiveness of the backsheet and exposes the PV modules to a higher risk of degradation and damage [36].

According to the previous study, cracking in the PV modules is more common in non-fluoropolymer backsheets with PET or PA external layers. For example, in [38], a novel testing method has been used to evaluate the cracking tendency and crack spreading of backsheets by using a monitoring crack structure during the deformation of backsheet after UV irradiation. The presented work endeavors to perform the acceleration aging by injecting an extra UV dose compared to field exposure on the backside of the PV module. An in situ surface cracking test was performed under a confocal microscope and the tensile deformation applied to the exposed samples was measured using a tensile apparatus. In [39], a simple test approach that combines weathering with a sequential fragmentation test is designed to rapidly and efficiently mimic the cracking behavior of PA-based backsheets in the field. The novel testing method has been used to analyze and forecast backsheet cracking propensity. For the investigation, the PA-based backsheets have been exposed to UV radiation for different testing hours. The exposed samples were then pulled in tension step by step while being monitored in real-time with laser scanning confocal microscopy. Detailed characterizations of the backsheets reveal considerable chemical deterioration and surface erosion and ductility loss. Furthermore, cracking was discovered on the surface of the UV-exposed samples at a low applied strain of 1%. The study in [40] has tested field-deployed modules with broken PA backsheets. Material changes in field-aged backsheets and those aged in combined-accelerated stress testing (C-AST) have been evaluated using planar and cross-sectional optical microscopy, as well as Fourier-Transform Infrared Spectroscopy (FTIR). The materials analysis assisted in elucidating the underlying mechanisms that contribute to failure and validating that the alterations observed in C-AST are indicative of those observed in the field. The study in [41] has tested the material properties of a novel PA-based, Co-extruded backsheet film. The proposed backsheet was evaluated using C-AST and demonstrated greater durability than the traditional backsheet. However, the proposed backsheet failed by forming a crack. A different set of materials and tools have been used in the proposed study to identify changes induced by aging and their relation with backsheet cracking.

7.3 Temperature Impact on PV System Operation

The climatic and environmental parameters to which the PV modules are subjected have a massive impact in terms of degradation. The PV module performance is usually affected by solar irradiation and type of radiation, temperature, moisture, mechanical factors, and electrical operational parameters. Other regional climate factors to consider include snow, hail, wind, salt, sand, dust, and pollutants/gases, some of which are potentially corrosive such as NH_3, NO_2, Cl_2, etc. Furthermore, when PV modules are exposed to solar radiation, the temperature of the surface rises owing to the heating effect. This has a considerable influence on the temperature differential between the PV module surface and the ambient air. The temperature difference may create a visible temperature gradient between the PV, backsheet, and the encapsulation layers as well as a force known as thermophoresis force. The temperature significantly affects the degradation of the PV modules, particularly hot spots, adhesive bleaching, delamination failure on interconnections, etc. The majority of the chemical reactions that occur during module degradation are caused by temperature [42].

The importance of operating temperature in relation to the electrical efficiency of a PV device, whether it is a simple module, a thermal collector, or the PV modules that are integrated with the building, is well established and documented, as indicated by the interest of researchers and previous studies. Several correlations exist that express PV cell temperature (T_c) as a function of meteorological factors such as ambient temperature (T_a), the wind speed (V_w), the global solar radiation (I_T), and solar irradiation (G), with material and system-dependent attributes as parameters. A substantial number of correlations express the PV module's thermal properties' electrical efficiency. The majority of correlations include a reference state and the corresponding values of the relevant variables [43]. However, many of them take the acquainted linear form, varying only in the numerical values of the relevant parameters which are expected are dependent on the material and system. In terms of the relevant weather variables, it was discovered that the PV cell temperature rise over the ambient is extremely sensitive to wind speed, less sensitive to wind direction, and practically insensitive to atmospheric temperature.

However, it obviously depends heavily on the impinging irradiation, for example, the solar radiation flux on the cell or module. By theoretical, the correlations for the PV-operating temperature are either explicit or implicit. If the PV operating temperature correlations are implicit, an iteration procedure is required.

The discrepancies across the T_{pv}, T_c, T_b, and T_f, which stand for module temperature, cell semiconductor layer temperature, and back and front side temperatures of the module, are discussed in the experiments, T_b is measured and is commonly referred to as T_{pv} or T_m. The T_c is given by:

$$T_c = T_b + \frac{I_T}{I_{ref}} \Delta T_{c-b} \tag{7.5}$$

where

$$\Delta T_{c-b} = T_c - T_b \tag{7.6}$$

I_T is the intensity of the global solar radiation on the PV module. I_{ref} is the reference intensity of the solar radiation which is 10^3 W/m^2. The value of the difference between T_c and T_b is given to be 2–3 °C. The following equations were utilized to obtain the T_c from T_b, which is the temperature generally determined.

$$T_c = T_b + \dot{Q} \sum (\delta x_i/k_i) = T_b \left(1 - \frac{\sum(\delta x_i/k_i)}{U_b^{-1}} \right) - T_\alpha \frac{\sum(\delta x_i/k_i)}{U_b^{-1}} \tag{7.7}$$

$\dot{Q}$ is considered the best approximation as half of the heat rate generated in the semiconductor. $\sum(\delta x_i/k_i)$ is the global resistance per m^2 because of the heat transfer between the layers. U_b is the heat loss coefficient from the PV module. Using basic equations, the influence of temperature on the electrical efficiency of a PV cell/module is given by the following equations:

$$P_m = I_m V_m = FF.I_{sc} V_{sc} \tag{7.8}$$

In Eq. (7.8), the FF is the fill factor, I_{sc} and V_{sc} are the short current and the open-circuit voltage, respectively. The impact leads in the PV module is given as a linear relation:

$$\eta_c = \eta_{ref} [1 - \beta_{ref}(T_c - T_{ref}) + \gamma \log I(t)] \tag{7.9}$$

where η_{ref} is the module's electrical efficiency at the reference temperature. T_{ref} is the temperature at 1000 W/m^2. β_{ref} and γ are the temperature and the solar irradiation coefficients, which have the values of 0.004 and 0.12, respectively. The quantities η_{ref} and β_{ref} are normally given by the PV manufacturer. The traditional linear expression for the PV electrical efficiency is given in Eq. (7.10).

$$\eta_c = \eta_{ref} [1 - \beta_{ref}(T_c - T_{ref})] \tag{7.10}$$

The actual value of the temperature coefficient depends on the PV materials and the T_{ref}, and it is given by the ratio as follows:

$$\beta_{ref} = \frac{1}{T_0 - T_{ref}} \tag{7.11}$$

A set of equations is defined in the group of physics-based models for T_c prediction, including the Energy Balance Equation (EBE) under steady or transient circumstances, combined with equations on heat propagation from the semiconductor layer to the front and back surfaces, and then subsequently to the environment. It is very important to consider the radiated heat exchanged between the PV surface and the air. Those sets of equations that consider all environmental statuses include a general simulation model. Nevertheless, the heat convection and radiated heat coefficients utilized do not adequately cover the entire range of climatic circumstances. Additionally, grey models for electric and heat transfer parameters connected with the functioning PV module have been integrated into an EBE, as well as regression analysis of recorded data from monitored quantities, which are applied for the creation of semiempirical models for T_c prediction.

7.4 Irradiance Impact on PV System Operation

As described in the previous subsections, the maximum power that a PV module can be collected is determined by many natural conditions such as ambient temperature, solar radiation, moisture, and dust deposition. However, the influence of these factors leads to a reduction in PV module performance and efficiency. As a result, these factors must be addressed and taken into account in order to deal with changing weather conditions while also achieving the maximum output power from the PV module. The partial shadowing condition (PSC), defined as a circumstance in which a portion of the PV module gets nonuniform solar radiation, is one of the critical factors. PSC is the primary factor that decreases and minimizes the PV system's output power. PSC plays a crucial role in the output power of the PV system. When PSC happens, numerous maximum power points (MPPs) on the PV curve exist, resulting in additional difficulties on the PV features. The irradiance that contributes to creating short-circuit current in a PV device is referred to as effective irradiance. It is crucial to remember that effective irradiance is also affected by PV modules' possibly varying angular responsivity and pyranometers. Hence, the term "spectrally effective" is employed. Reflection losses must be included in addition to the spectral impact when calculating energy ratings or yield projections. It may not always be able to separate spectral and angular effects [44].

The impact of broadband irradiance on PV power is well-analyzed excessively, as will be discussed more later. In contrast, despite several studies completed throughout the world, data quantifying the impact of solar spectrum irradiance on PV performance have remained elusive. [45] investigated the impact of spectral irradiance variation and spectral response on energy production by PV modules. The experimental work has been conducted outdoors and used optical filters to find various zones of the solar irradiance spectrum. In [46] proposed that a minutely solar irradiance forecasting approach based on a real-time surface irradiance mapping model be used to improve solar power forecasting accuracy. The authors investigated the mapping relationship between sky pictures and solar irradiance and extracted the red-green-blue (RGB) values and position information of pixels in sky images after background removal and distortion rectification. Then, a real-time sky image-irradiance mapping model is developed, trained, and updated using real-time sky photos and solar irradiance. A study in [47] investigated the lack of information that impacted PV power estimation and designed a PV model that considers the spectrum distribution of irradiance and the spectral response of the panels. The developed model was evaluated using PV power measurements for a monocrystalline Si module. This model is used to investigate the impact of solar zenith angle and clouds on the performance of PV modules. As a result, the PV performance may be increased by 5% in overcast situations by spectral filtering near-infrared irradiance and by 18% when only useable irradiance is used to compute performance. Usually, the results are difficult to compare since they focus on different locations and time periods for energy effect (instantaneous, monthly, and annual), employ different indicators, and often do not account for measurement error. However, they agree on one critical point: the effect of different irradiance on the performance of the PV modules is largely determined by its spectral response.

The irradiance influences the value of the short-circuit current since it is directly proportional to Eq. (7.12), where I_{sh} and G are the short-circuit current and the irradiance, respectively. I_{sh}^* and G are the short-circuit current and the irradiance, both under STC, respectively.

$$I_{sh} = \frac{I_{sh}^* \, G}{G^*} \tag{7.12}$$

The open-circuit voltage is influenced by solar irradiance because it decreases as the temperature rises. This occurs because temperature increases affect the gap energy, which, in turn, affects the diode's saturation current.

The prospect of the absorbed photon describes the spectral response of a PV module that will generate a photogenerated current I_{ph} through PV cell, the spectral response is obtained by the bandgap, the thickness, and characteristics of the material. The spectral response is defined as the short-circuit current, $I_{sh}(\lambda)$, resulting from a single wavelength of light normalized by the maximum possible current [48].

$$SR(\lambda) = \frac{I_{sh}(\lambda)}{qAf(\lambda)} \tag{7.13}$$

where q is the electronic charge which is 1.6×10^{-19} C, A is the surface area of the PV module and $f(\lambda)$ is the incident photon flux. The degree to which the spectral response and incident irradiance spectrums coincide varies with spectrum change, resulting in a spectral effect on device current and efficiency. In terms of spectral influences, the situation is reversed: methods for measuring a module's spectral response exist and are used as cutting-edge procedures despite the lack of a valid international standard (national standards do exist, however); whereas data on spectral irradiance is rare, and models to account for the influence of spectral effects under all-sky conditions are not readily available. The spectral factor (SF) of non-concentrating PV modules is given by:

$$SF = \frac{\int E_G(\lambda) \, SR(\lambda) d\lambda \int E_G^*(\lambda) d\lambda}{\int E_G^*(\lambda) SR(\lambda) d\lambda \int E_G(\lambda) d\lambda} \tag{7.14}$$

where E_G and E_G^* are spectral irradiance that effectively teaches the module under the STC. When SF is more than one value, it has spectral power gains. In contrast, it will lose some values in comparison to the STC. The short-circuit current of a PV module is proportional to the product of the spectral irradiance and the integrated SR along the wavelength [45].

$$I_{cc} \propto \int E_G(\lambda) \, SR(\lambda) d\lambda \tag{7.15}$$

The spectral effective responsivity (*SEF*) is defined as:

$$SEF = \frac{\int E_G(\lambda) \, SR(\lambda) d\lambda}{\int_{\lambda < \lambda_0} E_G(\lambda) d\lambda} \tag{7.16}$$

λ_0 is the present wavelength corresponding to the bandgap energy of the material. The index represents the ratio between the short-circuit current simply accounting for irradiance and spectral effects and the spectrum power available for PV conversion in units of A/W.

The IEC 60904-7 standard defines the spectral mismatch (*MM*) factor as a method of assessing the relative spectral effect of a sample PV device and a reference PV device for utilization.

$$MM = \frac{\int E_G(\lambda) \, SR_{sample}(\lambda) d\lambda \int E_G^*(\lambda) \, SR_{ref} d\lambda}{\int E_G^*(\lambda) SR_{sample}(\lambda) d\lambda \int E_G(\lambda) \, SR_{ref} d\lambda} \tag{7.17}$$

if MM > 1 : Spectral Gain compared to STC

if MM < 1 : Spectral Loss compared to STC

where: SR_{sample} and SR_{ref} represent the spectral response of the sample PV module and the reference PV module, respectively.

The above explanations show that not considering the spectrum impact is problematic since the spectral impact is sensitive to the PV modules, introducing a bias in the expected energy output depending on the technology. Without spectrally irradiance data for random locations, or reliable and generally applicable prediction models for forecasting the spectral impacts of various module technologies, the only possible data sources are local spectral irradiance measurements. Based on these measurements, energetic losses or gains for different technologies can be computed with respect to different periods. Moreover, the spectral impact can be determined using the measurement of the short-circuit current of the different PV modules. The drawback of this approach is that influences resulting from the temperature, solar irradiance, and reflection must be removed. Thus, the different angular responsivity of flat PV panels and pyranometers with glass domes will impact the findings, especially when a pyranometer is employed as a reference device.

As an index for spectral irradiance and spectral impact, many studies have used the spectral *MM* factor according to IEC6094-3, as it is largely applied in the PV community. However, the other factors, such as average photon energy (*APE*) or useful fraction (*UF*) directly indicate whether a specific spectral distribution causes gains or losses compared to the reference condition.

7.5 Capacitor ESR Impact on PV System Operation

The power grid integrated with PV stations is impacted by the performance of the PV panels, inverter conversion losses, and power extraction efficiency (PEE). The PV panel's performance is determined by the material used and the production method. The second issue is inverter conversion losses, which include switching and conduction losses [49]. The solar inverter's PEE is the third factor that may influence output power. This factor is described as the proportion of the inverter's average power extracted from the PV panel to the maximum power available from the PV panel. The extraction efficiency may be extremely reduced due to the PV voltage oscillations in the MPP. These oscillations are caused by the algorithms used to obtain maximum power point tracking (MPPT), and DC-link voltage ripple (of twice the grid frequency) in the case of a single-phase inverter. Generally, the extraction efficiency of various MPPT algorithms ranges from 90 to 99.9% [50]. Furthermore, the oscillations in DC-link voltage at twice the grid frequency are considered one of the factors that affect extraction efficiency in single-phase PV inverters. The capacitors are popularly applied to reduce these oscillations in the power system due to their efficiency and low cost. The capacitor in DC-link

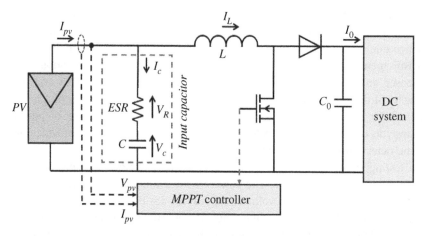

Figure 7.4 The configuration of the capacitor system in a single-phase inverter.

represents a power decoupling element, absorbing this oscillating power and balancing the instantaneous power difference between the inverter input and output. Thus, the voltage and current of the DC-link capacitor will invariably show as a second-order harmonic ripple. However, the capacitors degrade due to their efficiency loss with time, which leads to a rise in equivalent series resistance (ESR) and a decrease in capacitance ability. An increase in ESR value and a decrease in capacitance leads to an increase in power oscillations around its MPP [51]. When the value of ESR increases and the aging capacitance reduces, the PEE and the power dispatch from the distributed energy resources will reduce. This leads to a decrease in the profits from the power system. Hence, the optimal capacity of the capacitor and timely replacement are desired. The degradation of capacitors is due to temperature, electrical, mechanical, and environmental stresses, and there are several root causes and failure modes. The lifetime of capacitors can be estimated using the ambient temperature and self-heating temperature. The characteristics of capacitors are typically evaluated using a sinusoidal current waveform, which is specified in the datasheet. However, the actual excitation currents of a capacitor in a power electronic converter are complex square and/or triangular waveforms [52]. Figure 7.4 shows a sample of the configuration of the capacitor system in a single-phase inverter.

Various techniques are suggested in the literature to improve the PV system operation using the capacitor ESR. In [53], this work proposed a novel sensor-less approach for monitoring the state of inverter DC-link using the aluminum electrolytic capacitors (AECs) based on the estimations of the ESR and capacitance. The suggested method is to estimate the ESR and capacity of the capacitor using the inverter whenever the inductive loads are turned off. The parameters are determined without the use of additional sensors and utilize the inverter's DC-link voltage and stator current measurements. Because parameter change due to aging is most noticeable at low temperatures, automatic offline measurements of ESR and C are collected before inductive load initiation. The proposed approach is a simple, low-cost solution for estimating ESR and the capacity of the capacitor status evaluation. In [54], The authors present a novel method for online life-cycle monitoring and voltage management of power electrolytic capacitors in DC-voltage link converters. The circuit provides an online detection of the capacitor's ESR by using the converter's current ripple as a testing signal. The suggested concept's basic detection principle is based on estimating capacitor losses. The power loss is estimated while the AC ripple components of the capacitor voltage and current are measured. The measured values correspond to the ESR value. Thus, the specific frequency compensation filters required in alternative concepts can be avoided to a large extent. In [55], the authors presented a quasi-online method for condition monitoring of the AECs in inverters in a single-phase grid-connected PV system by estimating the capacitance and ESR. This approach operates during nighttime when there is no solar irradiation. In this scenario, the inverter injects current at various odd harmonic frequencies, and DC-link capacitor impedances at various frequencies are measured. Using these impedance data, the least mean square approach is applied to predict the ESR and capacitance values. In [56] the authors proposed a three-dimensional piecewise linear model based on the circuit parameter-related bifurcation behaviors and control weight-related dynamical distributions.

The results show that when placing a small output capacitor ESR, the inductor current weight must be increased to provide a stable operation as the feedback gain or output capacitor ESR increases. In [57], a two-stage diagnostic approach is proposed to obtain the optimal operation of the DC-link capacitor in a single-phase grid-connected PV plant. The values of ESR and C have been used as indicators in the estimation of the degradation stage. The authors proposed electrochemical impedance spectroscopy (EIS) to estimate the impedance curve of the DC-link capacitor. Moreover, the multi-fitting algorithm has been used to determine the ESR and C parameters. The fault severity was determined by comparing the estimated values of the ESR and C to the nominal values. The EIS was shown to be capable of determining capacitor impedance independent of the solar generator's real operating conditions, such as during irradiance variations and with the MPPT algorithm is off. In [58], proposed a simple and cost-efficient online approach for monitoring the health of the DC-link capacitor using AECs. The suggested approach measures the AEC impedance at twice the grid frequency. A sampling of AEC, PV voltage, PV current, and inductor current based on mathematical models is applied to obtain and estimate the second harmonic.

Table 7.3 Comprehensive comparison of the existing methodologies.

Ref.	Method	Advantages	Disadvantages
[53]	An inverter is modulated to inject a DC in the motor just before starting the motor	No need for injection of external current/voltage	Health monitoring is possible only at the starting of motor
[57]	ESR or capacitance value is estimated by injecting low-frequency voltage in DC-link	Low sampling rate	May violate the limits of subharmonic line currents mentioned in IEEE 519-2014 standard
[54]	ESR is calculated using average power loss and RMS of current flowing through the capacitor	Less computational complexity	Use of extra current sensor in series with the capacitor
[55]	ESR or capacitance is estimated using the recursive least square method.	Elimination of current sensor in series with the capacitor	Require special operating conditions.

Table 7.3 shows a comprehensive comparison of the existing methodologies, advantages, and disadvantages. When the ESR or capacitance value reaches a critical point, most existing techniques consider electrolytic capacitors to be unsuitable.

When the AC component of the current flows through the capacitor, the presence of ESR leads to heat generation in the capacitor. ESR represents all of the ohmic losses of the capacitor and is connected in series with the capacitance C. The ESR of the capacitor is expressed as:

$$ESR = R_f + R_{elec} + R_{constant} \tag{7.18}$$

where R_f is the frequency-dependent part, R_{elec} is the resistance of electrolyte, which depends on temperature also, and $R_{constant}$ represents the constant part of ESR, which is equal to the summation of foil resistance, tabs resistance, lead wires resistance, and ohmic contact resistance.

The voltage across ESR is given by:

$$v_{ESR}(t) = ESR \left\{ I_{pv} - \frac{V_{pv}}{L} t \right\} \tag{7.19}$$

where V_{pv} and I_{pv} are the average solar PV voltage and the current, respectively. L is the inductance.

The voltage across C is given by:

$$v_C(t) = \frac{1}{C} \int_0^t i_c(t)\, dt + V_c(0) \tag{7.20}$$

The PV voltage is given by:

$$v_{pv}(t) = ESR \left\{ I_{pv} - \frac{V_{pv}}{L} t \right\} + \frac{1}{C} \int_0^t i_c(t)\, dt + V_c(0) \tag{7.21}$$

The main stress factors are the hot-spot temperature T_h and operating voltage V_{dc}. The T_h is estimated based on the capacitor power losses, which are calculated as:

$$P_{C,losses} = \sum_i ESR\,(T_h, iw_n)\, I_{i\,(RMS)}^2 \tag{7.22}$$

where $I_{i\,(RMS)}$ is the RMS value of each ith harmonic component of the capacitor current and ESR is the capacitor's equivalent series resistance.

$$T_h = T_a + R_C \cdot P_{C,losses} \tag{7.23}$$

where T_a and R_C correspond to the ambient temperature and the capacitor-equivalent thermal resistance, respectively.

The relation between the volume of electrolyte present in a capacitor and its ESR is given by:

$$\frac{ESR}{ESR_0} = \left(\frac{V_0}{V} \right)^2 \tag{7.24}$$

where V_0 and ESR_0 are the initial volumes of electrolyte and ESR, respectively. Ripple current passing through the capacitor leads to heating of the capacitor. As the temperature rises, the evaporation rate of the electrolyte rises, increasing the rate of electrolyte loss. The capacitor's heat dissipation increases as the ESR rises, resulting in additional electrolyte loss. Manufacturers recommend replacing the capacitor if the volume of electrolyte has decreased to 40% of its initial value.

7.6 Data Acquisition for Failure Modes in PV System

The optimum design, size, and analysis of the PV system performance are considered essential to reduce the cost and obtain the economic startup of the PV stations. To achieve these objectives, detailed information about operational output system data, temperature, humidity, and solar irradiance must be collected over a long period [59]. Therefore, the data acquisition system (DAS) has been deemed necessary for monitoring and collecting PV system data to evaluate its performance. In most cases, the distinctive behavioral patterns of the PV system can be recognized within the respective quantities in the supervision and the DAS. Using the DAS assists in disseminating PV systems and primarily in developing countries [60]. Moreover, the DAS with a granularity of 15 minutes is recommended by standards such as IEC 61724. These collected data are commonly used to calculate standardized performance indicators like the performance ratio or technical availability. Based on these indicators, statistical approaches, simulations, and intelligent methods have all been applied in recent years as computational tools for fault diagnosis [61].

In addition, the high cost of commercial automatic DAS is able to control and monitor the PV solar plants, as well as measuring, storing, and statistically processing the variables required for sizing and evaluating the performance of PV system, is a major impediment to the development of PV system projects in developing countries. Currently, DASs serve as a link between electrical and meteorological environments. The system is widely used in research laboratories for measuring and analyzing scientific and engineering experiments, as well as for PV plant assessment in a variety of institutions and industries [60]. Several DAS techniques have been developed for operating in a wide range of implementations, for instance, monitoring and evaluating the performance of PV plants, monitoring the status of batteries in water pumping integrated with PV systems, and measuring operational parameters of hybrid PV-CHP systems [62].

DASs are typically integrated into an island or grid-connected PV systems to measure, record, and display data in the absence of human or computer mediations. The system aims to receive many input parameters using components created in the signal conditioning unit and to record data from a variety of sensors and PV systems simultaneously. DASs can adapt the signals and settings to fit the changing situations under varying weather conditions [63]. The DAS process may fail transiently or permanently. Transient interruptions in data capture are caused by communication failures, outages, outliers, and incorrect operations. On the other hand, cyber-attacks on component failures are sources of long-term disruption. Communication failure and continuous data loss cause existing forecasting models to reach their optimal performance due to most of these approaches are not designed to handle the situation where new temporal measurements are not provided in the PV system. Outliers and compromised data must be updated with more accurate values, and missing information must be provided. Advances in PV system measurement and control and the introduction of new cyber threats and vulnerabilities need the use of short-time forecasting approaches to address partial observability in training datasets. This problem has lately received attention in the field of power system forecasting research [64].

In this sense, a lot of DAS designs for fault diagnosis have been established in the literature such as in [65], a proposed failure diagnostic system developed and experimentally demonstrated using actual data acquired from a test PV system installed. The operation of the PV system and the meteorological factors have been recorded and stored according to the IEC-61724 conditions by using a DAS platform. The meteorological factors that have been measured include solar irradiance, wind speed, temperature, maximum point DC current, the maximum point DC voltage, and the MPP. In [61], the authors presented an integrated approach to model-based fault detection through physical and statistical models and failure diagnosis based on the physics of failure. Both approaches obtain the optimal PV system operation and maintenance based on typically available data from the DASs. A case study based on six years of data from an outdoor PV system was used to show the failure detection and diagnostic capabilities. Underperforming values of the PV plant's inverters were reliably recognized in this case study, and various underlying causes were identified. [66] proposed the ANN model per failure mode and enhanced a practical implementation in DASs for different PV plants. This technique was easier to enhance decision-making processes in condition-based maintenance and risk modeling, allowing for savings in direct and indirect corrective maintenance costs or demonstrating residual life until ultimate equipment failure. When enough data for significant training is available, a better implementation of the proposed approach will operate to decrease costs and increase understanding of the plant's life cycle when subjected to nonhomogeneous operational and environmental situations.

Several studies used the data analysis techniques for failure detection based on DAS, such as in [67] presented a design and development of a grid-connected PV virtual instrumentation system (GCPV-VIS) that uses statistical approaches to simplify monitoring and fault detection of a grid-connected. The approach has been validated using Hub 4 communication manager, Mate3 device, and the sensor block, which support DASs and database transmission of environment and electrical parameters. The fault detection is based on a statistical comparison of measured and theoretical output power using the *T*-test. The study results show that the suggested technique can detect grid-connected PV system problems and be utilized to inspect PV system conditions regularly. In [66], the authors adapted a reliable model to combine monitoring data on operating assets, and the information on environmental conditions. The proposed technique shows a logical choice tool based on two ANN models. This application allows updating the dependability analysis based on changes in operating and/or environmental circumstances and easily be automated within a supervisory control and DAS. In [68] proposed a GISTEL (solar radiation by tele tele-detection) model to predict the solar irradiation profile of the PV system. The difference measurement between actual and simulated DC power was utilized as an indicator for faults. Simultaneously, the current and voltage ratios are applied to discriminate shading faults from string faults.

To summarize, the techniques to diagnose the faults based on DASs automatically extract knowledge from a given PV dataset and can accurately estimate the projected energy production, which may successfully identify defects in PV systems. However, many approaches such as ANN, ML, support vector machine (SVM), decision tree, and fuzzy logic require a large amount of training data and short- and long-term training processes. Furthermore, to decide on fault conditions, experimental data (current, voltage, irradiance, ambience, and temperature) containing both healthy and faulty operating conditions in a PV system are required.

7.7 Fault Classifier Development and Monitoring

In island mode or grid-connected PV system, a failure (or fault) is defined as an event that reduces the performance of the power systems. The failures may lead to a partial decrease in the performance of the power system or a complete shutdown in some cases where the power output is completely null. Grid-connected PV system failures can occur on either the AC or DC side of the system and are classified primarily by location and configuration. Therefore, the power system are required for fault detection and operation of the protection equipment in the minimum possible time so that the power system could remain in a stable condition. The faults in transmission lines in electrical power systems are required to be detected first, then classified accurately, and then cleared as quickly as feasible. An effective fault detection system provides relaying operations more reliable, fast, and secure [69].

The faults in the grid-connected PV system may occur on the DC or AC side. There are two types of faults on the AC side of the PV system: complete blackout and grid outage (lightning and unbalanced voltage). Instead, faults on the DC side of the PV system are mostly caused by the balance of the system or PV array problems. The balance of system component faults involve MPPT errors and cabling problems. The cabling problem may issue three types of faults in the power system: bridging, open-circuit, and earth faults [70].

Generally, PV system failures are caused by PV module faults, e.g. either defect or aging during their operational lifetime. PV system failures can be permanent or temporary, which can occur at the cell, module, or array level where the electrical properties of some PV modules are significantly deviating considerably from the others. Temporary failures are due to partial shading (cloud and other light-blocking obstacles) or soiling (dust, dirt, and other particles that cover the PV cell) on PV modules. Thus, the dissipation of the power occurring in the damaged area results in overheating, which leads to an elevated operating temperature of the defective PV cells. Figure 7.5 shows the classification of the failure that occurs in the PV plants.

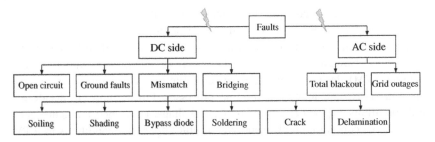

Figure 7.5 Classification of faults in PV systems.

The experimental studies have presented the effect of each individual failure on PV system performance, focusing on the impact of the failures changing system operational characteristics such as on the DC side, the faults that occur in the inverters and MPPT lead to very low output power. Furthermore, the shading of PV modules within a string reduces voltage, current, and power. Consequently, when two electrically parallel PV cells inside a module are shaded, the temperatures of the cells will increase. The soldering faults lead to a decrease in power generation. Similar effects are noted in PV modules that are affected by the rest of the faults. The next section briefly presents the most common types of faults peculiar to PV systems.

7.7.1 Types of PV System Faults

7.7.1.1 Ground Faults
The metallic elements of the PV array are often grounded using earth-grounding cables (EGCs) to protect users from any electric shock. Any unintended interconnection between a current-carrying conductor and an EGC that results in current flow to the ground is referred to as a "ground fault" [71]. Ground faults must be defined and isolated to ensure safety and prevent permanent failure in the PV systems. PV systems are divided into two categories: grounded PV systems and ungrounded PV systems. Usually, a residual current monitoring device (RCMD) and the offline measurements of DC insulation resistance such as insulation monitoring devices (IMDs) are employed to detect ground faults [72].

7.7.1.2 Line-Line Faults
A line-line fault is an unintentional short circuit between two points in a PV array that result in differing voltage potential. In a PV system, two types of line-line faults commonly occur: intra-string and cross-string faults. These types of faults are very complicated to detect and distinguish compared to other faults and are frequently classified as short-circuit faults in grounded PV systems. PV system protection devices may fail to recognize line-line faults, particularly in real-time low

irradiance conditions, the level of *MM* is below 20%, and when the MPPT gets defected [72].

7.7.1.3 Arc Faults

Arc faults in PV systems are a phenomenon caused by the frequent high-power discharge of electricity through an air gap, which cause serious fire threats for combustible materials and safety hazards. An arc fault protection device must be able to detect the irregular types very fast to isolate the fault before it causes a fire. Arc faults are difficult to detect due to their randomness and intermittent nature. In a PV system, two types of arc faults may occur: series arc faults and parallel arc faults. Contrary to the other faults, the arc faults slightly impact the electrical characteristics of a PV module [72].

7.7.1.4 Shading Faults

Shading faults occur when obstacles such as trees, surrounding buildings, and clouds create shadows on PV modules. The shading in the PV module can be either homogeneous, with balanced decreased irradiation across the panels, or nonhomogeneous, with unbalanced reduced irradiation across the panels [72]. In addition to the reduction in output power, the shading leads to an accelerating aging process of the PV panels.

7.7.2 Failure Detection and Classification in PV Systems

Different diagnostic approaches are applied to diagnose the faults in grid-connected PV systems using failure detection techniques. Moreover, failure classification techniques include various methods utilized to detect the type of failure in the PV system. The classification of the failure routine process is the most important because it allows quantifying the factors that cause different failure mechanisms quickly. The power losses due to faults could be reduced by alerting operators to the failure parts and operating to take corrective action. Generally, the failure diagnosis approaches are based on signature analysis, electrical numerical methods, and advanced statistical approaches.

7.7.2.1 Signature Analysis Approaches

Signature analysis approaches are based on the analysis of electrical parameters to identify and classify failures in PV systems. These approaches required direct electrical, solar irradiation, meteorological measurements, $I-V$ characteristic curves, signal generators, circuits, and simulation models.

The electrical characteristics information of the PV modules are acquired by the $I-V$ curves, where valuable information about the quality of the power in the PV modules can provide. These characteristics can provide detailed information for diagnosing fault conditions. Following the authors in [73]used a simple equation, which corresponds to the normalized error by the comparison between the $I-V$ curve in normal operation and the $I-V$ curve in shading conditions. This study has shown that it is possible to detect and identify the avalanche effect of the shaded PV cell through the presented analysis of $I-V$ curves. In [74] investigated a novel fault diagnostic method for a PV system according to three steps. The optimal fault characteristics are analyzed using the $I-V$ curves. Then, the trust-region reflective (TRR) algorithm is proposed to calibrate the fault characteristics under the STC. In the last step, a multi-class adaptive boosting (AdaBoost) algorithm is proposed to optimize the fault diagnostic model. In [75] proposed a very simple fault detection and classification (FDC) method for PV module shading faults using real electrical measurements $I-V$. At first, the features for different experimental tests under healthy and shading conditions are obtained to create the database. Then, the actual data of $I-V$ are analyzed using a principal component analysis (PCA). The accuracy of the data classification is evaluated using the confusion matrix as a metric of class separability.

Furthermore, in [76], a method for online diagnosis of PV module crack through $I-V$ curve is proposed. In order to extract the fault characteristics of the cracked PV modules, the authors investigated and collected the data from different cracked PV modules. The faults are detected by the derivative approach, and the convex function is detected to determine the crack fault. In [77], the impedance characteristics of the PV modules have been measured as Nyquist plots. The $I-V$ curves compare the characteristics among the failure modes and demonstrate the applicability of impedance spectroscopy as a potential PV diagnosis tool. The equivalent circuit parameters of the impedance characteristics have been compared to obtain failure and degradation of the PV modules. In [78] has proposed the use of two variables for distinguishing between normal operation and fault conditions in an experimental PV array. Based on these two variables, a fault detection algorithm was proposed. This algorithm is proven to be successful in distinguishing between three possible scenarios: normal operation, partial shading, and permanent faults. The proposed algorithm can be operated accurately under fast-moving cloud conditions and does not require any training data from different conditions. However, this method is not efficient in large-scale PV systems, where the solar radiation on the different modules is varying.

In this sense, the author in [79] presented an automatic method based on power loss analysis for detecting faults on both the DC and AC sides of a PV system. This technique was graphically developed and could correctly recognize and obtain many types of failures, such as faulty modules, string or inverter, false alarm, and a set of other problems, such as partial shading, PV aging, and MPPT error. In [80] proposed an approach to detect PV plant faults through the generation of fault indicator signals called "residuals" for each string and the comparison of residuals with a threshold value. Furthermore, a regression-based approach is proposed to estimate fault location as a function of fault current and irradiance level measurements. The proposed approach is demonstrated by specifically focusing on intra-string line-line faults. Various line-line fault case studies are verified through simulations and validated on an experimental setup in a solar PV plant. In [81], a novel fault diagnosis methodology in a PV system is presented that can efficiently identify and locate the open-circuit and short-circuit faults. The developed diagnosis approach is based on a metaheuristic algorithm that optimizes the string current considering the effect of nonuniform irradiance and the temperature of the PV modules. In [82], the cable fault characteristics in DC microgrids are analyzed in detail to find and diagnose the pole-to-ground faults and the pole-to-pole faults. The proposed method of diagnostic for fault is divided into several stages. The fault circuits and main equations have been presented in each stage by using a simulation model based on the power systems computer aided design (PSCAD). In [83], the fault detection and isolation (FDI) algorithm was developed to detect the occurrence of the fault in line current sensors, grid voltage sensors, and DC-line voltage based on residual generation defined by the difference between the measurement data and the data created. The model reference adaptive system (MRAS) proposed to reconstitute the information of three-phase line currents and DC-link voltage sensors and the reconstituted information is injected back into the FDI algorithm.

7.7.2.2 Numerical Methods

Most of the numerical methods used in failure detection and diagnosis adopt artificial intelligence (AI) models, such as MLs, SVMs, wavelets, neural network, fuzzy logic, decision tree, long- and short-term memory network, linear regression, etc. The dataset obtained from an experimentally accurate PV system using data acquisition and measurement tools will be split into training and test datasets and used to create the models and the algorithms trained to learn the relationships between the input and output parameters of PV systems. These techniques should be able to detect, identify, and locate the position of the fault [70].

The previous works include many detection tools based on AI models to identify failures in the PV system. In [84], the authors proposed string-level fault detection and diagnosis technique for PV systems based on the k-nearest neighbors (KNN) method that can detect and classify open-circuit faults, line-line faults, partial shading with and without bypass diode faults, and partial shading with inverted bypass diode faults in real-time. Comprehensive modeling of PV systems based on experimental data required data from the manufacturer's datasheet reported of the PV panel under STC and normal operating cell temperature (NOCT). The proposed model has been evaluated using temperature-dependent variables that impact the electrical characteristics such as junction thermal voltage, diode quality factor, and series resistance. In [85] proposed a fault detection algorithm based on multi-resolution signal decomposition (MSD) for feature extraction, and two-stage support vector machine (SVM) classifiers for decision making. This detection method only requires data on the total voltage and current from a PV array and a limited amount of labeled data for training the SVM. In this study, both simulation and experimental case studies verify the accuracy of the proposed method. The proposed model is low-cost and does not require numerous costly sensors where only the measurements of the overall voltage and current of a PV array are used.

Dhibi et al. [86] proposed enhanced fault detection and diagnosis model for a grid-connected PV system using random forest (RF) technique based on data reduction structure. The main objective of this work is to minimize the large operation data collected in grid-connected PV systems using Euclidean distance without degrading the prediction model's performance. In addition, the reduced data is used to create the kernel PCA (KPCA) model for feature extraction and selection purposes. The authors applied the RF classifier to the extracted and selected features to address the fault detection and diagnosis problem. In [87], a novel PV fault diagnosis method can automatically extract the features from raw data for PV module fault classification, which feeds into a soft-max regression classifier model for fault diagnosis. The proposed method has proven extremely effective for high fault diagnosis accuracies on both noisy and noiseless data compared to other techniques such as SVM, ANN, and probabilistic neural networks (PNNs). In [88] presented a statistical algorithm for classification of fault causes on power transmission lines. The proposed algorithm is based upon the root mean square (RMS) current duration, voltage dip, and discrete wavelet transform (DWT) measured at the sending end of a line and the decision tree method, a commonly accessible measurable method. Fault duration of RMS current signal, voltage dip, and DWT gives concealed data of a fault signature as a contribution to decision tree calculation which is utilized to classify various fault causes.

The authors in [89] presented a comprehensive framework of novel data acquisition. A Bayesian belief network (BBN)-based fault detection and diagnostic system has been developed, which analyzes the obtained data to detect faults and the possible causes of the existing faults. The NASA Exploration Missions and Navigation (FINEMAN) system has been developed and integrated as a package that allows for linking the interface of the DAS, processing the measurement values, fault injector for failure testing, and presenting the BBN interface engine. In [90] proposes a new fault detection algorithm for PV systems based on artificial neural networks (ANN) and fuzzy logic system interface. There are a few instances of machine learning techniques that have been applied in fault detection algorithms in PV systems. Meanwhile, this work proposed a system capable to detect possible faults in PV systems using the radial basis function (RBF) ANN. Improved wavelet neural networks, wavelet neural networks, and backpropagation neural networks were used to improve the model provided in [91] for the defect diagnosis of the PV system. The Gaussian function is used as an activation function, an additional momentum mechanism, and an adaptive learning rate method in the training technique. The final results show that the proposed technique can efficiently diagnose the PV system problem with high-performance accuracy, convergence time, and stability under critical conditions. Additionally, a supervised learning decision tree model was proposed in [92] for detecting different diagnostic types of faults such as inverter failure, partial shading, and bypass diode failure. The developed algorithm accurately detected the failures, and the trained models showed high accuracy when classifying each type of failure used for benchmarking during the testing phase. The results show 98.7% for the inverter failure, 95.3% for bypass diode fault, and 96.6% for partial shading.

7.7.2.3 Statistical Analysis

Over the years, statistical fault detection methods in PV systems have been proposed and developed to monitor PV plant performance by accurately detecting failures on the DC side of the system's power. The most common statistical analysis methods use thresholds for each monitored characteristic of the PV and compare observed values to threshold limits (lower and upper average) to determine whether the statuses are normal operation or fault conditions.

A statistical fault detection approach in PV plants was proposed [93], which allows for defining suitable thresholds on the AC and DC power in order to indicate the fault that occurs. The proposed models are developed and implemented in a 1-MW power plant using a SCADA system to interface and operate the data on a Web page. To begin with, the daily adjusted performance ratio of the PV system was evaluated. Then, multiple methods for estimating the PV plant's AC and DC power were tested in an outdoor system. The study in [6] aims to create a novel fault detection and diagnostic technique for PV systems on the DC side. The authors proposed a statistical approach that exploits the features of the one-diode model and those of the univariate and multivariate exponentially weighted moving average (EWMA) charts to improve fault detection. The residual's capture of the electrical characteristics is used to find the difference between the measurements and the predictions of the MPP of the current, voltage, and power of the one-diode model and utilize them as fault factors. In [94] proposed a statistical approach for detecting the faults on the DC side in PV systems. The proposed approach combines the sensitivity of the multivariate cumulative SUM algorithm (MCUSUM) for identifying incipient changes with the flexibility and accuracy of a model simulation based on a single diode model. The method has been implemented in a real-time application with an actual 20 MWp grid-connected PV system. In [95], an effective monitoring framework using the main component analysis approach and multivariate monitoring schemes is designed to control the PV systems. The residuals for anomaly detection are generated using the PCA approach. These residuals are examined for defect detection by computing the monitoring schemes. The main objective of the proposed methodology is to bridge the gaps from the traditional schemes by designing assumption-free PCA-based schemes. Actual measurement data was collected from a 9.54 kW grid-connected PV system, and different cases of study have been concerned to evaluate the fault detection capabilities of the proposed schemes.

In this sense, the study in [96] proposed a novel design using the statistical analysis approach to address the unpredictable arcing phenomenon and to derive an arc fault detector, which experimental results have investigated. The experimental results show that the prototype arc fault detector can achieve 99.99% of fault detection accuracy in given arcing conditions and 0.027% of false detection in given nonarcing conditions while avoiding malfunctions from the switching noise of the DC side. The work in [97] deals with the nonsymmetric faults in a hybrid system of PV-wind by analyzing the system parameters using these techniques and an algorithm and considering the current characteristics, DWT's multi-resolution analysis, and Stockwell transforms statistical coefficient analysis. Depending on the results obtained from the proposed methodology based on the statistical analysis approach, the optimal solution was detected. An algorithm was proposed for fault prediction in PV-wind hybrid systems. In [98], the proposed statistical fault detection method is based on different outlier detection levels, which are 3-Sigma, Hampel identifier, and Boxplot rule. According to the results obtained from the experimental work, the outlier detection methods can implement to identify line-to-line, open-circuit, partial shading, and degradation problems. Despite this, the highest performing outlier detection method for short-circuit faults was shown to generate false alarms during normal conditions, both after and before partial shading conditions.

7.7.3 Monitoring System for Failure Detection in PV Systems

An automatic fault diagnosis technique in a monitoring system plays a critical role in detecting energy production causes. With automatic fault diagnosis techniques, the monitoring system plays a critical key in ensuring the reliable and detecting causes affecting energy production [99]. The PV-monitoring systems are classified as ground-based or space-based monitoring systems. Monitoring in the PV system enhances the performance and provides rapid healing of the power system by detecting the possible energy losses from changes before the faults occur, which can affect energy production [100].

To create an optimum system performance, a suitable monitoring system is required to assess the PV system evaluation, whether it operates according to anticipation, analyzes contractual restrictions, and performs immediate healing if the failure occurs [101]. The aim of integrating the PV system's monitoring system is to have high-quality simultaneous measurement values of different environmental conditions such as the solar irradiance, temperature, and instant electrical parameters data of the PV system. The sensors and weather measurement stations are integrated onsite to measure the temperature and solar radiation, the speed and direction of the wind, and additionally electrical sensors for obtaining the PV operational measurements [70].

The signal conditioning unit is considered one of the important components in the PV-monitoring system. This unit performs signal amplification and sensor measurement filtering for further data measurement processing. This unit consists of a microcontroller that transfers the outputs of the signal conditioning unit to a computer in real-time using a specific protocol. The computer can analyze, display, and store the collected data and send commands to the system control unit for future actions based on internal analysis and external commands from users [99].

7.8 Conclusion

PV systems are increasingly gaining worldwide attention due to the requirement to serve growing electricity demands combined with the desired green energy solutions for power production development. PV systems frequently have various factors that significantly degrade their performances and efficiency such as aging and degradation of the PV cell, temperature, solar irradiance, the capacitor ESR, and the faults that occur in the PV system. Faults such as short circuit, open circuit, shading, and arc faults in PV systems can reduce solar energy production, leading to economic losses. Fault detection and diagnosis in a large-scale PV system are considered a major issue as more and more PV plants with increasingly large capacities continue to come into existence. In this paper, a comprehensive review of existing factors that reduce the PV system performance reported in the literature has been presented in terms of degradation and aging, weather impacts, and faults as well as the tools being used to enhance the performance of the PV modules. Moreover, this paper describes the performance monitoring requirements for detecting failures in monitored PV systems and the various data analytic techniques for detecting and classifying failures in PV systems. As described in this paper, depending on the measuring equipment, the quality of the data, and the technique used for failure diagnosis, each technique produces different results with varying classification accuracy. According to previous studies, the developed data analysis methods are the most promising diagnostic tools due to accurately detecting faults as they occur using existing equipment/monitoring systems and then determining the probability of specific failures. In addition, the existing data analytics-based failure diagnosis approaches are divided into three categories: statistical, numerical, and electrical signature analysis. Future works are required to determine the most effective and appropriate fault diagnosis solution for PV systems and to benchmark the performance of the different failure diagnosis methodologies under different conditions and fault scenarios.

References

1 Zahraoui, Y., Alhamrouni, I., Mekhilef, S. et al. Energy management system in microgrids : a comprehensive review. *Sustainability* 13 (19): 10492.

2 Madeti, S.R. and Singh, S.N. (2017). A comprehensive study on different types of faults and detection techniques for solar photovoltaic system. *Sol. Energy* 158 (June): 161–185. https://doi.org/10.1016/j.solener.2017.08.069.

3 Dhoke, A., Sharma, R., and Saha, T.K. (2020). A technique for fault detection, identification and location in solar photovoltaic systems. *Sol. Energy* 206 (February): 864–874. https://doi.org/10.1016/j.solener.2020.06.019.

4 REN21 (2017). *Renewables 2017 Global Status Report.* REN21 Secretariat: Paris http://www.ren21.net/gsr-2017.

5 Maaløe, L., Winther, O., Spataru, S., and Sera, D. (2020). Conditional monitoring in photovoltaic systems by semi-supervised machine learning. *Energies* 13 (3): 1–14. https://doi.org/10.3390/en13030584.

6 Harrou, F., Sun, Y., Taghezouit, B. et al. (2018). Reliable fault detection and diagnosis of photovoltaic systems based on statistical monitoring approaches. *Renew. Energy* 116: 22–37. https://doi.org/10.1016/j.renene.2017.09.048.

7 Zahraoui, Y., Basir Khan, M.R., AlHamrouni, I. et al. (2021). Current status, scenario, and prospective of renewable energy in Algeria: a review. *Energies* 14 (9): 2354. https://doi.org/10.3390/en14092354.

8 Hu, Y., Zhang, J., Cao, W. et al. (2015). Online two-section PV array fault diagnosis with optimized voltage sensor locations. *IEEE Trans. Ind. Electron.* 62 (11): 7237–7246.

9 Fazai, R., Abodayeh, K., Mansouri, M. et al. (2019). Machine learning-based statistical testing hypothesis for fault detection in photovoltaic systems. *Sol. Energy* 190 (July): 405–413.

10 Garoudja, E., Chouder, A., Kara, K., and Silvestre, S. (2017). An enhanced machine learning based approach for failures detection and diagnosis of PV systems. *Energy Convers. Manag.* 151 (September): 496–513. https://doi.org/10.1016/j.enconman.2017.09.019.

11 Huerta Herraiz, Á., Pliego Marugán, A., and García Márquez, F.P. (2020). Photovoltaic plant condition monitoring using thermal images analysis by convolutional neural network-based structure. *Renew. Energy* 153: 334–348. https://doi.org/10.1016/j.renene.2020.01.148.

12 Samara, S. and Natsheh, E. (2019). Intelligent real-time photovoltaic panel monitoring system using artificial neural networks. *IEEE Access* 7: 50287–50299. https://doi.org/10.1109/ACCESS.2019.2911250.

13 Procopiou, A.T. and Ochoa, L.F. (2017). Voltage control in PV-rich LV networks without remote monitoring. *IEEE Trans. Power Syst.* 32 (2): 1224–1236. https://doi.org/10.1109/TPWRS.2016.2591063.

14 Procopiou, A.T., Long, C., and Ochoa, L.F. (2015). On the effects of monitoring and control settings on voltage control in PV-rich LV networks. *IEEE Power Energy Soc. Gen. Meet.* 2015: 2014–2015. https://doi.org/10.1109/PESGM.2015.7285791.

15 Buerhop, C., Schlegel, D., Niess, M. et al. (2012). Reliability of IR-imaging of PV-plants under operating conditions. *Sol. Energy Mater. Sol. Cells* 107: 154–164. https://doi.org/10.1016/j.solmat.2012.07.011.

16 Zou, Z., Hu, Y., Gao, B. et al. (2014). Temperature recovery from degenerated infrared image based on the principle for temperature measurement using infrared sensor. *J. Appl. Phys.* 115 (4): https://doi.org/10.1063/1.4863783.

17 Takashima, T., Yamaguchi, J., Otani, K. et al. (2009). Experimental studies of fault location in PV module strings. *Sol. Energy Mater. Sol. Cells* 93 (6–7): 1079–1082. https://doi.org/10.1016/j.solmat.2008.11.060.

18 Takashima, T., Yamaguchi, J., and Ishida, M. (2008). Disconnection detection using Earth capacitance measurement in photovoltaic module string. *Prog. Photovoltaics Res. Appl.* 20 (16): 669–677. https://doi.org/10.1002/pip.860.

19 Garoudja, E., Kara, K., Chouder, A., and Silvestre, S. (2015). Parameters extraction of photovoltaic module for long-term prediction using artifical bee colony optimization, *3rd International Conference on Control, Engineering and Information Technology, CEIT 2015.* https://doi.org/10.1109/CEIT.2015.7232993.

20 Hu, Y., Cao, W., Wu, J. et al. (2014). Thermography-based virtual MPPT scheme for improving PV energy efficiency under partial shading conditions. *IEEE Trans. Power Electron.* 29 (11): 5667–5672. https://doi.org/10.1109/TPEL.2014.2325062.

21 Kumar, B.P., Ilango, G.S., Reddy, M.J.B., and Chilakapati, N. (2018). Online fault detection and diagnosis in photovoltaic systems using wavelet packets. *IEEE J. Photovoltaics* 8 (1): 257–265. https://doi.org/10.1109/JPHOTOV.2017.2770159.

22 Murtaza, A.F., Bilal, M., Ahmad, R., and Sher, H.A. (2020). A circuit analysis-based fault finding algorithm for photovoltaic array under L-L/L-G faults. *IEEE J. Emerg. Sel. Top. Power Electron.* 8 (3): 3067–3076. https://doi.org/10.1109/JESTPE.2019.2904656.

23 Chouder, A. and Silvestre, S. (2010). Automatic supervision and fault detection of PV systems based on power losses analysis. *Energy Convers. Manag.* 51 (10): 1929–1937. https://doi.org/10.1016/j.enconman.2010.02.025.

24 Younes, Z., Alhamrouni, I., Mekhilef, S., and Reyasudin, M. (2021). A memory-based gravitational search algorithm for solving economic dispatch problem in micro-grid. *Ain Shams Eng. J.* 12 (2): 1985–1994.

25 National Renewable Energy Laboratory, (NREL) (2017). Researchers at NREL Find Fewer Failures of PV Panels and Different Degradation Modes in Systems Installed after 2000 https://www.nrel.gov/news/program/2017/failures-pv-panels-degradation.html (accessed 10 April 2017).

26 Jordan, D.C., Silverman, T.J., Wohlgemuth, J.H. et al. (2017). Photovoltaic failure and degradation modes Dirk. *Prog. photovoltaics Res. Appl.* 25 (4): 318–326.

27 Mansour, D.E., Barretta, C., Pitta Bauermann, L. et al. (2020). Effect of backsheet properties on PV encapsulant degradation during combined accelerated aging tests. *Sustainability* 12 (12): 5208.

28 Hocine, L. and Mounia Samira, K. (2019). Optimal PV panel's end-life assessment based on the supervision of their own aging evolution and waste management forecasting. *Sol. Energy* 191 (April): 227–234. https://doi.org/10.1016/j.solener.2019.08.058.

29 Agresti, A., Berionni Berna, B., Pescetelli, S. et al. (2020). Copper-based Corrole as thermally stable hole transporting material for perovskite photovoltaics. *Adv. Funct. Mater.* 30 (46): 2003790.

30 Chundi, N., Kesavan, G., Ramasamy, E. et al. (2021). Ambient condition curable, highly weather stable anti-soiling coating for photovoltaic application. *Sol. Energy Mater. Sol. Cells* 230 (November): 111203. https://doi.org/10.1016/j.solmat.2021.111203.

31 Repins, I.L., Kersten, F., Hallam, B. et al. (2020). Stabilization of light-induced effects in Si modules for IEC 61215 design qualification. *Sol. Energy* 208: 894–904. https://doi.org/10.1016/j.solener.2020.08.025.

32 Ottersböck, B., Oreski, G., and Pinter, G. (2017). Comparison of different microclimate effects on the aging behavior of encapsulation materials used in photovoltaic modules. *Polym. Degrad. Stab.* 138: 182–191. https://doi.org/10.1016/j.polymdegradstab.2017.03.010.

33 Husain, A.A.F., Hasan, W.Z.W., Shafie, S. et al. (2018). A review of transparent solar photovoltaic technologies. *Renew. Sustain. Energy Rev.* 94 (June): 779–791. https://doi.org/10.1016/j.rser.2018.06.031.

34 de Oliveira, M.C.C., Diniz Cardoso, A.S.A., Viana, M.M., and de Lins, V., F.C. (2018). The causes and effects of degradation of encapsulant ethylene vinyl acetate copolymer (EVA) in crystalline silicon photovoltaic modules: a review. *Renew. Sustain. Energy Rev.* 81 (March): 2299–2317. https://doi.org/10.1016/j.rser.2017.06.039.

35 Adothu, B., Bhatt, P., Zele, S. et al. (2020). Investigation of newly developed thermoplastic polyolefin encapsulant principle properties for the c-Si PV module application. *Mater. Chem. Phys.* 243 (August): 122660. https://doi.org/10.1016/j.matchemphys.2020.122660.

36 Omazic, A., Oreski, G., Halwachs, M. et al. (2019). Relation between degradation of polymeric components in crystalline silicon PV module and climatic conditions: a literature review. *Sol. Energy Mater. Sol. Cells* 192 (September): 123–133.

37 Julien, S.E., Kempe, M.D., Eafanti, J.J. et al. (2020). Characterizing photovoltaic backsheet adhesion degradation using the wedge and single cantilever beam tests, Part II: accelerated tests. *Sol. Energy Mater. Sol. Cells* 215: 110669.

38 Lin, C.C., Lyu, Y., Jacobs, D.S. et al. (2019). A novel test method for quantifying cracking propensity of photovoltaic backsheets after ultraviolet exposure. *Prog. Photovoltaics Res. Appl.* 27 (1): 44–54.

39 Lyu, Y., Kim, J.H., Fairbrother, A., and Gu, X. (2018). Degradation and cracking behavior of polyamide-based backsheet subjected to sequential fragmentation test. *IEEE J. Photovoltaics* 8 (6): 1748–1753. https://doi.org/10.1109/JPHOTOV.2018.2863789.

40 Owen-Bellini, M., Miller, D.C., Schelhas, L.T. et al. (2019). Correlation of advanced accelerated stress testing with polyamide-based photovoltaic backsheet field-failures. In: *2019 IEEE 46th Photovoltaic Specialists Conference (PVSC)*, 1995–1999. IEEE.

41 Ulicna, S., Sinha, A., Springer, M. et al. (2021). Failure analysis of a new polyamide-based fluoropolymer-free backsheet after combined-accelerated stress testing. *IEEE J. Photovoltaics* 11 (5): 1197–1205.

42 Kawajiri, K., Oozeki, T., and Genchi, Y. (2011). Effect of temperature on PV potential in the world. *Environ. Sci. Technol.* 45 (20): 9030–9035. https://doi.org/10.1021/es200635x.

43 Kaplanis, S., Kaplani, E., and Kaldellis, J.K. (2022). PV temperature and performance prediction in free-standing, BIPV and BAPV incorporating the effect of temperature and inclination on the heat transfer coefficients and the impact of wind, efficiency and ageing. *Renew. Energy* 181: 235–249. https://doi.org/10.1016/j.renene.2021.08.124.

44 Dirnberger, D., Blackburn, G., Müller, B., and Reise, C. (2015). On the impact of solar spectral irradiance on the yield of different PV technologies. *Sol. Energy Mater. Sol. Cells* 132: 431–442. https://doi.org/10.1016/j.solmat.2014.09.034.

45 Leitão, D., Torres, J.P.N., and Fernandes, J.F.P. (2020). Spectral irradiance influence on solar cells efficiency. *Energies* 13 (19): https://doi.org/10.3390/en13195017.

46 Wang, F., Xuan, Z., Zhen, Z. et al. (2020). A minutely solar irradiance forecasting method based on real-time sky image-irradiance mapping model. *Energy Convers. Manag.* 220: 113075.

47 Lindsay, N., Libois, Q., Badosa, J. et al. (2020). Errors in PV power modelling due to the lack of spectral and angular details of solar irradiance inputs. *Sol. Energy* 197 (December): 266–278. https://doi.org/10.1016/j.solener.2019.12.042.

48 Eke, R., Betts, T.R., and Gottschalg, R. (2017). Spectral irradiance effects on the outdoor performance of photovoltaic modules. *Renew. Sustain. Energy Rev.* 69 (December): 429–434. https://doi.org/10.1016/j.rser.2016.10.062.

49 Zahraoui, Y., Alhamrouni, I., Mekhilef, S. et al. (2021). A novel approach for sizing battery storage system for enhancing resilience ability of a microgrid. *Int. Trans. Electr. Energy Syst.* June: 1–19. https://doi.org/10.1002/2050-7038.13142.

50 Agarwal, N., Arya, A., Ahmad, M.W., and Anand, S. (2016). Lifetime monitoring of electrolytic capacitor to maximize earnings from grid-feeding PV system. *IEEE Trans. Ind. Electron.* 63 (11): 7049–7058. https://doi.org/10.1109/TIE.2016.2586020.

51 de Paula Silva, R., Da Silveira, D.B., De Barros, R.C. et al. (2021). Third-harmonic current injection for wear-out reduction in single-phase PV inverters. *IEEE Trans. Energy Convers.* 37 (1): 120–131.

52 Matsumori, H., Urata, K., Shimizu, T. et al. (2018). Capacitor loss analysis method for power electronics converters. *Microelectron. Reliab.* 88–90 (July): 443–446. https://doi.org/10.1016/j.microrel.2018.07.049.

53 Lee, K.W., Kim, M., Yoon, J. et al. (2008). Condition monitoring of DC-link electrolytic capacitors in adjustable-speed drives. *IEEE Trans. Industry Appl.* 44 (5): 1606–1613.

54 Vogelsberger, M.A., Wiesinger, T., and Ertl, H. (2011). Life-cycle monitoring and voltage-managing unit for DC-link electrolytic capacitors in PWM converters. *IEEE Trans. Power Electron.* 26 (2): 493–503. https://doi.org/10.1109/TPEL.2010.2059713.

55 Agarwal, N., Ahmad, M.W., and Anand, S. (2018). Quasi-online technique for health monitoring of capacitor in single-phase solar inverter. *IEEE Trans. Power Electron.* 33 (6): 5283–5291. https://doi.org/10.1109/TPEL.2017.2736162.

56 Zhang, X., Zhang, Z., Bao, H. et al. (2021). Stability effect of control weight on multiloop COT-controlled Buck converter with PI compensator and small output capacitor ESR. *IEEE J. Emerg. Sel. Top. Power Electron.* 9 (4): 4658–4667. https://doi.org/10.1109/JESTPE.2020.3014523.

57 Plazas-Rosas, R.A., Orozco-Gutierrez, M.L., Spagnuolo, G. et al. (2021). Dc-link capacitor diagnosis in a single-phase grid-connected pv system. *Energies* 14 (20): 1–23. https://doi.org/10.3390/en14206754.

58 Arya, A., Ahmad, M.W., Agarwal, N., and Anand, S. (2017). Capacitor impedance estimation utilizing dc-link voltage oscillations in single phase inverter. *IET Power Electron.* 10 (9): 1046–1053. https://doi.org/10.1049/iet-pel .2016.0603.

59 Younes, Z., Alhamrouni, I., Mekhilef, S., and Khan, M.R.B. (2021). Blockchain applications and challenges in smart grid. In: *2021 IEEE Conference on Energy Conversion (CENCON)*, 208–213. IEEE.

60 Rezk, H., Tyukhov, I., Al-Dhaifallah, M., and Tikhonov, A. (2017). Performance of data acquisition system for monitoring PV system parameters. *Meas. J. Int. Meas. Confed.* 104: 204–211. https://doi.org/10.1016/j .measurement.2017.02.050.

61 Gradwohl, C., Dimitrievska, V., Pittino, F. et al. (2021). A combined approach for model-based pv power plant failure detection and diagnostic. *Energies* 14 (5): 1261.

62 Forero, N., Hernández, J., and Gordillo, G. (2006). Development of a monitoring system for a PV solar plant. *Energy Convers. Manag.* 47 (15–16): 2329–2336. https://doi.org/10.1016/j.enconman.2005.11.012.

63 Meyer, E.L., Apeh, O.O., and Overen, O.K. (2020). Electrical and meteorological data acquisition system of a commercial and domestic microgrid for monitoring pv parameters. *Appl. Sci.* 10 (24): 1–18. https://doi.org/10.3390/ app10249092.

64 Heidari Kapourchali, M., Sepehry, M., and Aravinthan, V. (2019). Multivariate spatio-temporal solar generation forecasting: a unified approach to deal with communication failure and invisible sites. *IEEE Syst. J.* 13 (2): 1804–1812. https://doi.org/10.1109/JSYST.2018.2869825.

65 Livera, A., Florides, M., Theristis, M. et al. (2018). Failure diagnosis of short- and open-circuit fault conditions in PV systems. In: *2018 IEEE 7th World Conference on Photovoltaic Energy Conversion (WCPEC) (A Joint Conference of 45th IEEE PVSC, 28th PVSEC & 34th EU PVSEC)*, 0739–0744. IEEE.

66 Olivencia Polo, F.A., Ferrero Bermejo, J., Gómez Fernández, J.F., and Crespo Márquez, A. (2015). Failure mode prediction and energy forecasting of PV plants to assist dynamic maintenance tasks by ANN based models. *Renew. Energy* 81: 227–238. https://doi.org/10.1016/j.renene.2015.03.023.

67 Dhimish, M., Holmes, V., and Dales, M. (2016). Grid-connected PV virtual instrument system (GCPV-VIS) for detecting photovoltaic failure. In:

2016 4th International Symposium on Environmental Friendly Energies and Applications (EFEA), 1–6. IEEE.

68 Tadj, M., Benmouiza, K., Cheknane, A., and Silvestre, S. (2014). Improving the performance of PV systems by faults detection using GISTEL approach. *Energy Convers. Manag.* 80: 298–304. https://doi.org/10.1016/j.enconman.2014.01.030.

69 Jamil, M., Sharma, S.K., and Singh, R. (2015). Fault detection and classification in electrical power transmission system using artificial neural network. *Springerplus* 4 (1): https://doi.org/10.1186/s40064-015-1080-x.

70 Livera, A., Theristis, M., Makrides, G., and Georghiou, G.E. (2019). Recent advances in failure diagnosis techniques based on performance data analysis for grid-connected photovoltaic systems. *Renew. Energy* 133: 126–143. https://doi.org/10.1016/j.renene.2018.09.101.

71 Abubakar, A., Almeida, C.F.M., and Gemignani, M. (2021). Review of artificial intelligence-based failure detection and diagnosis methods for solar photovoltaic systems. *Machines* 9 (12): 328. https://doi.org/10.3390/machines9120328.

72 Pillai, D.S. and Rajasekar, N. (2018). A comprehensive review on protection challenges and fault diagnosis in PV systems. *Renew. Sustain. Energy Rev.* 91 (July): 18–40. https://doi.org/10.1016/j.rser.2018.03.082.

73 Bressan, M., El Basri, Y., Galeano, A.G., and Alonso, C. (2016). A shadow fault detection method based on the standard error analysis of I-V curves. *Renew. Energy* 99: 1181–1190. https://doi.org/10.1016/j.renene.2016.08.028.

74 Huang, J.M., Wai, R.J., and Gao, W. (2019). Newly-designed fault diagnostic method for solar photovoltaic generation system based on IV-curve measurement. *IEEE Access* 7: 70919–70932. https://doi.org/10.1109/ACCESS.2019.2919337.

75 Fadhel, S., Delpha, C., Diallo, D. et al. (2019). PV shading fault detection and classification based on I-V curve using principal component analysis: application to isolated PV system. *Sol. Energy* 179: 1–10.

76 Ma, M., Zhang, Z., Xie, Z. et al. (2020). Fault diagnosis of cracks in crystalline silicon photovoltaic modules through I-V curve. *Microelectron. Reliab.* 114 (October): 113848. https://doi.org/10.1016/j.microrel.2020.113848.

77 Katayama, N., Osawa, S., Matsumoto, S. et al. (2019). Degradation and fault diagnosis of photovoltaic cells using impedance spectroscopy. *Sol. Energy Mater. Sol. Cells* 194 (September): 130–136. https://doi.org/10.1016/j.solmat.2019.01.040.

78 Hariharan, R., Chakkarapani, M., Saravana Ilango, G., and Nagamani, C. (2016). A method to detect photovoltaic array faults and partial shading in PV systems. *IEEE J. Photovoltaics* 6 (5): 1278–1285. https://doi.org/10.1109/JPHOTOV.2016.2581478.

79 Chine, W., Mellit, A., Pavan, A.M., and Kalogirou, S.A. (2014). Fault detection method for grid-connected photovoltaic plants. *Renew. Energy* 66: 99–110. https://doi.org/10.1016/j.renene.2013.11.073.

80 Dhoke, A., Sharma, R., and Saha, T.K. (2019). An approach for fault detection and location in solar PV systems. *Sol. Energy* 194 (November): 197–208. https://doi.org/10.1016/j.solener.2019.10.052.

81 Hazra, A., Das, S., and Basu, M. (2017). An efficient fault diagnosis method for PV systems following string current. *J. Clean. Prod.* 154: 220–232. https://doi.org/10.1016/j.jclepro.2017.03.214.

82 Zhang, Z., Chen, Q., Xie, R., and Sun, K. (2019). The fault analysis of PV cable fault in DC microgrids. *IEEE Trans. Energy Convers.* 34 (1): 486–496. https://doi.org/10.1109/TEC.2018.2876669.

83 Ben Youssef, F. and Sbita, L. (2017). Sensors fault diagnosis and fault tolerant control for grid connected PV system. *Int. J. Hydrogen Energy* 42 (13): 8962–8971. https://doi.org/10.1016/j.ijhydene.2016.11.147.

84 Madeti, S.R. and Singh, S.N. (2018). Modeling of PV system based on experimental data for fault detection using kNN method. *Sol. Energy* 173 (June): 139–151. https://doi.org/10.1016/j.solener.2018.07.038.

85 Yi, Z. and Etemadi, A.H. (2017). Line-to-line fault detection for photovoltaic arrays based on multi-resolution signal decomposition and two-stage support vector machine. *IEEE Trans. Ind. Electron.* 64 (11): https://doi.org/10.1109/TIE.2017.2703681.

86 Dhibi, K., Fezai, R., Mansouri, M. et al. (2020). Reduced kernel random forest technique for fault detection and classification in grid-tied PV systems. *IEEE J. Photovoltaics* 10 (6): 1864–1871.

87 Appiah, A.Y., Zhang, X., Ayawli, B.B.K., and Kyeremeh, F. (2019). Long short-term memory networks based automatic feature extraction for photovoltaic array fault diagnosis. *IEEE Access* 7: 30089–30101. https://doi.org/10.1109/ACCESS.2019.2902949.

88 Asman, S.H., Aziz, N.F.A., Amirulddin, U.A.U., and Kadir, M.Z.A.A. (2021). Decision tree method for fault causes classification based on rms-dwt analysis in 275 kv transmission lines network. *Appl. Sci.* 11 (9): https://doi.org/10.3390/app11094031.

89 Coleman, A. and Zalewski, J. (2011). Intelligent fault detection and diagnostics in solar plants. In: *Proceedings of the 6th IEEE International Conference on Intelligent Data Acquisition and Advanced Computing Systems*, vol. 2, 948–953. IEEE.

90 Dhimish, M., Holmes, V., Mehrdadi, B., and Dales, M. (2018). Comparing Mamdani Sugeno fuzzy logic and RBF ANN network for PV fault detection. *Renew. Energy* 117: 257–274. https://doi.org/10.1016/j.renene.2017.10.066.

91 Li, X., Yang, P., Ni, J., and Zhao, J. (2014). Fault diagnostic method for PV array based on improved wavelet neural network algorithm. In: *Proceeding of the 11th World Congress on Intelligent Control and Automation*, 1171–1175. IEEE.

92 Livera, A., Phinikarides, A., Makrides, G., and Georghiou, G.E. (2017). Advanced failure detection algorithms and performance decision classification for grid-connected PV systems. In: *33rd European Photovoltaic Solar Energy Conference and Exhibition*.

93 Ventura, C. and Tina, G.M. (2015). Development of models for on-line diagnostic and energy assessment analysis of PV power plants: the study case of 1 MW Sicilian PV plant. *Energy Procedia* 83: 248–257. https://doi.org/10.1016/j.egypro.2015.12.179.

94 Harrou, F., Sun, Y., and Saidi, A. (2018). Model-based fault detection algorithm for photovoltaic system monitoring. In: *2017 IEEE Symposium Series on Computational Intelligence (SSCI)*, 1–5. IEEE.

95 Taghezouit, B., Harrou, F., Sun, Y. et al. (2020). Multivariate statistical monitoring of photovoltaic plant operation. *Energy Convers. Manag.* 205 (November): 112317. https://doi.org/10.1016/j.enconman.2019.112317.

96 Seo, G.S., Ha, J.I., Cho, B.H., and Lee, K.C. (2016). Series arc fault detection method based on statistical analysis for dc Microgrids. In: *2016 IEEE Applied Power Electronics Conference and Exposition (APEC)*, 487–492. IEEE.

97 Ray, D.K. and Chattopadhyay, S. (2020). Fault analysis in solar–wind microgrid using multi-resolution analysis and Stockwell transform-based statistical analysis. *IET Sci. Meas. Technol.* 14 (6): 639–650. https://doi.org/10.1049/iet-smt.2019.0279.

98 Zhao, Y., Lehman, B., Ball, R. et al. (2013). Outlier detection rules for fault detection in solar photovoltaic arrays. *Conf. Proc. – IEEE Appl. Power Electron. Conf. Expo. –APEC* 2913–2920. https://doi.org/10.1109/APEC.2013.6520712.

99 Madeti, S.R. and Singh, S.N. (2017). Online fault detection and the economic analysis of grid-connected photovoltaic systems. *Energy* 134: 121–135. https://doi.org/10.1016/j.energy.2017.06.005.

100 Madeti, S.R. and Singh, S.N. (2017). Monitoring system for photovoltaic plants: A review. *Renew. Sustain. Energy Rev.* 67: 1180–1207. https://doi.org/10.1016/j.rser.2016.09.088.

101 Zahraoui, Y. et al. (2021). Self-healing strategy to enhance microgrid resilience during faults occurrence. *Int. Trans. Electr. Energy Syst.* November: 1–16. https://doi.org/10.1002/2050-7038.13232.

8

Fault-Tolerant Converter Design for Photovoltaic System

Azra Malik and Ahteshamul Haque

Advance Power Electronics Research Lab, Department of Electrical Engineering, Jamia Millia Islamia
(A Central University), New Delhi, India

8.1 Introduction

The growing demand for electricity has motivated the adoption of renewable energy in a gradually increasing manner. The government's ambitious targets and environmental and climatic conditions are also propelling more penetration of renewable energy sources in electrical power system [1]. This has led to more acceptance of micro-grids, smart-grids, transfer of more power with fewer losses, etc. The last decade has seen tremendous growth in a grid-tied photovoltaic (PV) system (GTPVS). In these systems, power electronics plays a vital role in fulfilling the high electricity demand. These systems require a power electronics converter, i.e. inverter, which is essentially a DC/AC converter. The inverter is utilized for effective power transfer from the PV side (DC side) to the grid side (AC side) [2]. It is also responsible for providing better power quality and maintaining efficient power control. Further progress in an inverter involves proper inverter topology, identifying voltage and power range, modularity, improved power quality, reliability requirements etc. However, since these power electronics converters utilize switching devices in a dominant manner, they may be affected by degraded reliability [3]. The switching device circuit increases complexity, which leads to an increased likelihood of device failure. During the investigation, it is found that insulated gate bipolar transistors (IGBTs) are the most preferred power-switching devices for power electronic converters. Over the last decade, they have witnessed significant performance improvement in terms of manufacturing, compact structure, better voltage and current handling capability. However, they can still suffer from failures due to thermomechanical and electrical overstress events. These failures may be characterized by two classes- one class consists of failure

Fault Analysis and its Impact on Grid-connected Photovoltaic Systems Performance, First Edition.
Edited by Ahteshamul Haque and Saad Mekhilef.

due to the operation of a device for longer periods typically known as wear-out failures, and the other class consists of failure due to overstress events, also known as catastrophic failures [4]. Several potential causes lead to the wear-out failures including the bond wire heel crack, bond wire lift-off, solder joint failure, substrate crack, etc. The catastrophic failure may occur in IGBTs due to the overstress events [5]. Besides, these failures can be characterized through bond wire melting, latch-up failure, high voltage breakdown, electrostatic discharge (ESD), secondary breakdown, etc.

In general, the reliability of a power converter is mainly characterized by the most vulnerable device in its topology. Particularly, if one critical switching device has experienced a fault, it may eventually lead to the collapse of the whole converter operation instigating disastrous consequences [6]. Studies have also revealed that in practical GTPVS, inverters are the weakest components. In other words, they are most susceptible to fault occurrences. Therefore, devising proper CM techniques and fault-tolerant solutions is of paramount importance to ensure the reliable operation of GTPVS. The discussion regarding fault-tolerant power systems has become a research hotspot in critical applications like hospitals, health care systems, military, aircraft, etc. Some metrics have also been defined to quantify reliability including failure rate, mean time to repair (MTTR), mean time between failures (MTBF) etc. [7]. A hierarchical and advanced health monitoring, diagnostic and fault-tolerant strategy can be implemented for inverters. It would significantly help in improving the reliability and availability of the overall system [8]. The general fault-tolerant design approach for GTPVS applications is presented in Figure 8.1. Considering this approach as a three-level fault-tolerant solution, the device health can be enhanced and the overall device lifetime can be improved. At the first level, online health CM can be utilized for predicting device failures and further measures can be taken to avoid failure events. The second level consists of the fault-diagnostic technique, which helps in fault identification and localization [9]. At the third level, the fault-tolerant operation is desired as a post-fault reconfiguration step. It will help restore optimum operation conditions after the fault has occurred [10]. This comprehensive assortment is presented in Figure 8.2, which shows the hierarchy involving fault prognosis, health monitoring, fault diagnosis, and fault-tolerant strategy implementation.

A fault-tolerant technique for inverters should be capable to handle potential device fault events like open-circuit, short-circuit faults, etc. Other than that, it should be able to fulfil the following criteria:

1) The considered fault-tolerant inverter should have a fewer number of redundant power devices so as to offer a cost-effective solution. Otherwise, the market acceptance of the solution may be prevented.

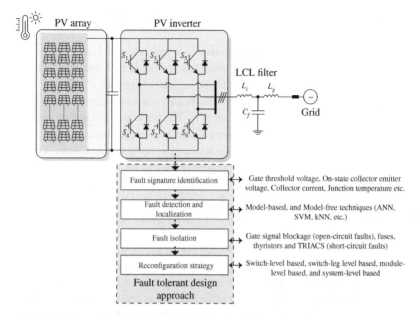

Figure 8.1 Fault-tolerant design approach for GTPVS.

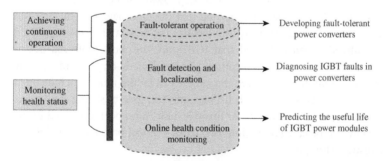

Figure 8.2 A three-step process for ensuring reliable and fault-tolerant operation of GTPVS.

2) The fault-tolerant strategy should help enhance the overall lifetime of the inverter. It should ensure improved performance of the inverter under normal operating conditions.

3) The design criteria for the fault-tolerant inverter should consider both dynamic and steady-state performance in normal as well as post-fault operation of the system.

4) It is desired that the fault-tolerant inverter is able to provide rated currents and voltages throughout the post-fault stage. This feature is of utmost importance to ascertain the optimum performance of GTPVS.

8.2 Fault Signature Identification and Fault Diagnosis

The switching device has some parameters that indicate the fault occurrence in that device. Those parameters may also indicate the health condition of the device. These device parameters are termed "Fault Signatures." These fault signatures are found to be beneficial for the health condition monitoring (CM) of the switching devices in the converters [11]. The basic understanding behind this conclusion is that the output voltage or current characteristics under faulty conditions are different from that obtained under normal conditions. Direct monitoring of the switching device characteristics can be carried out through the allocated embedded sensors. The severity level due to a fault can also be assessed with their help. For instance, bond-wire fault in IGBT devices can be detected with electric resistance measurement [12]. Further, IGBT device parameters like threshold gate voltage ($V_{G(TH)}$), on-state resistance ($R_{(ON)}$), on-state collector-emitter voltage ($V_{CE(ON)}$), collector current (I_C), junction temperature (T_J), internal thermal resistance ($R_{(TH)}$), device turn-on time ($T_{(ON)}$), turn-off time ($T_{(OFF)}$), etc., can be utilized for monitoring the device condition. It has been observed that a sudden increment of 3% in on-state collector-emitter voltage indicates the presence of a bond-wire fault in an IGBT [13]. Once proper fault signature has been decided, the subsequent step toward achieving fault-tolerant operation is the fault diagnosis. It is desired that the fault detection mechanism should be able to accurately identify and localize the fault within the stipulated time. It has already been indicated that IGBTs are the most preferred power switching devices for GTPVS applications. These devices may be affected by numerous kinds of failures. However, the majority of failure modes include open circuit fault, short-circuit fault, high leakage current, gate misfiring faults, etc. Various fault-diagnostic techniques for inverter faults have been proposed in the literature. They can be categorized into two types – model-based and model-free fault-diagnostic techniques.

8.2.1 Model-based Fault-Diagnostic Techniques

Model-based techniques incorporate the mathematical and analytical information related to the system. These techniques entail residual generation based on the in-depth understanding of their physical characteristics. The residue obtained is the difference between the true value and predicted value of the considered model parameter [14]. Residual generation can be manifested through algorithms like state observers, parity equations, parameter estimations, etc. [15]. They have the advantage that they do not demand additional hardware units; therefore, they are cost-effective. However, their shortcoming is that defining an accurate mathematical model is sometimes cumbersome. Moreover, precise physical and

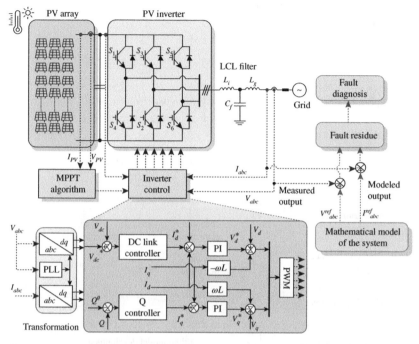

Figure 8.3 A general model-based fault-diagnosis approach.

mathematical characteristics and knowledge about system behavior are difficult to obtain. A general process for model-based fault-diagnostic techniques has been presented in Figure 8.3.

8.2.2 Model-free Fault Diagnostic Techniques

Model-free fault-diagnostics techniques are free from these shortcomings, and they are also known as "data-driven fault-diagnostic techniques". Essentially, they utilize the immense capability of artificial intelligence (AI) to solve real-world problems like that of fault diagnosis and protection applications [16]. In recent years, fault-diagnostic techniques based on AI have performed exceptionally with higher accuracy and efficiency [17]. In practical applications, AI is manifested through its branch universally known as machine learning (ML). ML works through employing the ability to imitate human learning with regular progress till it reaches the desired outcomes [18]. The model-free fault-diagnostic procedure is elaborated in Figure 8.4. Further, it is classified into supervised learning, unsupervised learning, and semi-supervised learning. These techniques utilize a significant amount of data and, hence, the very first step in fault detection is collecting proper data, its cleaning, and preprocessing. The next step consists of

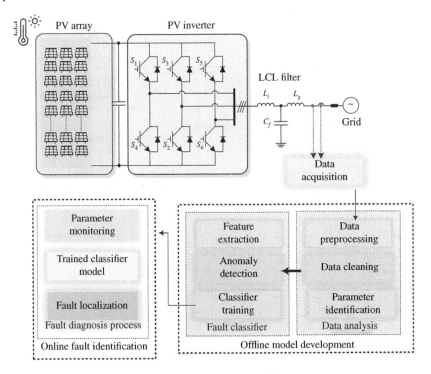

Figure 8.4 A general model-free fault-diagnostic technique.

feature extraction from the relevant data, and the final step uses the acquired features for training based on ML techniques [19]. The training process for ML techniques generally adopts classification approaches for fault diagnosis in GTPVS. There are various categories under ML-based classification techniques including artificial neural network (ANN), k-nearest neighbor (kNN), fuzzy logic, Bayesian network, decision trees, random forest, support vector machine (SVM), etc. The detailed comparison among these algorithms is summarized in Table 8.1.

ANN is a supervised learning-based technique, which takes inspiration from the human brain structure [20]. It includes a biological neural network-like configuration, and functions through emulating the human brain for learning. It has been utilized widely for fault diagnosis in GTPVS applications [21, 22]. Another common technique is decision trees, which comes under supervised learning-based methods. It incorporates an upside decision arrangement similar to a tree, having a top with a root, further splitting into many branches. At a point, when the branch discontinues splitting, it is the final outcome for the desired purpose. It is seldom utilized for fault diagnosis applications since it has a shortcoming of giving

Table 8.1 Summary of model-free fault classification techniques.

ML fault-diagnostic technique	Advantages	Disadvantages
Artificial neural network (ANN)	• Network configuration is simple • Insensitive to noise in data • Capable of doing more than one assigned tasks	• Computationally intensive • Cannot handle high-dimensional data • Structure of network is not defined properly
K-nearest neighbor (KNN)	• Simple and easy in implementation • Adaptable to a new dataset • Appropriate for classification of more than one class	• Finding optimum "k" is difficult • Computational complexity • Noise sensitive
Decision trees	• Tree structure is simple • Insensitive to outliers • Not affected by noise in data	• Trees with a large number of branches may suffer from overfitting • Decision outcome may be biased
Random forest	• Capable to handle missing values • Robust to overfitting • Insensitive to noise	• Computationally intensive • Time for training is more • Not adaptable to new data
Bayesian networks	• Capable of generalization • Features can be utilized optimally • Applicable to reason in forward and backward directions	• High computational efforts • Network construction is difficult • Cannot handle high dimensional data
SVM (support vector machine)	• Memory efficient • Robust to high dimensional data • Ability to utilize non-linear data • Strong generalization capability	• Noise sensitive • Interpreting the outcome is challenging • High training time
Deep learning	• Highly accurate and efficient • Minute details of data are utilized • Outstanding feature extraction	• High computational efforts • Huge training data

biased results in a few circumstances [23]. A supervised learning-based classification approach, particularly known as random forest works as a combination of decision trees [24]. Numerous decision trees together are arranged to form a structure similar to a forest. These combined decision trees are uncorrelated and are able to provide better predictions in terms of accuracy [25]. kNN is based on the idea that there exists association or some relationship among the closely located data points. This proximity or association information is utilized for the fault classification purpose through quantifying with procedures like Euclidean distance etc. [26]. Another category under supervised learning is the Bayesian network, which comes under probabilistic information depiction [27]. They aim to study conditional dependence by employing inference over the respective variables. It incorporates edges and nodes in a network, in which each node provides a unique variable and each edge denotes a conditional dependency. Through the optimal implementation of features, it provides better estimation for classification applications. SVM is a popular approach, which aims to create a hyperplane based on the available data [28]. Further, it constructs a barrier or boundary between the data points corresponding to different classes. It has been widely applied for fault diagnosis applications in GTPVS [29, 30].

An advanced field under ML, known as deep learning consists of deeper neural networks [31]. It is a promising technology and is considered to be more accurate and efficient as compared to the traditional ML techniques. Furthermore, they are free from hand-engineered feature extraction processes, which helps in saving time and effort [32]. They consist of powerful inherent feature extraction ability, which enhances its overall performance as a classification method. There are many divisions under deep learning including convolutional neural networks (CNNs), Boltzmann machines, recurrent neural networks (RNNs), deep belief network (DBN), etc. These divisions have been successfully applied for fault detection and classification in GTPVS applications [33, 34].

8.3 Fault Isolation Strategies

After the fault diagnosis step, it is required that the faulty area should be isolated from the remaining healthy part of the system. Essentially, the fault event should not influence the rest of the system and its propagation should be limited [35]. A fault isolation circuitry compels the affected switch to be electrically segregated so that its impact on the system performance is eliminated. The design consideration should take the following into account – (i) precision, (ii) speediness, (iii) fault treatment, (iv) effect on the system operation, (v) simplicity, and (vi) cost. Open-circuit faults can be isolated simply by stopping their gate signals. However, short-circuit faults require proper isolation through elements like fuses, triode for alternating current (TRIACs), etc.

The fault isolation strategy for an open-circuit fault consists of gate signal blockage as shown in Figure 8.5a. For a short-circuit fault, the inverter pole is isolated through silicon-controlled-rectifier (SCR) triggering as can be seen in Figure 8.5b [36]. Further, it will blow up the corresponding fuses. Capacitor value and size are critical to the fault isolation duration. Its value should be such that the energy

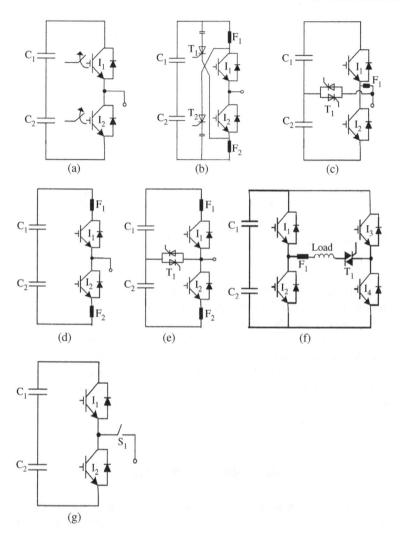

Figure 8.5 Fault isolation strategies proposed for fault-tolerant operation of GTPVS.

transfer is adequate for blowing the fuses in the stipulated time [37]. The disadvantage associated with this particular technique is that the cost is more due to more number of components. Another fault isolation strategy for short-circuit faults is proposed in [38], as described in Figure 8.5c. Here, if short-circuit fault occurs in switch S1, then complementary switch S2 is opened and TRIAC is triggered. Consequently, the associated phase current will increase leading to the blowing of the fuse. The drawback of this strategy is that the whole leg is isolated, even with one switch failure. In addition, it is unable to manage the fault, where switches in the same leg fail concurrently. With the help of fast fuses, the implementation of fault isolation can be carried out in a simple manner as shown in Figure 8.5d [39]. In this strategy, when a short circuit is detected in one or two switches of a single leg, the switch command is obstructed. It results in the fuse blowing out and, hence, the respective switch leg is isolated.

A simpler version of the fault isolation technique [36] is proposed in [40]. It consists of only one TRIAC and two fuses as shown in Figure 8.5e. Once a short-circuit fault occurs in S1, S2 is opened, and the TRIAC T1 is fired. Further, it causes fuse F1 to blow leading to the isolation taking place. The disadvantage here is that this technique might lead to an increase in parasitic inductances. Another fault-isolation strategy specific to the neutral leg topology is proposed as shown in Figure 8.5f [41]. Note that the load impedance preferably should be low, and the TRIAC current rating should be sufficient to endure the shoot-through current. In the event of a short-circuit fault in the upper or lower switch in a leg, the gating signal for the complementary switch is blocked. This logic implementation is obtained through hardware. Further, the respective TRIAC is activated along with the corresponding switch in the auxiliary leg. Consequently, it leads to the formation of a shoot-through loop composing a switch in the auxiliary leg, the affected faulty switch, and the DC-link capacitors. The subsequent high inrush current in the loop will result in clearing the fuse of the respective faulty leg. A fuse-independent approach for fault isolation is implemented [42]. It consists of a controllable switch, which can be TRIAC, relay, or bi-IGBT. Here, the switch is TRIAC as depicted in Figure 8.5g, and in case of a short-circuit fault, TRIAC and complementary switch are turned off. Here, the drawback is that this approach is unable to handle multiple switch failures.

8.4 Post-Fault Reconfiguration Techniques

Reconfiguration strategy may usually be based on hardware redundancy or corresponding control mechanism. Many fault-tolerant techniques for grid-tied PV inverters have been proposed in literature. These can be characterized based on level of redundancy, inverter topology, switching criteria, etc. The classification based on hardware redundancy consists of four types – switch-based, switch leg-based, module-based, and system-based.

8.4.1 Switch-based Reconfiguration Strategies

Switch-based fault-tolerant technique may further be divided into two methods. One method realizes the inherent redundancy present in the switching states of a converter. In the majority of cases, this applies to multilevel inverters like neutral-point-clamped (NPC), flying-capacitor, etc., since these multilevel inverters have more number of switching devices as compared to generally two- or three-level converters. The other method utilizes the connection of the faulty converter to the DC-bus midpoint through adding extra switches as per the requirement. Another solution involves adding switches either in series or parallel to main switches. These may work both in online and offline modes depending on the application.

In case of flying capacitor multilevel inverters, an inherent redundant switching states-based solution is proposed [43]. It deals with the short-circuit fault tolerance for a few switching cycles. The design of the measurement and monitoring circuit is presented and real-time results manifest remedial reconfiguration of the pulse width modulation (PWM) modulator such that the fault is withstood for a very short period. However, the disadvantage is that voltage waveforms obtained are non-optimized causing higher harmonics. Another unique approach for fault tolerance in flying capacitor inverter topology is developed [44]. This work utilizes the reconfiguration in the gate control signal when a single-switch fault occurs. It has the benefit that the power conversion quality does not suffer in the post-fault operation. The experimental results suggest a unique capacitor-balancing method works well and is able to validate the effectiveness of the presented approach. The demerits of this approach include complicated capacitor control and not being extendable to higher levels. A predictive control-based fault-tolerant technique for a three-phase three-cell flying capacitor converter is proposed [45]. It consists of adopting a finite set-constrained method, such that it is able to achieve robustness to semiconductor device faults. Through a meticulous design of switching sequences and utilizing the phase voltage measurements, the faulty area can be isolated. Further, restructuring occurs in such a way that the original operation of the system is restored. However, this method is restricted to only three-cell converters and is difficult to apply for a higher number of cells. In [46], the model-based predictive control method is considered for identifying faulty switching devices in a flying capacitor multilevel inverter. After fault detection, the faulty switch leg is isolated and the impacted phase is joined at the DC-bus midpoint. The finite-set cost function is utilized for obtaining the optimum quality of the inverter output. It involves the reduction in voltage level during the reconfiguration step as compared to that attained during normal operation. In addition, it comprises oversized capacitors for fault diagnosis. In [47], a simple fault handling approach for flying capacitor-based inverter is proposed. Fault identification is done using model-based predictive control. This approach works by using the switching sequence details and capacitor charging state information. When the faulty switch is detected, it bypasses the switch and processes the reconfiguration step. In this approach, capacitor voltage balancing becomes critical as the number of levels increase.

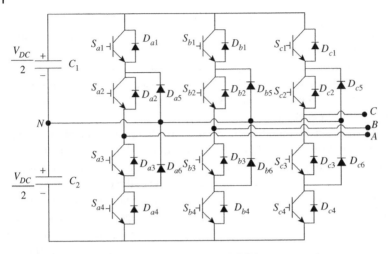

Figure 8.6 A general neutral-point-clamped (NPC) inverter topology.

Fault-tolerant strategy for NPC multilevel inverter is proposed [48]. NPC topology for a three-phase inverter is provided in Figure 8.6. The strategy is based on open-circuit fault diagnosis without any additional hardware or complex diagnostic circuits. It has the benefit of utilizing redundant non-faulty switching states for faulty phases such that the normal operation of the system is restored. The inverter is able to transfer maximum power with proper sinusoidal currents during the post-fault operation. The method has a shortcoming in that it covers only open-circuit faults. Strategies for three-phase NPC inverter fault-tolerant operation are proposed [49]. The fault detection scheme is executed through a digital signal processor (DSP), which takes into account the gating signals and measured voltage across the single switching device. It is observed that for a unique switch fault, the pertaining voltage vectors become unavailable and for the system to continue operating, PWM modulation has to be reformed to stop utilizing the unavailable voltage vectors. The only demerit of this method is that it results in reduced output power, derating the whole system. Another work on NPC converter with detailed fault analysis and two fault-tolerant solutions is presented [50]. These solutions are then compared with the original NPC converter operation in terms of various parameters. The first solution is attained by replacing the clamping diodes based on IGBTs. However, it is found that this solution would result in increased conduction losses and the cost of the converter. The second solution is based on improving the converter behavior through replacing clamping diodes with active switches. It will help in allowing the output to be connected to the neutral point under all operating conditions. This solution slightly increases the cost of the system along with enhancement in the nominal power by 20%, which is a great advantage achieved with this solution. Consequently, two solutions can be applied as per the requirement and with a few improvements, they can serve to make the system more robust and reliable. A hybrid fault-tolerant converter combining a three-level NPC converter with a three-level flying capacitor leg is developed [51]. The flying capacitor leg presents an ability to balance the neutral-point voltage of the NPC converter. Additionally, it has redundant hardware having two fuses and one thyristor in each leg. A filter with a dedicated mathematical design is adopted for operating in normal mode. After the fault is detected, the hardware modification is employed such that the faulty leg is isolated and a neutral point connection takes place. The reconfigured converter is controlled through a field-programmable gate array (FPGA) and the system continues to operate. Additional hardware increases the cost of the system.

A redundant parallel switch-based fault-tolerant matrix converter (MC) is proposed [52]. MC is a promising technology as it does not depend on a DC link capacitor and has a compact design. However, there stand certain barriers in the way of its huge adoption. A fault-tolerant MC would play a significant role in making this technology more reliable and available. A novel MC is developed in this work, which has redundant parallel switches allowing flexible reconfiguration of the converter. With fault occurrence in any of the switches, it is immediately replaced by the additional switch in parallel, retaining optimum power, and overall performance. An analysis of the fault-tolerant operation for multilevel active clamped (MAC) converters is presented [53]. For MAC converters, more number of conduction paths exist, which can be utilized well for redundancy. However, during fault events, switches might have to withstand increased blocking voltage. To maintain normal device blocking voltage, a scheme is proposed, which however compromises a few voltage levels. To further improve the fault-tolerant capability, three hardware solutions are proposed adding auxiliary switches in different configurations. These solutions are able to tackle simultaneous multiple device fault occurrences. Furthermore, these solutions are compared for different fault tolerance performance parameters including conduction losses, cost, blocking voltage capability, etc. A fault-tolerant approach based on the idea that after fault events healthy phase has to endure overcurrent necessitating the use of oversized switching devices is proposed [54]. The approach stresses the faulty phase output voltage for VSI to become zero and further adjusts the phase angles of the remaining phase voltages. It will help in retaining balanced voltages during the post-fault operation. Already existing modulation strategy can contribute well to the post-fault operation performance. In case of applying a similar strategy for multilevel converters, bidirectional switches (e.g. TRIACs) can be utilized for joining the midpoint to the output. An improved three-level NPC topology known as active-NPC (ANPC) utilizing IGBTs to replace the clamp diodes for fault-tolerant operation is presented [55]. The major advantage of this method

Table 8.2 Summary of switch-level fault-reconfiguration strategies.

References	Topology	Fault type	Advantages	Disadvantages
[43]	FC	Short circuit	• Able to handle faults in a short time	• High harmonics
[44]	FC	Open, short circuit	• Great post-fault performance • Better efficiency due to capacitive balance approach	• Complicated control • Cannot extend to high levels
[45]	FC	Open, short circuit	• Robustness to semiconductor device faults	• Difficult to apply for a high number of levels
[46]	FC	Open, short circuit	• Quality of inverter output is good	• Need oversized capacitors
[47]	FC	Open, short circuit	• Simple fault handling • Better capacitor charging state information	• Difficult to apply for high number of levels
[48]	NPC	Open circuit	• Does not require extra hardware • Utilizes redundant non-faulty switching states	• Can only work for open circuit faults
[49]	NPC	Open, short circuit	• Robustness to single semiconductor device faults	• Leads to reduced output power in post-fault operation
[50]	NPC	Open, short circuit	• Improved solution provides better output power in post-fault operation	• Cost might increase
[51]	NPC with FC	Open, short circuit	• Optimum post-fault performance	• Cost might increase with extra hardware
[52]	MC	Open, short circuit	• Retains optimum power in post-fault operation	• Cost incurred is high
[53]	MAC	Open, short circuit	• Increase in number of conducting path provides better redundancy • Can handle simultaneous multiple device faults	• Switches may withstand increased blocking voltage during faults
[54]	VSI	Open, short circuit	• Retains balanced voltage in after-fault operation	• Generalization ability is not good
[55]	ANPC	Open, short circuit	• Does not require additional hardware	• Reduced output voltage in after-fault operation
[56]	ANPC	Short circuit	• Improved output voltage in post-fault operation	• Excessive burden on device may lead to reduced device lifetime
[57]	VSI	Short circuit	• Improved reliability • Better converter operation	• Increased conduction losses

is that it does not involve additional switching devices and faulty phase output can be connected to the midpoint. However, the output voltage is reduced during the post-fault operation. As a solution to this, some thyristors and fast fuses are incorporated for a better fault-tolerant performance [56]. The short-circuit fault event enables the associated thyristor to clear the corresponding fuse. Therefore, configuring it to a two-level structure in the post-fault reconfiguration. The disadvantage is that a few devices have to endure overvoltages affecting the device life. Another fault-tolerant approach based on the redundant device in series is developed [57]. It consists of additional devices in series with the main switching devices to counter the short-circuit failures. Therefore, it enhances the overall reliability and the converter operation effectiveness. However, it also involves increased conduction losses in the no-fault operation. In addition, the voltage sharing concern needs to be addressed properly. Various switch-level reconfiguration strategies discussed here are summarized in Table 8.2.

8.4.2 Switch-leg-based Reconfiguration Strategies

This kind of fault-tolerant operation requires connecting an additional switch-leg or phase-leg such that it is able to replace the faulty leg during the post-fault operation. The switch-leg may be connected both in series and in parallel arrangements. A redundant parallel-leg-based inverter fault-tolerant topology as shown in Figure 8.7 is proposed [41]. It also contains additional bidirectional switches to facilitate the isolation process once the fault is diagnosed. Further, the reconfiguration process connects the redundant switch-leg such that the original topology is restored. This technique has the merit of improved output power and voltage. However, the demerit is that it can only handle single open switch/phase faults. A similar fault-tolerant inverter is presented in [58, 59]. Another NPC

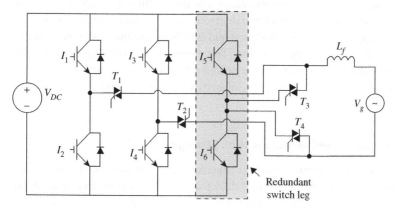

Figure 8.7 Redundant switch-leg-based fault-reconfiguration technique.

converter-based fault-tolerant topology is proposed consisting of an additional fourth leg [60]. This leg provides a flying capacitor converter structure enabling fault-tolerant operation. An important benefit

offered by this technique is that it eliminates the low-frequency oscillations appearing at the midpoint or neutral point. It results in making the modulation strategy to be independent of ensuring the voltage balance. Therefore, the modulation technique can be modified to attain optimum output power and reliability of the converter. A similar work for wind generation applications is proposed in [61]. An improvement of the above technique based on a resonant structure minimizing the losses encountered due to additional leg is presented in [62]. It also incorporates the comparison between the two strategies and presents further analysis with their merits and demerits.

Redundant series leg-based fault-tolerant strategies are also proposed for various applications. A fault-tolerant H-Bridge power converter in an electric vehicle for a three-phase permanent magnet synchronous motor (PMSM) [63]. The power circuit consists of three H-bridges and each H-bridge supplies a distinct PMSM phase to perform fault-tolerant functionality. It does not depend on fault isolation relays and it works without a neutral point. The advantages of proposed method includes reduced switching losses, allowing complete DC link voltage for each stator winding, and optimum drive performance. A comparison is also made with standard space vector PWM (SVPWM) control methods for different parameters. This technique has a major drawback in that the voltage is reduced during the after-fault operation. Another fault-tolerant method for a single DC-link dual inverter in induction motor drive is proposed to tackle the switching device incipient faults [64]. For fault diagnosis, device degradation is monitored over time and further, a mechanism is enabled to trigger the corrective action depending on the detected fault. For fault signature, large-frequency components are introduced during switching and the machine response is detected. This response is utilized to track the degradation in the power switching device. The control approach is well designed and it utilizes additional switching states for proper operation of the machine. The experimental outcomes validate the efficiency of the fault-tolerant technique and manifest improvement in the overall reliability of the system. A summary of switch leg-based solutions is provided in Table 8.3.

8.4.3 Module-based Reconfiguration Strategies

Module-based fault-tolerant solutions consider the reconfiguration in the remaining modules once a module is found defected. Typically, it is utilized in modular multilevel converters (MMCs), Figure 8.8 and cascaded multilevel converters (CMCs), Figure 8.9. For these converters, the arrangement of modules consists

Table 8.3 Summary of switch-leg level fault-reconfiguration strategies.

References	Topology	Fault Type	Leg configuration	Advantages	Disadvantages
[41, 58, 59]	H-Bridge	Open circuit	Parallel	• Improved output performance in after-fault operation	• Can only handle open circuit faults
[60, 61]	NPC	Open, short circuit	Parallel	• Eliminates low frequency oscillations at neutral point	• Increased number of switching devices
[62]	NPC	Open, short circuit	Parallel	• Less switching losses • Reduced stress on devices	• Cost is high
[63]	H-Bridge	Open, short circuit	Series	• Less switching losses • Optimum output performance	• Reduced voltage in after-fault operation
[64]	H-Bridge	Open, short circuit	Series	• Improved reliability	• Increased thermal stress

of either series or parallel arrangement. Redundant module-based fault-tolerant solutions have become quite popular recently. For redundant devices connected in series in two-level converters, the converter works well even with one device shorted. It does not even need any modification in the control design. This is device-level redundancy. Modular redundancy is also applied in a similar manner by adding an additional module in series or parallel depending upon the configuration. Cascaded H-bridge (CHB) multilevel converter (CHMC) has a high number of power switching devices and it is considered a promising technology specifically for applications like static synchronous compensator (STATCOM) [65]. A fault-tolerant technique adopting an additional H-bridge building block (HBBB) for CHMC application is presented. It consists of bypassing the faulty HBBB block as soon as the fault is detected and further, reconfiguring the redundant block in the system. In this process, the converter level degrades from $(2N+1)$ to $(2N-1)$, which is suitable for STATCOM application. The disadvantage is that with the voltage degrading, large transients in current are observed and, therefore, it needs to be addressed properly for fault-tolerant design. A fault-tolerant MMC topology is presented having redundant modules with proper design and control strategy [66]. The spare modules are connected to the system once a faulty module is detected. It presents a dual sorting strategy for implementing voltage balance control and minimizing the switching commutations for an individual

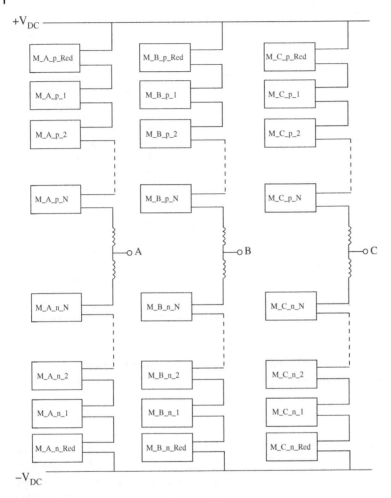

Figure 8.8 A general modular multilevel converter (MMC) topology for fault reconfiguration.

module. Further, a novel process is developed, which can handle multiple module failures including faults in the spare modules. A detailed investigation is incorporated for ensuring fault-proof energy transfer. The technique is able to reduce the undesired switching losses as well along with proper power balance. The work mainly focuses on its application in high-voltage direct current (HVDC) and presents comprehensive simulation studies for both normal and faulty conditions. Another MMC topology-based fault-tolerant method implementing modulation reconfiguration is developed [67]. It incorporates fault management through modifying the reference sinusoidal modulation with the carrier rotation

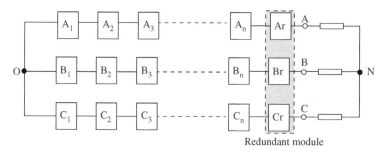

Figure 8.9 A general CMC topology for fault reconfiguration.

method, such that it can maintain regulated output voltages. The multicarrier PWM is able to achieve voltage balancing for the capacitors in sub-modules. The experimental results demonstrate effective flexibility as compared to other methods proposed in this domain. The major advantage is that it improves the overall reliability and robustness of the system without any additional hardware cost. In general, for CHB multilevel converters, the geometric method is applied for the after-fault operation of the system [68]. In this method, common mode voltage is determined by the after-fault control approach. Utilizing this approach, inverter phase voltages may be expressed as the addition of output phase voltages and common mode voltages. This common-mode voltage introduces high-order harmonics leading to unbalance in output voltages. Therefore, two solutions are proposed to improve the post-fault functioning of the converter [69]. The first solution considers finding optimum after-fault states from the available possible states resulting in increased available voltage. The main aim of this solution is to minimize the fundamental component of common-mode (FCCM) voltage to ensure the balanced operation of the converter. The second solution finds effective common-mode voltage by calculating a reduction factor, which is essentially a ratio of required output amplitude to the available output. Hence, the common-mode voltage is minimized using this approach. These two solutions can mutually be applied for the after-fault control operation of the converter to ensure balanced output. Another fault-tolerant solution for a three-phase CHB converter with an auxiliary module, i.e. two-level voltage source inverter is proposed [70]. The auxiliary module is connected in series with the converter system and it comprises six power-switching devices and one capacitor. This approach can handle both open-switch and short-switch faults. After fault diagnosis, the auxiliary module is joined to the system and it is able to attain maximum output power even during fault conditions. Space vector modulation (SVM) is employed to find the reference vector for the inverter and the auxiliary module's voltage is retained as constant. The major benefit of the proposed approach is the utilization of a DC link capacitor, which helps in decreasing the system complexity and overall size. However, various voltage-regulation modes are presented to regulate the capacitor voltage. Another fault-tolerant topology for CHB inverter, based on an extra cross-coupled CHB (X-CHB) unit, is explored [71]. This additional unit along with other units present in the system improves the system's reliability, availability and robustness, even during the open or short circuit fault occurrence. The proposed topology is a five-level inverter with a self-voltage balancing feature provided by the attached flying capacitor. It is also able to boost the voltage by a factor of two without introducing an additional boost stage. Various switching states describing different operation modes at different levels of the converter are elaborated with a unique modulation strategy divided into unity and non-unity power factor functionality. The topology is able to attain full power factor operation. It is considered appropriate for applications like uninterrupted power supplies (UPSs) and battery energy storage systems (BESS).

In MMCs and CMCs, particularly for variable-speed drive applications, the converter works in a reduced mode during fault events. After the fault has occurred, the number of module cells are not equal and it results in deviated phase shifts in the corresponding voltages causing voltage to unbalance. Therefore, the neutral-shift method is applied to handle that shift and it modifies the phase-voltage references such that the equivalent neutral point is changed so as to maintain proper balance. It can be achieved by shunting the same number of module cells in other phases. This kind of operation leads to reduced output. Nonetheless, it also provides continuous operation of the system with balanced output. However, it is not a desirable solution since it results in non-optimized operation as the healthy cells in modules remain inactive. Therefore, a solution is desired, which helps in maintaining continuous operation and simultaneously utilizing all the healthy module cells to maximize the output. A method is proposed utilizing voltage command signals, which are produced for delivering peak voltage decline [72]. These signals can be modified for compensating for the identification of faulty modules. This approach is particularly suited for PWM converters. Consequently, the designed circuit provides balanced output line-to-line voltage and optimum utilization of healthy modules in the converter system. Another neutral-shift fault-tolerant approach is presented, which takes benefits from the fact that module star-point is floating and not joined with the motor neutral [73]. Thus, it moves away from the neutral and module voltage phase angles can be rearranged. It helps in obtaining balanced line voltages for the motor. This approach is similar to providing a zero-sequence component to the voltage line vectors. In this situation, the module star point is no longer aligned with the motor neutral and still, motor voltage is stable and balanced. This method helps in utilizing 80% of the healthy modules with 70% of the available module voltage. However, some precautions are of concern, for instance, the power factor control should be briefly considered specifically at high speeds. Other than that, due to neutral shift, loading will be unequal leading to degraded input power quality and output power quality is also a major concern. A similar fault-tolerant approach that allows the adjustments in the module voltage phase in order to obtain balanced motor voltages, utilizing the neutral-shift method is proposed [74]. Considering the problem to solve for the angles to generate balanced output voltages, the phase voltage is characterized by the vectors. These are located in the real axis. Therefore, the imaginary part of the vector is found to be zero. Representing these voltage vectors further in equation forms and solving for the corresponding angles using software, the angles are found to produce balanced line voltages. Furthermore, this solution results in 30% more output voltage. A post-fault technique based on the neutral-shift method for MMC is proposed, which, after bypassing the defected module, works on improving the output line-to-line voltage [75]. The analysis regarding the neutral-shift method works on finding its impact on MMC operation along with building an internal variable analysis strategy. The post-fault strategy functions by modifying the magnitude and phase angles of the MMC phase voltages. This helps in shifting the neutral point and subsequently achieving balanced line voltages. Line-to-line voltages constitute an equilateral triangle and to maximize the triangle or achievable output voltage, output phase voltages are properly employed and center of the triangle is moved away from the middle point. Further, initial modulation indices and phase angles are modified and correspondingly, reference voltages of all the phases are changed. For subsequent performance improvement, an adjusted circulating current suppression (CCS) method is adopted to overcome the double-frequency components.

An alternate method for optimizing the output voltage during post-fault operation is the rise in the input DC-bus voltage. This will help in keeping the optimum output voltage levels. DC-bus voltage post-fault reconfiguration differs from the fact that overvoltage is handled by only the defected phase or all three phases [76]. Here, in CMCs, H-bridge modules are utilized as active rectifiers. With the fault event, the DC-link voltages of the healthy modules are enhanced to maintain optimum voltage levels. A similar principle is utilized in [77], where neutral-shift is adopted to share the overvoltage equally among the healthy modules. This further assists in escalation in the DC-link voltage of the non-faulty modules by 35%. However, the disadvantage here is that the growing DC-bus voltage may overstress the devices with increased voltages, compelling oversized device design. Another work that utilizes simultaneous neutral-shift on AC and DC, both sides for MMC is proposed [78]. The strategy controls DC-side and respective circulating currents. Further, the voltage in faulted nonredundant modules is supported by the redundant module voltage. The major idea is to explore the voltage sharing in a way that the over-/under-voltage introduced by

Table 8.4 Summary of module-level fault-reconfiguration strategies.

References	Module topology	Fault Type	Reconfiguration method	Advantages	Disadvantages
[65]	CMC	Open, short circuit	Redundant module	• Improved output voltage and power in after-fault operation	• Large transients in current need to be addressed
[66]	MMC	Open, short circuit	Redundant module	• Minimizes switching losses • Can handle multiple faults	• Large transients in system may be experienced
[67]	MMC	Open circuit	Redundant module	• Optimum voltage balance is achieved	• Can only handle open circuit faults
[68]	CMC	Open, short circuit	Redundant module	• Capable of avoiding the region of over-modulation	• Imbalance in output phase voltages
[70]	CMC	Open, short circuit	Redundant module	• Improved output performance • Reduced system complexity and size	• Capacitor voltage regulation is critical
[71]	CMC	Open, short circuit	Redundant module	• Reliability is enhanced	• Sensitive to DC-link capacitor failure
[73]	MMC	Open, short circuit	Neutral shift method	• Optimum utilization of healthy modules • Balance output voltage	• Degraded output power quality
[74]	CMC	Open circuit	Neutral shift method	• Increase in output voltage by 30%	• Input current quality is deteriorated
[75]	MMC	Open, short circuit	Neutral shift method	• Optimum post-fault performance	• Circulating current suppression is critical
[76]	CMC	Short circuit	DC-bus voltage approach	• Enhanced output voltage during after-fault operation	• Reduced output current magnitude
[77]	CMC	Open, short circuit	DC-bus voltage with neutral shift approach	• Balanced and optimum phase voltages	• Oversized device design
[78]	MMC	Open, short circuit	DC-bus voltage approach	• Complexity is reduced for increased module number	• Difficulty in handling energy variations
[79]	MMC	Open, short circuit	DC-bus voltage approach	• Good dynamic behavior • Output line voltage is increased	• Different phases experience unequal power factor

the faulty nonredundant modules is compensated by the non-faulty redundant modules. In addition, the power imbalance on the DC as well as AC side is recompensed by the dedicated power feed-forward control. An improvement over this technique, which works on optimum utilization of healthy capacity of MMC is presented [79]. The major contribution of this work includes employing the optimum DC component and finest neutral-point location. The combined method offers advantages like fewer limitations in fault numbers and fewer limitations in power factors. Overall, the capability of the proposed fault-tolerant strategy is found to be better as compared to the discussed techniques since it provides a comprehensive and articulated formulation, maximum achievable output line voltage in post-fault operation and performance evaluation under different fault conditions. The module-based fault-tolerant solutions discussed earlier are detailed in Table 8.4.

8.4.4 System-based Reconfiguration Strategies

The system-based fault-tolerant solutions consist of two hardware configurations. One of them is series configuration and the other is parallel configuration. For both hardware solutions, further categories may include keeping the redundant converter in operation during the normal converter function or connecting the redundant converter after fault occurrence. Parallel converter solutions help in improving overall output power during after-fault operation. These are utilized in many applications like UPSs [80], active power filters [81] etc. An important point to consider here is that parallel converter operation may induce circulating currents in the system if the converters are not properly synchronized [82]. Therefore, improved control strategies are proposed like circular chain control (3C) technique [83], master-slave control [84], droop control [85], etc.

The series inverter topology for fault-tolerant functionality in case of single-phase converter application is described in Figure 8.10a. A cascaded inverter topology for motor drive application is presented [86]. Three TRIACS are connected in series with the three output phases. The topology presented is able to mitigate single open and short switch faults, and phase open circuit faults. However, the overall after-fault performance is degraded since the output power achieved is lower than that in the normal operation.

A distributed control is adopted for the input series output parallel (ISOP) inverter system [87]. This system is quite appropriate for applications based on high input voltage requirements and high output current requirements. This work utilizes a decentralized control scheme to attain proper power balance with a redundant converter, which is in contrast to the centralized control that restricts the working of a redundant converter-based system. Hot-swap technique is realized along with the distributed control for a complete modular system,

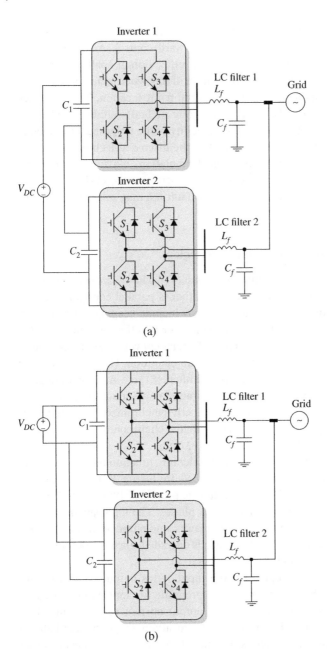

Figure 8.10 Series and Parallel inverter topology for fault-tolerant operation.

which helps in significantly improving the reliability of the redundant ISOP system. A quantitative analysis is done to determine the number of redundant inverters such that maximum reliability is achieved with minimum cost [88]. A mathematical model is derived such that a proper trade-off between the cost and reliability is obtained. Various reliability indices are discussed including MTTR, MTTF, MTBF, failure rate, etc. With dynamic power distribution scheme implementation, a reduction in system cost is achieved.

The parallel inverter-based fault-tolerant solution for a single-phase converter is shown in Figure 8.10b. For a parallel inverter system, it is required that the load current should be shared among them equally. An instantaneous average current-sharing approach for a parallel inverter system is proposed, which allows for equivalent current sharing among different units [89]. The approach models the deviation in the inverter current as the source of disturbance and the objective is to minimize this disturbance. It helps achieve stability enabling the parallel inverters to share an equal load among them. Another work on a parallel converter system works on replacing a single converter unit with smaller six units, such that if fault occurs in one of the modules, the remaining five converters are able to supply 5/6 of the rated power to the grid [90]. This solution helps in optimizing efficiency in comparison to the single converter-based solution. Further, a small shift in the PWM switching results in the PWM-harmonics elimination. The adopted control mechanism is effective as the decision regarding the addition or subtraction of active converter modules is taken in such a way that overall thermal stress is minimized. In conclusion, this solution improves the system's reliability along with reducing the total cost of energy. For fault-tolerant functionality, a parallel redundant multiple-inverter system is developed, which aims to keep the output current constant even in fault conditions [91]. When one converter has a fault, its current reduces to zero, then the remaining healthy converters are partially charged to maintain a constant load current. Consequently, it enhances the reliability and availability of the system. A self-checking monitor-based fault-tolerant scheme is proposed for the parallel inverter system [92]. It detects the faults affecting the inverter and generates the corresponding alarm. With the alarm generation, the affected inverter is disconnected and the redundant inverter unit is configured to the system. The main benefit of the diagnostic mechanism is that the monitor is self-checking when the monitor itself has a fault. Thus, this kind of fault-tolerant operation improves the reliability of GTPVS. Another fault-tolerant design for parallel inverters for IGBT short-circuit faults is presented [93]. It carries out proper stage analysis during the IGBT short circuit failure and further surge current and fault isolation duration are explored. The fault-tolerant mechanism is designed in such a way that it is able to deal with post-fault surge currents and load voltage deviations. In addition, a mathematical model is assessed for identifying these issues and surge currents are further suppressed. The discussed

Table 8.5 Summary of system-level reconfiguration strategies.

References	Fault type	System configuration	Advantages	Disadvantages
[86]	Open, short circuit	Series	• Can handle multiple faults	• Degraded after-fault performance
[87]	Open, short circuit	Input series output parallel	• Reliability is enhanced	• Switching device may experience overstress
[89]	Open, short circuit	Parallel	• Equal load sharing among different parallel units	• Voltage regulation may be a concern
[92]	Open, short circuit	Parallel	• Improved reliability and fault-tolerant ability	• Efficiency might be affected
[93]	Short circuit	Parallel	• Deals with load voltage deviations • Surge current suppression	• Can only handle short circuit faults

system-level reconfiguration strategies along with their advantages and disadvantages are detailed in Table 8.5.

8.5 Summary

This chapter details the design process for a fault-tolerant converter for GTPVS applications. It is very crucial to understand the fault-tolerant design process to make the system more robust, reliable and available. It broadly consists of three steps – fault diagnosis, fault isolation, and post-fault reconfiguration. These steps have been studied thoroughly in order to offer an optimum fault-tolerant converter design structure. Various fault-diagnostic techniques have been discussed along with their merits and demerits. Fault isolation is an essential part of this whole design criteria and therefore, plenty of isolation techniques have been detailed for enhancing system reliability. These techniques employ fast-acting fuses and different bidirectional switches like TRIAC, etc. Furthermore, the remedial actions after fault events consist of software or control, and hardware solutions at different levels. Switch-level solutions majorly comprise internal modifications in switching sequences or control algorithms to achieve optimum output performance during the reconfiguration process. They are widely utilized for multilevel converters including NPC, FC, etc. Leg-level solutions require an auxiliary leg for reconfiguration and are found to attain a balance between cost and reliability. However, module-level solutions including MMC and CMC topologies can provide reconfiguration structures without auxiliary switches or switch legs. System-level solutions are found to be more mature and widely appropriate for GTPVS applications. Since the majority of the fault-tolerant topologies are found to be suitable for single device open or short circuit faults, there are only a few strategies, which can be utilized for multiple device faults. Further investigation is required to devise methods, which are capable to handle multiple faults. It is observed that there is a lack of proper quantitative analysis for a transition from the faulty condition to the post-fault condition. It is required to study how to accomplish fast and seamless transient progression from faulty to post-fault state. Overall, less complex and cost-effective fault-tolerant solutions are still being researched and great progress in this area is highly anticipated in the near future.

References

1 Alagh, Y.K. (2006). India 2020 energy policy review. *J. Quant. Econ.* 4 (1): 1–14. https://doi.org/10.1007/bf03404634.

2 Kurukuru, V.S.B., Haque, A., Tripathi, A.K., and Khan, M.A. (2021). Condition monitoring of IGBT modules using online TSEPs and data-driven approach. *Int. Trans. Electr. Energy Syst.* 31 (8): 1–24. https://doi.org/10.1002/2050-7038.12969.

3 Malik, A., Haque, A., Bharath, K.V.S., and Jaffery, Z.A. (2021). Transfer learning-based novel fault classification technique for grid-connected PV inverter. *Innov. Electr. Electron. Eng.* https://doi.org/10.1007/978-981-16-0749-3_16.

4 Malik, A., Haque, A. and Bharath, K. V. S. (2021). Deep learning based fault diagnostic technique for grid connected inverter. *2021 IEEE 12th Energy Conversion Congress & Exposition – Asia (ECCE-Asia)*, May 2021, pp. 1390–1395. doi: 10.1109/ECCE-Asia49820.2021.9479371

5 Wu, R., Blaabjerg, F., Wang, H. et al. (2013). Catastrophic failure and fault-tolerant design of IGBT power electronic converters - An overview. *IECON Proc. (Industrial Electron. Conf.)* 507–513. https://doi.org/10.1109/IECON.2013.6699187.

6 Bharath, K. V. S., Haque, A., and Khan, M. A. (2018). Condition monitoring of photovoltaic systems using machine learning techniques. *2018 2nd IEEE International Conference on Power Electronics, Intelligent Control and Energy Systems. ICPEICES 2018*, Delhi, India (22–24 October 2018), pp. 870–875. doi: https://doi.org/10.1109/ICPEICES.2018.8897413.

7 Abuelnaga, A., Narimani, M., and Bahman, A.S. (2021). A review on IGBT module failure modes and lifetime testing. *IEEE Access* 9: 9643–9663. https://doi.org/10.1109/ACCESS.2021.3049738.

8 Pradeep Kumar, V.V.S. and Fernandes, B.G. (2017). A fault-tolerant single-phase grid-connected inverter topology with enhanced reliability for solar PV applications. *IEEE J. Emerg. Sel. Top. Power Electron.* 5 (3): 1254–1262. https://doi.org/10.1109/JESTPE.2017.2687126.

9 Kumar, V.V.S.P. and Fernandes, B.G. (2017). A fault-tolerant single-phase grid-connected inverter topology with enhanced reliability for solar PV applications. *IEEE J. Emerg. Sel. Top. Power Electron.* 5 (3): 1254–1262. 10.1109/JESTPE.2017.2687126.

10 Malik, A., Haque, A., and Bharath, K. V. S. (2022). Fault tolerant inverter for grid connected photovoltaic system. *2022 IEEE International Conference on Power Electronics, Smart Grid, and Renewable Energy (PESGRE)* (January 2022), pp. 1–6. 10.1109/PESGRE52268.2022.9715825..

11 Keller, J. and Kroposki, B. (2010). *Understanding Fault Characteristics of Inverter-Based Distributed Energy Resources*. Golden, CO (United States): https://doi.org/10.2172/971441.

12 Choi, U.M. and Blaabjerg, F. (2018). Separation of wear-out failure modes of IGBT modules in grid-connected inverter systems. *IEEE Trans. Power Electron.* 33 (7): 6217–6223. https://doi.org/10.1109/TPEL.2017.2750328.

13 Oh, H., Han, B., McCluskey, P. et al. (2015). Physics-of-failure, condition monitoring, and prognostics of insulated gate bipolar transistor modules: a review. *IEEE Trans. Power Electron.* 30 (5): 2413–2426. https://doi.org/10.1109/TPEL.2014.2346485.

14 Khan, S.S. and Wen, H. (2021). A comprehensive review of fault diagnosis and tolerant control in DC-DC converters for DC microgrids. *IEEE Access* 9: 80100–80127. https://doi.org/10.1109/ACCESS.2021.3083721.

15 Poon, J., Jain, P., Konstantakopoulos, I.C. et al. (2017). Model-based fault detection and identification for switching power converters. *IEEE Trans. Power Electron.* 32 (2): 1419–1430. https://doi.org/10.1109/TPEL.2016.2541342.

16 Ramachandran, S. (2020). Applying AI in power electronics for renewable energy systems. *IEEE Power Electron. Mag.* 7 (3): 66–67. https://doi.org/10.1109/MPEL.2020.3012009.

17 Zhao, S., Blaabjerg, F., and Wang, H. (2021). An overview of artificial intelligence applications for power electronics. *IEEE Trans. Power Electron.* 36 (4): 4633–4658. https://doi.org/10.1109/TPEL.2020.3024914.

18 Chen, Y., Tan, Y., and Deka, D. (2018). Is machine learning in power systems vulnerable? *2018 IEEE Internation Conference on Communication, Control, and Computing Technologies for Smart Grids, SmartGridComm*

2018, Aalborg, Denmark (29–31 October 2018). 1–6. doi: 10.1109/SmartGrid-Comm.2018.8587547.

19 Bharath Kurukuru, V. S., Haque, A., Kumar, R., Khan, M. A., and Tripathy, A. K. (2020). Machine learning based fault classification approach for power electronic converters. *9th IEEE International Conference on Power Electronics, Drives andEnergy Systems. PEDES 2020*, Jaipur, Rajasthan, India (16–19 December 2020). doi: https://doi.org/10.1109/PEDES49360.2020.9379365.

20 Chowdhury, D., Bhattacharya, M., Khan, D., Saha, S., and Dasgupta, A. (2017). Wavelet decomposition based fault detection in cascaded H-bridge multilevel inverter using artificial neural network. *RTEICT 2017 – 2nd IEEE International Conference on Recent Trends in Electronics,. Information and Communication Technology Proceedings*, Piscataway, New Jersey (19–20 May 2017), vol. 2018, pp. 1931–1935. https://doi.org/10.1109/RTEICT.2017.8256934.

21 Dhumale, R.B. and Lokhande, S.D. (2016). Neural network fault diagnosis of voltage source inverter under variable load conditions at different frequencies. *Meas. J. Int. Meas. Confed.* 91: 565–575. https://doi.org/10.1016/j.measurement.2016.04.051.

22 Khomfoi, S. and Tolbert, L.M. (2007). Fault diagnostic system for a multilevel inverter using a neural network. *IEEE Trans. Power Electron.* 22 (3): 1062–1069. https://doi.org/10.1109/TPEL.2007.897128.

23 Nguyen, N.T. and Nguyen, H.P. (2017). Fault diagnosis of voltage source inverter for induction motor drives using decision tree. *Lect. Notes Electr. Eng.* 398: 819–826. https://doi.org/10.1007/978-981-10-1721-6_88.

24 Dhibi, K., Fezai, R., Mansouri, M. et al. (2021). A hybrid fault detection and diagnosis of grid-tied PV systems: enhanced random forest classifier using data reduction and interval-valued representation. *IEEE Access* 9 (Ml): 64267–64277. https://doi.org/10.1109/ACCESS.2021.3074784.

25 Dhibi, K., Fezai, R., Mansouri, M. et al. (2020). Reduced kernel random forest technique for fault detection and classification in grid-tied PV systems. *IEEE J. Photovoltaics* 10 (6): 1864–1871. https://doi.org/10.1109/JPHOTOV.2020.3011068.

26 Manohar, M., Koley, E., Kumar, Y., and Ghosh, S. (2018). Discrete wavelet transform and kNN-based fault detector and classifier for PV integrated microgrid. *Lect. Notes Networks Syst.* 38: 19–28. https://doi.org/10.1007/978-981-10-8360-0_2.

27 Cai, B., Zhao, Y., Liu, H., and Xie, M. (2017). A data-driven fault diagnosis methodology in three-phase inverters for PMSM drive systems. *IEEE Trans. Power Electron.* 32 (7): 5590–5600. https://doi.org/10.1109/TPEL.2016.2608842.

28 Kim, D.E. and Lee, D.C. (2009). Fault diagnosis of three-phase PWM inverters using wavelet and SVM. *J. Power Electron.* 9 (3): 377–385.

29 Yuan, W., Wang, T., and Diallo, D. (2019). A secondary classification fault diagnosis strategy based on PCA-SVM for cascaded photovoltaic grid-connected inverter, *IECON Proceedings (Industrial Electronics Conference, Lisbon, Portugal (14–17 October 2019), vol. 2019, pp. 5986–5991. https://doi.org/10.1109/IECON.2019.8927090.

30 Hu, Z.K., Gui, W.H., Yang, C.H. et al. (2011). Fault classification method for inverter based on hybrid support vector machines and wavelet analysis. *Int. J. Control. Autom. Syst.* 9 (4): 797–804. https://doi.org/10.1007/s12555-011-0423-9.

31 Gong, W., Chen, H., Zhang, Z. et al. (2019). A novel deep learning method for intelligent fault diagnosis of rotating machinery based on improved CNN-SVM and multichannel data fusion. *Sensors (Switzerland)* 19 (7): https://doi.org/10.3390/s19071693.

32 Haque, M., Shaheed, M. N., and Choi, S. (2018). Deep Learning Based Micro-Grid Fault Detection and Classification in Future Smart Vehicle. *2018 IEEE Transportation and. Electrification Conference and Expo, ITEC 2018*, Long Beach, CA, USA (3–15 June 2018), pp. 201–206. doi: 10.1109/ITEC.2018.8450201.

33 Sun, Q., Yu, X., and Li, H. (2020). Open-circuit fault diagnosis based on 1D-CNN for three-phase full-bridge inverter. *Proceedings – 11th International Conference on Prognostics System Health Management (PHM-Jinan 2020)*, Jinan, China (23–25 October 2020), pp. 322–327. doi: 10.1109/PHM-Jinan48558.2020.00064.

34 Gong, W., Chen, H., Zhang, Z. et al. (2020). A data-driven-based fault diagnosis approach for electrical power DC-DC inverter by using modified convolutional neural network with global average pooling and 2-D feature image. *IEEE Access* 8: 73677–73697. https://doi.org/10.1109/ACCESS.2020.2988323.

35 Hu, K., Liu, Z., Yang, Y. et al. (2020). Ensuring a reliable operation of two-level IGBT-based power converters: a review of monitoring and fault-tolerant approaches. *IEEE Access* 8: 89988–90022. https://doi.org/10.1109/ACCESS.2020.2994368.

36 Bolognani, S., Zordan, M., and Zigliotto, M. (2000). Experimental fault-tolerant control of a PMSM drive. *IEEE Trans. Ind. Electron.* 47 (5): 1134–1141. https://doi.org/10.1109/41.873223.

37 Mohammadi, D. and Ahmed-Zaid, S. (2016). Active common-mode voltage reduction in a fault-tolerant three-phase inverter. *Conference Proceedings – IEEE Applied Power Electronics Conference Exposition – APEC*, Long Beach, CA, USA (20–24 March 2016), vol. 2016-May, no. 1, pp. 2821–2825. doi: 10.1109/APEC.2016.7468264.

38 Wang, W., Zhang, J., and Cheng, M. (2017). Common model predictive control for permanent-magnet synchronous machine drives considering single-phase open-circuit fault. *IEEE Trans. Power Electron.* 32 (7): 5862–5872. https://doi .org/10.1109/TPEL.2016.2621745.

39 De Araujo Ribeiro, R.L., Jacobina, C.B., Cabral da Silva, E.R., and Nogueira Lima, A.M. (2004). Fault-tolerant voltage-fed PWM inverter AC motor drive systems. *IEEE Trans. Ind. Electron.* 51 (2): 439–446. https://doi .org/10.1109/TIE.2004.825284.

40 Moujahed, M., Ben Azza, H., Frifita, K. et al. (2016). Fault detection and fault-tolerant control of power converter fed PMSM. *Electr. Eng.* 98 (2): 121–131. https://doi.org/10.1007/s00202-015-0350-5.

41 Song, Y. and Wang, B. (2013). Analysis and experimental verification of a fault-tolerant HEV powertrain. *IEEE Trans. Power Electron.* 28 (12): 5854–5864. https://doi.org/10.1109/TPEL.2013.2245513.

42 Richardeau, F., Mavier, J., Piquet, H., and Gateau, G. (2007). Fault-tolerant inverter for on-board aircraft EHA. *2007 European Conference on Power Electronics and Applications*, pp. 1–9. doi: 10.1109/EPE.2007.4417537.

43 Richardeau, F., Baudesson, P., and Meynard, T.A. (2002). Failures-tolerance and remedial strategies of a PWM multicell inverter. *IEEE Trans. Power Electron.* 17 (6): 905–912. https://doi.org/10.1109/TPEL.2002.805588.

44 Kou, X., Corzine, K.A., Familiant, Y.L., and Member, S. (2004). A unique fault-tolerant design for flying capacitor multilevel inverter. *IEEE Trans. Power Electron.* 19 (4): 979–987. https://doi.org/10.1109/TPEL.2004.830037.

45 Aguilera, R.P., Quevedo, D.E., Summers, T.J., and Lezana, P. (2008). Predictive control algorithm robustness for achieving fault tolerance in multicell converters. *2008 34th Annual Conference of IEEE Industrial Electronics* (November 2008), pp. 3302–3308. doi: 10.1109/IECON.2008.4758489.

46 Druant, J., Vyncke, T., De Belie, F. et al. (2015). Adding inverter fault detection to model-based predictive control for flying-capacitor inverters. *IEEE Trans. Ind. Electron.* 62 (4): 2054–2063. https://doi.org/10.1109/TIE.2014 .2354591.

47 Amini, J. and Moallem, M. (2017). A fault-diagnosis and fault-tolerant control scheme for flying capacitor multilevel inverters. *IEEE Trans. Ind. Electron.* 64 (3): 1818–1826. https://doi.org/10.1109/TIE.2016.2624722.

48 Choi, U.M., Lee, J.S., Blaabjerg, F., and Lee, K.B. (2016). Open-circuit fault diagnosis and fault-tolerant control for a grid-connected NPC inverter. *IEEE Trans. Power Electron.* 31 (10): 7234–7247. https://doi.org/10.1109/TPEL.2015 .2510224.

49 Li, S. and Xu, L. (2006). Strategies of fault tolerant operation for three-level PWM inverters. *IEEE Trans. Power Electron.* 21 (4): 933–940. https://doi.org/10 .1109/TPEL.2006.876867.

50 Ceballos, S., Pou, J., Robles, E. et al. (2010). Performance evaluation of fault-tolerant neutral-point-clamped converters. *IEEE Trans. Ind. Electron.* 57 (8): 2709–2718. https://doi.org/10.1109/TIE.2009.2026710.

51 Abdelghani, A.B., Abdelghani, H.B., Richardeau, F. et al. (2017). Versatile three-level FC-NPC converter with high fault-tolerance capabilities: switch fault detection and isolation and safe post-fault operation. *IEEE Trans. Ind. Electron.* 0046 (c): https://doi.org/10.1109/TIE.2017.2682009.

52 Andreu, J., Kortabarria, I., Ibarra, E., and De Alegría, I. M. (2019). A new hardware solution for a fault tolerant matrix converter. *2009 35th Annual Conference of IEEE Industrial Electronics* (November 2009), pp. 4469–4474. doi: 10.1109/IECON.2009.5414858.

53 Nicolas-Apruzzese, J., Busquets-Monge, S., Bordonau, J. et al. (2013). Analysis of the fault-tolerance capacity of the multilevel active-clamped converter. *IEEE Trans. Ind. Electron.* 60 (11): 4773–4783. https://doi.org/10.1109/TIE.2012.2222856.

54 Lezana, P., Pou, J., Meynard, T.A. et al. (2010). Survey on fault operation on multilevel inverters. *IEEE Trans. Ind. Electron.* 57 (7): 2207–2218. https://doi.org/10.1109/TIE.2009.2032194.

55 Li, J., Huang, A.Q., Liang, Z., and Bhattacharya, S. (2012). Analysis and design of active NPC (ANPC) inverters for fault-tolerant operation of high-power electrical drives. *IEEE Trans. Power Electron.* 27 (2): 519–533. https://doi.org/10.1109/TPEL.2011.2143430.

56 Ceballos, S., Pou, J., Robles, E. et al. (2007). Three-leg fault-tolerant neutral-point-clamped converter. *IEEE Int. Symp. Ind. Electron.* 00 (2): 3180–3185. https://doi.org/10.1109/ISIE.2007.4375124.

57 Julian, A.L. and Oriti, G. (2007). A comparison of redundant inverter topologies to improve voltage source inverter reliability. *IEEE Trans. Ind. Appl.* 43 (5): 1371–1378. https://doi.org/10.1109/TIA.2007.904436.

58 Santos, E.C., Jacobina, J.C.B., and Rocha, J.A.A.D.N. (2011). Fault tolerant ac – dc – ac single-phase to three-phase converter. *IET Power Electron.* 4 (9): 1023. https://doi.org/10.1049/iet-pel.2010.0342.

59 Song, Y. and Wang, B. (2012). A hybrid electric vehicle powertrain with fault-tolerant capability. *Conf. Proc. – IEEE Appl. Power Electron. Conf. Expo. – APEC* 951–956. https://doi.org/10.1109/APEC.2012.6165933.

60 Ceballos, S., Pou, J., Robles, E. et al. (2008). Three-level converter topologies with switch breakdown fault-tolerance capability. *IEEE Trans. Ind. Electron.* 55 (3): 982–995.

61 Ceballos, S., Pou, J., Zaragoza, J. et al. (2006). Efficient modulation technique for a four-leg fault-tolerant neutral-point-clamped inverter. *IECON Proc. (Industrial Electron. Conf.)* 55 (3): 2090–2095. https://doi.org/10.1109/IECON.2006.348023.

62 Ceballos, S., Pou, J., Zaragoza, J. et al. (2011). Fault-tolerant neutral-point-clamped converter solutions based on including a fourth resonant leg. *IEEE Trans. Ind. Electron.* 58 (6): 2293–2303. https://doi.org/10.1109/TIE.2010.2069075.

63 Kolli, A., Bethoux, O., De Bernardinis, A. et al. (2013). Space-vector PWM control synthesis for an h-bridge drive in electric vehicles. *IEEE Trans. Veh. Technol.* 62 (6): 2441–2452. https://doi.org/10.1109/TVT.2013.2246202.

64 Restrepo, J.A., Berzoy, A., Ginart, A.E. et al. (2012). Switching strategies for fault tolerant operation of single DC-link dual converters. *IEEE Trans. Power Electron.* 27 (2): 509–518. https://doi.org/10.1109/TPEL.2011.2161639.

65 Song, W. and Huang, A.Q. (2010). Fault-tolerant design and control strategy for cascaded H-bridge multilevel converter-based STATCOM. *IEEE Trans. Ind. Electron.* 57 (8): 2700–2708. https://doi.org/10.1109/TIE.2009.2036019.

66 Son, G.T., Lee, H.-J., Nam, T.S. et al. (2012). Design and control of a modular multilevel HVDC converter with redundant power modules for noninterruptible energy transfer. *IEEE Trans. Power Deliv.* 27 (3): 1611–1619. https://doi.org/10.1109/TPWRD.2012.2190530.

67 Shen, K., Xiao, B., Mei, J. et al. (2013). A modulation reconfiguration based fault-tolerant control scheme for modular multilevel converters, *Conf. Proc. – IEEE Appl. Power Electron. Conf. Expo. – APEC* 3251–3255. https://doi.org/10.1109/APEC.2013.6520766.

68 Carnielutti, F., Pinheiro, H., and Rech, C. (2012). Generalized carrier-based modulation strategy for cascaded multilevel converters operating under fault conditions. *IEEE Trans. Ind. Electron.* 59 (2): 679–689. https://doi.org/10.1109/TIE.2011.2157289.

69 Zolghadri, S.O.M.R., Khodabandeh, M., Shahbazi, M. et al. (2017). Improvement of post-fault performance of a cascaded H-bridge multilevel inverter. *IEEE Trans. Ind. Electron.* 64 (4): 2779–2788. https://doi.org/10.1109/TIE.2016.2632058.

70 Salimian, H. and Iman-Eini, H. (2017). Fault-tolerant operation of three-phase cascaded H-bridge converters using an auxiliary module. *IEEE Trans. Ind. Electron.* 64 (2): 1018–1027. https://doi.org/10.1109/TIE.2016.2613983.

71 Mhiesan, H., Wei, Y., Siwakoti, Y.P., and Mantooth, H.A. (2020). A fault-tolerant hybrid cascaded H-bridge multilevel inverter. *IEEE Trans. Power Electron.* 35 (12): 12702–12715. https://doi.org/10.1109/TPEL.2020.2996097.

72 Marc, P.W. (2001). *(12s) United States Patent* 1 (12).

73 Hammond, P.W. (2002). Enhancing the reliability of modular medium-voltage drives. *IEEE Trans. Ind. Electron.* 49 (5): 948–954. https://doi.org/10.1109/TIE.2002.803172.

74 Rodríguez, J., Pontt, J., Musalem, R., and Hammond, P. (2004). Operation of a medium-voltage drive under faulty conditions. *Conference Proceedings – IPEMC 2004 4th International Power Electronics and. Motion Control Conference*, Xi'an, China (14–16 August 2004), vol. 2, no. 4, pp. 799–803.

75 Yang, Q., Qin, J., and Saeedifard, M. (2016). A postfault strategy to control the modular multilevel converter under submodule failure. *IEEE Trans. Power Deliv.* 31 (6): 2453–2463. https://doi.org/10.1109/TPWRD.2015.2449875.

76 Lezana, P., Ortiz, G., and Rodríguez, J. (2008). Operation of regenerative cascade multicell converter under fault condition. *11th IEEE Work. Control Model. Power Electron. COMPEL 2008* 2: 1–6. https://doi.org/10.1109/COMPEL .2008.4634667.

77 Maharjan, L., Yamagishi, T., Akagi, H., and Asakura, J. (2010). Fault-tolerant operation of a battery-energy-storage system based on a multilevel cascade PWM converter with star configuration. *IEEE Trans. Power Electron.* 25 (9): 2386–2396. https://doi.org/10.1109/TPEL.2010.2047407.

78 Kucka, J., Karwatzki, D., and Mertens, A. (2017). Enhancing the reliability of modular multilevel converters using neutral shift. *IEEE Trans. Power Electron.* 32 (12): 8953–8957. https://doi.org/10.1109/TPEL.2017.2695327.

79 Farzamkia, S., Iman-eini, H., and Member, S. (2019). Improved fault-tolerant method for modular multilevel converters by combined DC and neutral-shift. *Strategy* 66 (3): 2454–2462.

80 Guerrero, J.M., Member, S., Vásquez, J.C. et al. (2009). Control strategy for flexible microgrid based on parallel line-interactive UPS systems. *IEEE Trans. Ind. Electron.* 56 (3): 726–736. https://doi.org/10.1109/TIE.2008.2009274.

81 Asiminoaei, L., Aeloiza, E., Member, S. et al. (2008). Shunt active-power-filter topology based on parallel interleaved inverters. *IEEE Trans. Ind. Electron.* 55 (3): 1175–1189. https://doi.org/10.1109/TIE.2007.907671.

82 Qinglin, Z., Zhongying, C., and Weiyang, W. (2006). Improved control for parallel inverter with current-sharing control scheme. *2006 CES/IEEE 5th International Power Electronics and Motion Control Conference* (August 2006), pp. 1–5. doi: 10.1109/IPEMC.2006.4778252.

83 Wu, T., Member, S., Chen, Y., and Huang, Y. (2000). 3C strategy for inverters in parallel operation achieving an equal current distribution. *IEEE Trans. Ind. Electron.* 47 (2): 273–281. https://doi.org/10.1109/41.836342.

84 Li, D., Member, S., Ngai, C. et al. (2020). A delay-tolerable master-slave current-sharing control scheme for parallel- operated interfacing inverters with low-bandwidth communication. *IEEE Trans. Ind. Appl.* 56 (2): 1575–1586. https://doi.org/10.1109/TIA.2019.2961335.

85 Wu, T., Liu, Z., Liu, J. et al. (2016). Transactions on power electronics a unified virtual power decoupling method for droop controlled parallel inverters in microgrids. *IEEE Trans. Power Electron.* 31 (8): https://doi.org/10.1109/TPEL .2015.2497972.

86 Welchko, B.A., Lipo, T.A., Fellow, L. et al. (2004). Fault tolerant three-phase AC motor drive topologies: a comparison of features, cost, and limitations. *IEEE Trans. Power Electron.* 19 (4): 1108–1116.

87 Fang, T., Shen, L., He, W., and Ruan, X. (2017). Distributed control and redundant technique to achieve superior reliability for fully modular input-series-output-parallel inverter system. *IEEE Trans. Power Electron.* 32 (1): 723–735. https://doi.org/10.1109/TPEL.2016.2530809.

88 Yu, X., Member, S., Khambadkone, A.M., and Member, S. (2012). Reliability analysis and cost optimization of parallel-inverter system. *IEEE Trans. Ind. Electron.* 59 (10): 3881–3889. https://doi.org/10.1109/TIE.2011.2175670.

89 Sun, X., Lee, Y., Member, S., and Xu, D. (2003). Modeling, analysis, and implementation of parallel multi-inverter systems with instantaneous average-current-sharing scheme. *IEEE Trans. Power Electron.* 18 (3): 844–856. https://doi.org/10.1109/TPEL.2003.810867.

90 Birk, J. and Andresen, B. (2007). Parallel-connected converters for optimizing efficiency, reliability and grid harmonics in a wind turbine. *2007 European Conference on Power Electronics and Applications*, pp. 1–7. doi: 10.1109/EPE.2007.4417318.

91 Seto, T., Hada, M., Data, R. U. S. A., and Processes, P. (1996). United States Patent [19], pp. 1–6.

92 Omana, M., Fiore, A., Mongitore, M., and Metra, C. (2019). Fault-tolerant inverters for reliable photovoltaic systems. *IEEE Trans. Very Large Scale Integr. Syst.* 27 (1): 20–28. https://doi.org/10.1109/TVLSI.2018.2874709.

93 Zhang, W. and Xu, D. (2018). Fault analysis and fault-tolerant design for parallel redundant inverter systems in case of IGBT Short-circuit failures. *IEEE J. Emerg. Sel. Top. Power Electron.* 6 (4): 2031–2041. https://doi.org/10.1109/ JESTPE.2018.2844092.

9

IoT-Based Monitoring and Management for Photovoltaic System

Azra Malik, Ahteshamul Haque, and V S Bharath Kurukuru

Advance Power Electronics Research Lab, Department of Electrical Engineering, Jamia Millia Islamia (A Central University), New Delhi, India

9.1 Introduction

Energy is the soul of today's human progress. Global energy consumption has increased drastically to meet the needs and uplift human living standards. Through the decades, we have relied on fossil fuels to generate energy. But, meeting the needs of modern living practices with such resources requires a huge amount of energy that increases global warming. On the other hand, fossil fuels are depleting, which are the main sources of energy generation [1, 2]. Either way, it is seen that alternatives are to be explored on an urgent basis. Electricity-generating fuel or coal-based thermal power plants are considered one of the major carbon-emitting sources and, alternatively, solar energy, wind energy, tidal energy, and geothermal energy have come into focus. It has taken several years to understand the need and now the power industry is moving toward these renewable sources and a few large-scale deployments in capacities of megawatts can be seen today. Further, for satisfying the current power demands, the International Energy Agency (IEA) gave the statement that renewable energy sources [3] can be introduced as the fastest growing alternative energy sources at present time. Among the rest of renewable energy sources, solar photovoltaic (PV) energy finds a strong place. Both the solar PV and concentrating solar-thermal power plants are increasing. The easy-to-install, scalable, and flexible design of the solar PV power plants has brought this technology to the domestic level as well. The net metering provision as initiated by countries like Germany has enabled domestic consumers to contribute their part [4]. Now the domestic consumers are generating their own electricity and feeding the surplus to the grid. At the utility scale, PV power plants are now reaching the size of megawatts and growing further. Recent years have seen rapid progress in PV power plants.

Fault Analysis and its Impact on Grid-connected Photovoltaic Systems Performance, First Edition.
Edited by Ahteshamul Haque and Saad Mekhilef.
© 2023 The Institute of Electrical and Electronics Engineers, Inc. Published 2023 by John Wiley & Sons, Inc.

An increase in PV plants has resulted in various technological advancements in the field [5, 6]. With more advanced systems, there come complications along. The internal power electronics of the PV components have become complex. Also, the larger the size of the PV plant, the probability of occurrence of an error also increases. Errors can be minor and may be ignored as they may not be affecting the energy yield of the power plant [7–9]. However, faults or errors can be severe which, if not dealt with, may result in lower yield [10–12]. The lower yield not only reduces the financial return on investments but also hurdles to achieving the goals of green and clean energy. In light of these issues, the monitoring systems for solar power plants, both traditional and recent are discussed in detail in the chapter. The recent methods for monitoring, also called remote monitoring systems, are focused on in the proposed methodology.

9.2 Background Information

9.2.1 General Description

The generalized structure for the monitoring system for a PV power plant is given in Figure 9.1. PV-generation setup connected to the grid has two phases of power flow that are DC stage and AC stage and its schematic representation are delineated in Figure 9.1. The DC stage power flow consists of a DC–DC converter, PV array, MPPT tracker to trace peak power, and controller to regulate the output of the converter. The converters are constructed with energy storage elements, namely inductors, capacitors, switching elements, and diodes in various configurations. The converters have the capability to transform one level of energy into another level of energy by performing switching actions. The converter boosts the

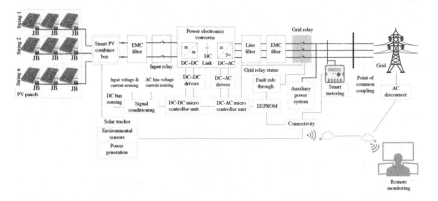

Figure 9.1 General structure of solar power plant-monitoring system.

voltage fed by PV since it dissents in pursuance to the irradiation and temperature. Further, the converter design varies with respect to that of requirements such as based on efficiency, input to the application such as high voltage or low voltage, and ripple content [13–16]. There are different configurations available for standalone PV or grid-connected PV. The applications usually require a stable output from the converter. The output voltage from the PV fed converter should be regulated before feeding it to the grid. The controllers are designed to tune the switching device duty cycle to overcome the non-linearity by yielding desired output voltage. It is necessary to design a PV system with advanced control strategies to increase the overall performance. The converter output is fed to the inverter and the output available from the inverter which is AC fed to the utility grid or the customer through the distribution panel. Energy meters are used to measure power based on voltage and current calculations. Graphical results and data are recorded with the help of the supervisory control and data acquisition (SCADA) system and stored for further use for analysis and monitoring purposes. In this aspect, the IoT-based remote monitoring and controlling system are briefed in upcoming sections. In recent years, IoT-based techniques have become a revolutionary technology for the remote sensing and monitoring of PV [17]. They are found to provide fast, reliable, and real-time information for a better understanding of PV fault diagnosis and prognosis [18]. Essentially, IoT is a progressive technology that is capable of exchanging information and sharing data among a wide scale of devices [19]. A more advanced version of IoT-based smart applications incorporates artificial intelligence (AI) to achieve enhanced adaptation and automation [20]. Generally, IoT is about sensors implanted into machines, which offer streams of data through the Internet connectivity. All IoT-related services inevitably follow five basic steps called "collect, aggregate, analyze, insight, and act." Over a network, IoT is deployed for the purpose of data exchange between distinct devices [21]. A brief overview of different concepts associated with IoT is shown in Figure 9.2.

9.2.2 Factors Influencing Optimum PV Yield

In solar arrays, reliability is vital for power-generation efficiency. The factors which reduce the efficiency of PV farms on DC side power flow are classified with respect to that of physical, environmental, and electrical fault occurrences on PV arrays. Other faults include converter switch fault, MPPT failure, and battery bank faults [11, 22–25]. The physical and environmental factors influencing the yield are exhibited in Figure 9.3. Based on the severity and the availability of devices for rectifying the electrical fault occurrences in the PV array. The AC stage consists of a DC–AC converter, transformer, and distribution panel, which feeds the utility grid or consumer. The fault occurrences which reduce the performance of the inverter are islanding operation and grid abnormalities [23, 25]. These

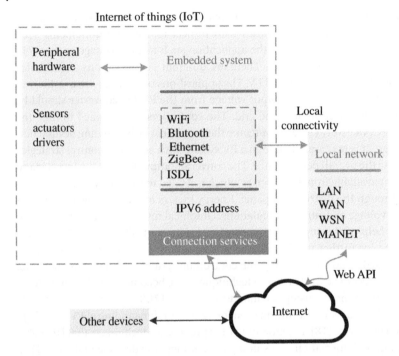

Figure 9.2 Internet of Things (IoT) system.

factors affecting the PV yield requires actions to be performed at the earliest in order to avoid interruption of power production and also to increase the PV performance.

Ground faults in the PV system can be either grounded or ungrounded [26–28]. It is due to the accidental connection between the earth's ground conductors and current-carrying conductors. Protection devices such as residual current monitoring devices and insulation-monitoring devices are available to protect the PV system. In a PV array, when two points are at different potentials, a short circuit occurs. This causes a line-line fault. It may occur in the same string or different string accordingly called an intra-string or cross-string respectively. An over-current protection device is available to overcome the fault.

Arc fault may be either series or parallel. Series arc occurs when PV producing current is interrupted [29–32]. Parallel arc occurs due to the breakdown of insulation. The protective measures for the series arc are arc fault circuit interrupter and arc fault detector. The parallel arc does not have any protective measures. Shading effects may be homogeneous or nonhomogeneous. The shaded panels are disconnected by bypassing diodes. Homogeneous or permanently shaded panels due to dust accumulation do not have any protective devices. Soiling decreases the

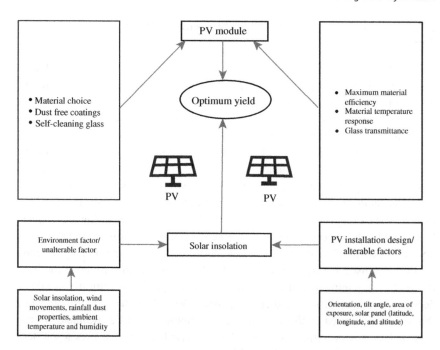

Figure 9.3 Environmental and physical factors influencing the PV yield.

quantum rays reaching the panel. It affects the power yield from PV. The accumulation of dust over the PV panels is influenced by factors such as orientation, tilt angle, wind velocity, ambient temperature, site characteristics, and the texture of the PV panel surface. If soiling is hard, it is called cementing. Cementing degrades the performance of the PV panel permanently. The dust intensity around the world is shown in the figure using different colors. Zones 3 and 4 using darker colors represent a higher level of dust deposited.

A loss of 1% of solar irradiation leads to 4.7% of degradation. From the recent surveys, it is found that the power-generation losses exceed 40% because of the deposition of coarse particles on the surface area of the PV panel systems. Near the equator, PV panels work in tropical conditions and the amount of dust intensity reduces the efficiency of the panel to 50% [33–36]. According to this study, shading effect due to the soiling factor cannot be prevented but can be reduced. In order to ensure that the PV yield altered not only due to temperature and solar irradiation but also due to soiling, a study was conducted in a company located near Chennai seashore in Tamil Nadu, India. The PV yield was recorded for the month of June–August. It was found that although irradiation and temperature were enough for energy conversion, the yield decreased. The decrease in yield is due to soiling. Once manual cleaning is c, the yield increases.

9.2.3 Methodologies to Determine the Factors Affecting the PV Performance

It becomes necessary to find out the reasons affecting the PV yield and take necessary actions to improve the PV performance and also to avoid interruption in power generation. The various techniques available to identify the factors are model-based and real-time difference measurement, output signal analysis, and learning techniques, namely machine learning (ML), deep learning, etc. [12, 37–42]. In real-time difference measurement method, the real-time parameters measured are studied with that of the threshold values to detect factors affecting the yield. In output signal analysis, the output obtained from the measured value of sensors is analyzed using the Fast Fourier transform [43–45] to identify the cause of reduction in the PV yield. In machine-learning techniques, the trained data are compared with the actual data to identify the reduction in power production. The drawback of this method is that model accuracy depends on the data used for training [46–49]. In the model-based difference method, the real-time parameters are compared with the predicted output to make a decision to act upon the predicted outcome [50–52]. In these methodologies, the sensor-measured real-time parameters should be made available to the user. IoT makes everything smarter. An IoT-based framework uses a wireless communication device to provide wireless access to end users.

Before the concept of IoT, most of the research and development for improving the operational efficiency of the PV system is focused on enhancing the tilt angle of the PV arrangement. Based on various surveys, there are multiple ways to choose the proper direction of the panels with respect to the sun. We will get more power generation only if our PV panels are in sync with the sun's direction. There are two major ways of the adjustment of PV panels – one is we can install panels in a fixed order and the other is we can have some amount of tilt with solar panels. If we consider the fixed panel method, the efficiency of the power generation is approximately 40% more in case of the summer season and it is 10% more in case of the winter season. The power-generation efficiency is also based on the location factor. If the power plant implies that the PV panels are located in the northern hemisphere, at that time, the direction of the solar panel should be facing the true south. And in case of solar panels located in the southern hemisphere, at that time, panels should be arranged in order to face the true north. The deviation in the true north and the magnetic north occurs due to the magnetic declination. In the winter season, the horizontal tilt angle should be at latitude plus fifteen degrees and in the summer season, it is at latitude minus fifteen degrees. We know it is always simple to have fixed panels [53] but to achieve maximum power, especially for power selling, manufacturers are opting for the method of tilting panels as per the season. Again, that is related to the sun's position in different seasons, as the sun is higher in summer and lower in winter.

Further, delta-type PLCs are utilized in [53] to track the solar panels automatically. In this method, the PLC monitors and tracks the solar array very effectively and efficiently. It has very easy and accurate controlling structures in all-weather situations. The exact maximum energy generation time is also evaluated using this method with the help of magnetic reed switches and light-dependent resistor (LDR) sensors interfaced with solar panels. These techniques help in getting high economic returns and compensating power demands within less time period. The work in [54] is more focused on how to maximize power generation with the use of sun tracking technique-based PLC and SCADA. In [55], the authors discussed how to implement and design the smart power system with the use of a personal area network. In [56], the authors demonstrated the Zigbee-based power plant-controlling technique being developed. This is a continuous wireless controlling and monitoring method, especially implemented in remote areas. The benefit of low power consumption by Zigbee is utilized in this paper. It has several other benefits also like a simple development structure, less delay, and increased data reliability with less pricing. In this method, temperature, voltage, current, humidity, and power are measured and sensed at the receiving end. After that, this data is forwarded to the central system for monitoring. SQL, python language, and DBMS are used for further processes. In [57], the monitoring of grid-tied solar hybrid systems using PLC and SCADA is demonstrated. Load shedding or power cuts are the major problems in many villages. To overcome this drawback, this method developed by the author can be very useful. In this system, major parts are the PLC controller, SCADA system, and PV cells with the inverter. This method aims to provide an uninterrupted power supply for industrial use or domestic use purposes. The SCADA system is used for visualizing the plant from start to end. This method requires less cost. A renewable energy-monitoring system is an instant monitoring system for renewable energy-generation systems that can be solar or wind power plants. This monitoring system is based on voltage and current measurements of the power plant. Microchip 18F4450 microcontrollers and sensors are utilized to measure related voltage and current values. These processed voltage and current parameters are transmitted to the PC over a USB interface and all data is stored in memory. Hence, the system can be easily monitored using this method. Distributed energy resources, load, storage, and grid agents are distributed generation (DG) agents. The monitoring unit agents are treated as communication mediums between higher agents and DG agents and they are worked as control agents. This method is processed with the help of Arduino Microcontroller [58, 59]. Instead of other methods for remote access for PV systems in rural areas especially, GSM modules can be efficiently used. There are some issues regarding solar panels like mean time to repair tends to difficult maintenance as well as inflexibility, and worst management ability. To mitigate this problem, the author proposed a model in which a gateway is embedded in nature and a GPRS-type Internet connection is provided to update all parameters in this smart system. IoT and wireless sensor network (WSN) can be used to perform these tasks smoothly. Nowadays smart grids as well as day-life applications like health, residential facilities, and safety applications can be effectively supported with the help of IoT and WSN. In smart or intelligent grid applications for renewable energy sources, what kind of issues occur that can be identified regarding nonintrusive load-type of monitoring and various appliances with load disaggregation? The problem of mismatching with load is caused due to the use of local generators that they are changing with respect to time. The development of renewable energy sources has been enhanced with the use of wireless remote-monitoring system [60]. For continuous tracking of energy performance measures and energy generation, the author proposed a low-cost data acquisition system. The output of this case study or project provides direct access to electrical energy yields at the rural area solar power plants with the use of test message transmission over mobile networks as well as WSN boards. The performance can be effectively measured at remote rural sites with the low-cost system.

9.3 Research Methodology

From the literature, it is observed that the traditional approaches are not capable of remotely accessing all the energy parameters and records of energy generation. To overcome this drawback, the advances in wireless technology allowed machine-to-machine M2M enterprise and industrial solutions for equipment monitoring and operation to become widespread [61]. Hence, in the proposed work, IoT is adapted to evolve the existing standards to form autonomous future communication architectures to support the intelligent exchange of data between millions of devices. The evolution of IoT from conventional approaches is shown in Figure 9.4.

This enhances the monitoring and management approach for the PV system to achieve its efficient operation. For developing a PV-monitoring approach with IoT, the general architecture consists of four layers: IoT hardware, communication layer, cloud infrastructure, and data analytics. The details of these layers are discussed in further sections.

9.3.1 Architecture of IoT-based Solar Monitoring System

IoT Hardware: The energy assets contain their own data source that requires hardware to interface with them. Low-cost IoT sensors can also be installed to

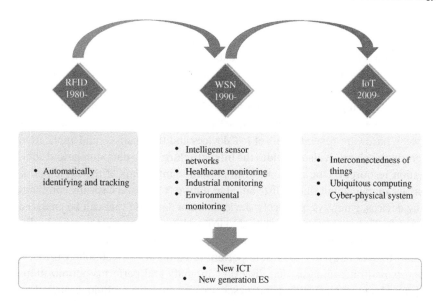

Figure 9.4 Evolution of the Internet of Things.

collect assets' status such as power, flow rate, pressure, etc. [18, 62, 63]. IoT devices include transducers such as sensors and actuators. With the low-cost and low-power IoT hardware, the traditional power equipment become connected, such as smart light bulbs, connected valves and pumps, smart meters, connected power plants, smart building components, etc. IoT devices are essentially capturing the data, which is the platform of IoT architecture and the foundation of connected and intelligent solutions. It enables IoT-related smart applications for upper layers.

Connectivity Layer: This is responsible for gathering data and transferring it over a network [64, 65]. The main components that complete the connectivity layer are sensors and devices. Sensors collect the information and send it off to the next layer where it is being processed. With the advancement of technology, semiconductor technology is used that allows for the production of micro-smart sensors that can be used for several applications.

Communication Layer: The communication layer is considered the backbone of the IoT systems. It is the main channel between the application layer and the IoT hardware layer. The IoT system is loaded with vast amounts of data and information that need to be shared within the network [64, 66]. Therefore, it is needed to set up a low-cost and low-bandwidth connection network among these nodes. The communication layer needs to interpret all industrial equipment communication protocols such as Modbus, DNP3, and transfer using either wired or wireless protocols, such as Lora and NB-IoT, and 5G. So, this is a broad layer with multiple devices, technologies, solutions (software and hardware), and functions. Typically, IoT gateways are used for connectivity aggregation, translation of various protocols, encryption and decryption of IoT data, the management of IoT devices, and some advanced edge computing. Moreover, networks are very vital components in IoT to connect things to the outside world. Cloud infrastructure needs to be able to manage a large number of the IoT endpoints in the fields, secure communication, authentication, and verification. With IoT platforms are software infrastructures between hardware-related layers of IoT devices and the business and application layers on the above. It also provides the infrastructure of big data storage and computation resources. One of the most important components of the IoT cloud is database management that is distributed in nature. The cloud basically combines many devices, gateways, protocols, devices, and a database that can be analyzed efficiently. These systems are essential to IoT architecture in order to provide efficient data analysis that can help improve the services and products.

Data Analytics Layer: The most important function of IoT technology is that it supports real-time analysis that discovers anomaly and performs optimizations [67]. This layer employs different data science and analytics techniques including ML algorithms to make sense of the data, can use data to find trends and gain actionable insights, evaluate the performance of devices, help identify inefficiencies, and create more efficient models for control applications. This layer is also considered the brain of the overall IoT system, which can help improve energy operations, efficiency, or even predict future events like a system failure.

9.3.2 IoT Components for Photovoltaic Systems

In this research, the IoT-based advanced monitoring scheme is employed to track the real-time performance and system progression and update the user regarding the system health periodically [68, 69]. Consequently, the understanding of the real-time working condition of the PV system enhances, and it helps in fast fault detection and maintenance for further analysis [70]. Adding IoT-based monitoring to a PV system makes it tremendously efficient, accurate, reliable, and cost-effective [71]. Further, it makes the system smart, flexible, and adaptable to various uncertainties. A general IoT-based PV-monitoring system comprises following components:

9.3.2.1 Sensor Devices

Sensors are the enabling assets of IoT systems. They collect and transmit data in real-time. In DER systems, there are pre-existing sensors and newly installed sensors to sense and collect data, for further processing, analytics, and decision-making [64, 67, 72]. There are the following commonly used sensors:

Circuit Sensors: These are commonly low-power IoT current and voltage sensors, which measure the basic performance of the device. Current transformer (CT) sensors measure alternating current (AC) flowing through. Some advanced circuit sensors include the capability to measure the harmonic and power factor for AC power.

Environment Sensors: These are used to detect the fluctuations in the environment, such as temperature, humidity, and solar radiation. Light sensors are used to measure luminance (ambient light level) or the brightness of a light, which are used for solar radiation. Humidity sensors are used to distinguish the amount of moisture and air's humidity and temperature sensors are used to detect the fluctuations on the surface. Both humidity and temperature have a direct impact on the PV panel performance.

Activity Sensors: These sensors are used to measure all human activities external to the IoT equipment themselves. Activities are often related to the load or demands. The common activities are passive infrared (PIR) sensors and proximity sensors, which provide feedback to control system output.

Mechanical Sensors: The mechanical sensors are used to measure variables, such as position, velocity, acceleration, force, pressure levels, and flow. For example, inertial sensors, such as accelerometers or gyroscopes can be used to measure position or motion. A popular application is vibration analysis, which can be used for earlier fault detection. Piezoresistive sensors respond to changes in pressure. Flow sensors can be used to measure liquid flow. Following is an example diagram to monitor solar panel (PV) performance using wireless IoT devices.

9.3.2.2 Other Components

Measurement Unit: The monitoring system consists of a measuring unit, with sensors utilized for continuous measurement of various PV parameters including temperature sensor, voltage sensor, current sensor, pyranometer, IR sensor, tilt sensor, humidity sensor, proximity sensor, etc. [73]. These sensors help in accurate and better-quality PV parameter measurement for further analysis in applications like maintenance, fault diagnostics, etc. It is required that sensors should be appropriately calibrated and accustomed as per the international standards [74].

Control and Communication Unit (CCU): Once the input is received from the respective sensors, the controllers modify it into numerous logical representations through controllers. It mainly consists of connecting a data acquisition unit (DAU) with a communication media. The DAU is employed for data collection from the measurement unit [75]. Moreover, it transforms the acquired raw information into an accessible and suitable form, in an optimal range, to be utilized in the further process, for instance, analog to digital conversion [76]. The communication media is responsible for transferring the updated information to the cloud. It can be in wired or wireless form, depending on the considered application. Commonly

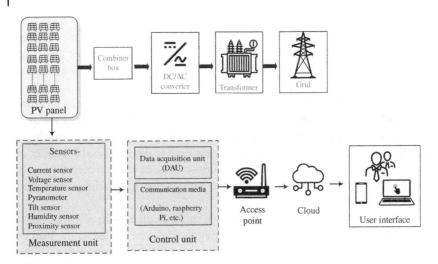

Figure 9.5 A typical IoT-based monitoring system for PV depicting IoT components.

used CCU include Arduino and Raspberry Pi, due to them being low-priced and user-friendly [73].

Cloud: The cloud is an important part of the PV-monitoring system, which is responsible for visualizing and analyzing the received data using a graphical web-page [77]. The cloud may be used for storing the relevant information in suitable form, e.g. spreadsheets. It creates a great impact in making the IoT a desirable and one of the most anticipated innovations in the near future since it can hold historical and present data in heavy amounts [78]. Hence, a record can be kept and it is highly beneficial in making the PV system robust and efficient. Another useful feature of the cloud is that it can send the status regarding the PV system to the concerned users via SMS, email, etc. [79]. These components of IoT in a typical PV system are depicted in Figure 9.5.

The earlier discussed components form a general IoT-based system for PV applications. Further, the obtained information by the concerned users can be utilized for many functions with respect to the PV system [80, 81]. It finds great importance in the area of PV power and energy forecasting since solar PV is intermittent and uncertain in nature [82]. For tracking the maximum power point (MPP) in the PV system, IoT-based algorithm may be very relevant and useful [83]. Another field where IoT finds relevance is the solar panel analysis with the help of thermal imaging [84]. Further, the real-time information provided may be utilized for finding the point, where the fault is initiated. Figure 9.6 shows the IoT-based PV monitoring used for anomaly detection, and it sends the information to the user when any anomaly is identified.

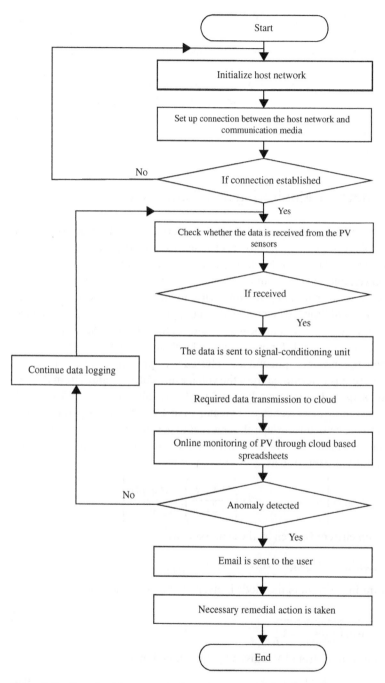

Figure 9.6 Flowchart for anomaly detection using IoT in a PV system.

Therefore, IoT has the potential to unite billions of devices for processing the available data, analyzing and further generating necessary instructions to be performed for enhancing efficiency without human intervention [85]. In the area of renewable energy, particularly PV systems, IoT can introduce a plethora of innovations to improve the comfort of a mass population through supplying continuous and reliable power at a highly reduced cost [86, 87]. Further, researchers are working on critical issues associated with solar panels including interoperability, storage system capacity, scalability, etc., for minimizing the carbon footprint with the help of IoT [88].

9.4 Remote PV Monitoring and Controlling

To inflate the yield and improve the PV farm performance from soiling losses, the methodology implemented consists of actuators, sensors, and controllers. The sensors are provided to audit the electrical substantial parameter, namely voltage, current, power, and environmental parameters such as irradiation and temperature. Since PV performance depends on environmental factors, it becomes difficult to set the threshold limits to determine the cause perturbing the PV performance. Because of this reason, the statistical parameters for that particular location, day, and time are made available for processing in the cloud server. It uses the PV model to determine the power output and it is referred to as predicted power. The actual PV power output is determined from the voltage and current values from the corresponding sensors in the string box and made available to the cloud server. If the difference in error percentage is greater than 15%, the cloud portal sends a command to the actuator through the Internet to perform the control action. The PV current is influenced by both solar irradiation and temperature. The output current from PV is expressed as

$$I_{(PV\ current)} = \left[I_{photon\ current} - I_{ds} \left(\frac{V + IR_s}{\frac{cKT}{q}} \right) - \frac{V + IRs}{R_P} \right] \tag{9.1}$$

The photon current is given by the expression as

$$I_{photon\ current} = [I_{sc(STC)} + K_i(T - T_{STC})] * \frac{\beta}{1000} \tag{9.2}$$

The saturation current of the diode is given by the expression as

$$I_{ds} = \frac{I_{sc(STC)}}{\exp\left(\frac{V_{oc}}{c*V_{tn}*ns} - 1\right)} \tag{9.3}$$

The output voltage from PV is given by the expression as

$$V_{pv} = V_{mp} + K_v(T - T_{STC}) \tag{9.4}$$

The PV power output is calculated using the expression

$$Power_{(PV\ model)} = V_{PV} * I_{PV} \tag{9.5}$$

$$\%Error = \frac{Power_{(PV\ model)} - Power_{(PV\ measured)}}{-Power_{(PV\ measured)}} \tag{9.6}$$

where I_{pv} is the PV output current, $I_{photon\ current}$ is the light-generated current, V_{pv} is the output voltage available at the PV panel, N_s is the number of solar cells, C is the ideality factor, I_{ds} is the diode saturation current, R_s is the series resistance.

The unknown parameters such as series resistance and shunt resistance are obtained from the Newton–Raphson method. The determined parameters evaluate the value of the PV current. The voltage influenced by the temperature and the value of voltage in pursuance to that of the current temperature is determined by the expression (9.4). For the given temperature and irradiation, the power generated is calculated from the PV model and that value is compared with the measured parameters. The difference in the percentage of error is calculated. For example, PV array KC200GT at 25 °C and 1000 W/m² have a maximum peak current of 7.61 A, a maximum peak voltage of 26.3 V and maximum power of 200.143 W. The short-circuit current is 8.21 A and the open-circuit voltage is 32.9 V with voltage and current coefficient as –0.1230v/k and 0.0032 with the number of cells as 54. Using the Newton–Raphson method, series and parallel resistance were calculated as 0.23 Ω and parallel resistance as 601.336 Ω. For a different value of the environmental parameter, the values of resistance, saturation current, PV voltage, current, and power are calculated. These determined values can also be used to determine the switching-device duty cycle to drive the converter to deliver the maximum power. If the power generated matches the calculated value, the actuator is at rest. If it varies above the fixed percentage, then the cleaning action is performed by the actuator for a time duration. After performing the action, if the power measured matches the power calculated, then there is no requirement for manual inspection. If the problem persists, then it requires manual inspection. In order to optimize PV productivity, the solar panel must face the sunlight directly. The optimized PV output for the year round is obtained by tilting the panel at that particular location latitude. Suppose, the Velachery location has a latitude of 12.976°N, then solar panels must be set around 13°N. To have the maximum yield during winter, the tilt angle is determined by adding 10–15° to the actual latitude value. But to have the maximum yield during summer, 10–15° must be subtracted from the actual latitude value. This information can be stored in the cloud portal. According to the seasonal variations, this information is processed and the stepper motor can be used to adjust the tilt angle of the panel to harness the maximum yield.

9.4.1 IoT in Monitoring and Maximizing PV Output

The way in which IoT interacts with the physical world resembles how people interact with the physical world. IoT devices enabled by the cloud meet the needs as well as respond to ameliorate the efficacy of the system. A smart PV system is programmed to monitor and perform intelligent action from a remote location. The main aim is to harness the maximum yield from PV. This chapter discusses the design of a smart PV farm based on cloud technologies. The automation system for harnessing the maximum power is more affordable and can be installed using the cloud. Cloud computing shares resources such as the storage unit that is the memory and software and delivers them over the Internet with the help of the Internet protocol. The applications are made available on the local machine or android and the service can be provided by the cloud and accessed by the clients. In order to deliver the clients with web-based applications, cloud computing provides software as a service (SAAS). The clients are allowed to develop, run, and manage applications without owning software and hardware, which is called platform as a service (PAAS). It shares hardware and software. Infrastructure as service (IAAS) is represented as virtual machine and storage. The client is allowed to run any application within a virtual machine. The constraint to run the application by a client is to hold a cloud license and it should be technically feasible. The proffered PV farm automation installation using the cloud approach provides monitoring as well as controlling over the Internet securely from any place without the need for hardware and large memory. A smart PV system can be established with a cloud setup. The cloud provides lots of services. It provides a large memory for storing data. The stored data can be processed and retrieved for analysis from the cloud. In this system, the cloud-processed data sends a command to the controller unit to perform the action using Wi-Fi. Cloud computing may be of various types, namely private cloud, public cloud, hybrid cloud, and community cloud. The private cloud is restricted to registered clients. It provides a distinct and secure environment to clients. The public model allows for the storage of applications and provides a virtual machine, and the users pay based on their usage model. Integration of private and public clouds makes a Hybrid cloud. The specific community from various organizations that share their infrastructure is called community cloud. Cloud storage for PV farms saves the cost spent on the operating systems. The automation system reduces the maintenance cost. Cloud storage, by virtualization, provides the entire setup in a powerful operating box. Thus, cloud storage allows end users to send files between cloud and local storage from anywhere. End users can access files using the Internet connection and data can be recovered whenever required by end users from the cloud storage.

9.4.2 IoT in Distributing Renewable Energy Harnessed

People nowadays are aware of the importance of energy management. To minimize the power consumption cost and to avoid interruption in power supply, people check out for equivalent sources to meet their demands. We have discussed previously how cloud technology helps in maximizing the PV yield. The maximum energy harnessed must be effectively distributed. IoT not only helps in disseminating the generated power but also allows changes between the energy sources available, that is, it helps in efficient power distribution for the electrical appliances such as lights and fans at home as well as commercial buildings.

In many nations, clean electrification contributors are integrated solar and wind. Solar and wind are equipped for domestic or commercial purposes. The hybrid energy is indispensable to meet power demand. Monitoring of parameters continuously becomes necessary to study the performance of the hybrid energy installed. IoT helps in the maximum utilization of the renewable energy source available by continuously monitoring the system and also providing a dedicated IP address and different parameter values of the system in detail.

This system focuses primarily on utilizing renewable energy, although supply from the grid is available. IoT monitors the power generated by solar and wind power and helps in changing over between the resources using a relay. The cloud portal receives the power generated from the hybrid energy source. The generated power is compared with that of the electrical appliance requirement.

There are many ways in which renewable power can be distributed to the load. Based on the installed capacity of solar and wind, some of the electrical appliances such as lights in the living room, porch, bedroom, kitchen, and bathrooms can be permanently fed by renewable power. By continuously monitoring the status of renewable power, in case of any interruption, the grid feeds the load. In other cases, the status of all the electrical appliances is monitored, whenever a change in status is identified and its power requirement is known. Accordingly, a decision is made to switch over between the renewable energy source or grid. Wind power is rectified using a rectifier and grid voltage is also rectified. The rectified voltage is selected using the relay unit. The inverter is fed either from the grid or from hybrid energy, i.e. solar and wind. The rectified voltage is fed to the flyback converter.

A flyback converter is an isolated DC-DC converter, which is used for boosting voltage available at the input side, i.e. the voltage fed from a renewable source. The yield available from the converter is fed to a multilevel inverter. The output of the inverter is fed to the load after filtering the harmonics using a filter circuit. The proposed scheme is delineated in Figure 9.7. The workflow of the proffered scheme is delineated in Figure 9.8.

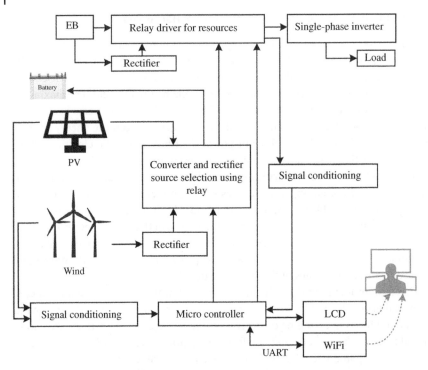

Figure 9.7 Proposed approach for distributing renewable energy harnessing.

9.4.3 Benefits of IoT in Distributed Energy Resources

IoT leads to a broad range of benefits for energy systems such as proactive actions and real-time intelligent automation, which boost the productivity and profitability of the renewable energy industry.

Boosting Operational Efficiency: Accurate and real-time data analytics are required to improve renewable energy-generation efficiency and capacity. IoT will help renewable energy companies manage the large capacity of assets located in remote and widely distributed areas, which can be very complex. IoT can analyze asset data in real-time with sophisticated algorithms. In addition, ML coupled with AI can predict the operation conditions. Technology can be used to ensure that energy transformation is evenly distributed. IoT can detect a change in demand and supply and automatically provides operators with information to react to these changes and increase efficiency. Hence, IoT offers an insight into which processes are redundant and time-consuming, and the productivity of renewable energy assets can be improved.

Better Production Management: Decisions about energy generation, load switching, and network configuration changes are constantly changing in distributed

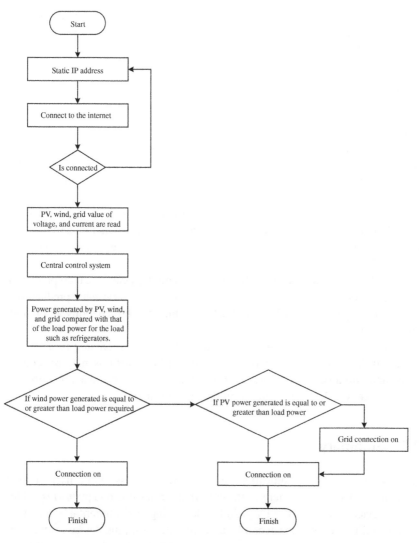

Figure 9.8 Workflow for power generation and management.

energy systems. Insightful information provided by IoT data analytics can provide insights to predict power production and asset reliability. For example, IoT offers real-time access to a solar farm's operation status, which can be turned into actionable insights, predict a broad range of risks, and automate for a prompt response.

Reduces Costly Downtime: Renewable energy includes solar panels, wind turbines, and transformer equipment which are remote and difficult to access.

Predictive maintenance with IoT can predict equipment failure in advance and schedule proactive maintenance procedures. Hence, with IoT, DERs can reduce unplanned downtime, resulting in lower operating costs.

9.4.4 AI Analytics of the IoT System

Traditionally, IoT devices were all about collecting data. Combining with AI algorithms, IoT can analyze data patterns, predict future energy generation and provide business intelligence to enhance operation. In some cases, IoT can control the system automatically, which can increase the system capability and reduce human involvement. Following are three areas in which IoT can generate values that will be described in the thesis in detail:

Fault Prediction: Using multisource-sensing data to predict grid and machinery failure, to achieve predictive maintenance of energy production and mechanical equipment can remarkably improve the distributed energy resource grid and energy reliability.

Service Life Prediction: AI algorithm and ML technology can take fault prediction one step further. With the remaining useful life (RUL) or service life prediction of critical grid components, DER can drastically improve the reliability and reduce the cost of maintenance.

AI-Enabled Optimization: AI-enabled algorithms can automate processes and make real-time system optimization. Not only can AI algorithms make better decisions than humans, they are also self-learning, and improve and adapt automatically to whatever the grid or energy market environments.

9.5 Summary

IoT-based wireless communication devices provide access to end users, reduce power loss in PV, and also increase the performance of PV. Cloud platform enabled by IoT devices increases the PV yield by identifying environmental occurrences such as soiling. It performs action by actuating the pumping device through the cloud portal. IoT not only increases the PV yield but it can also play an efficient role in distributing maximum energy harnessed by PV to electrical appliances at home or in a commercial building. Thus, IoT helps in achieving the automation and intelligent operation of renewable energy farms and increases the efficiency and reliability of farms to meet our future power demand.

In addition to providing real-time updates, utilizing IoT in alternative energy will cause enhancements in relation to metering. Energy suppliers will install OEM instrumentation in grids that job aboard metering parts to trace knowledge. The cellular property also can be utilized in star arrays to accommodate 2G,

LPWA, and different LTE networks. Thus, connections are going to be a lot of reliability still as these satisfy low power needs. Smarter meters would not solely increase overall potency. However, additionally these scale back prices for each energy supplier and customer. Improved metering potency is often achieved by employing link entrance or by putting a module before a star electrical converter in every grid.

References

1 Yıldız, İ. (2018). 1.12 fossil fuels. In: *Comprehensive Energy Systems*, 521–567. Elsevier https://doi.org/10.1016/B978-0-12-809597-3.00111-5.

2 Moriarty, P. and Honnery, D. (2017). Sustainable energy resources. In: *Clean Energy for Sustainable Development*, 3–27. Elsevier https://doi.org/10.1016/B978-0-12-805423-9.00001-6.

3 Agency, I.E. (2014). *Technology Roadmap*. SpringerReference https://doi.org/10.1007/SpringerReference_7300.

4 Roux, A. and Shanker, A. (2018). *Net Metering and PV Self-consumption in Emerging Countries*. Photovoltaic Power Systems Programme, Task 9, subtask 4. IEA-PVPS T9-18:2018. International Energy Agency (IEA).

5 Rakhshani, E., Rouzbehi, K., Adolfo, J.S., and Tobar, A.C. (2019). Integration of large scale PV-based generation into power systems: a survey. *Energies* 12 (8): 1–19. https://doi.org/10.3390/en12081425.

6 IRENA (2019). Future of solar photovoltaic: deployment, investment, technology, grid integration and socio-economic aspects. A Global Energy Transformation: Paper, vol. November.

7 Reise, C. and Müller, B. (2018). *IEA. Uncertainties in PV System Yield Predictions and Assessments*, vol. 21. Photovoltaic Power Systems Programme, Task 13, Subtasks 2.3 & 3.1. IEA-PVPS T13-12:2018. International Energy Agency (IEA).

8 Pfenninger, S. and Staffell, I. (2016). Long-term patterns of European PV output using 30 years of validated hourly reanalysis and satellite data. *Energy* 114: 1251–1265. https://doi.org/10.1016/j.energy.2016.08.060.

9 Vinod, K.R. and Singh, S.K. (2018). Solar photovoltaic modeling and simulation: as a renewable energy solution. *Energy Rep.* 4: 701–712. https://doi.org/10.1016/j.egyr.2018.09.008.

10 Chine, W., Mellit, A., Pavan, A.M., and Kalogirou, S.A. (2014). Fault detection method for grid-connected photovoltaic plants. *Renew. Energy* 66: 99–110. https://doi.org/10.1016/j.renene.2013.11.073.

11 Harrou, F., Sun, Y., Taghezouit, B. et al. (2018). Reliable fault detection and diagnosis of photovoltaic systems based on statistical monitoring approaches. *Renew. Energy* 116: 22–37. https://doi.org/10.1016/j.renene.2017.09.048.

12 Kurukuru, V.S.B., Blaabjerg, F., Khan, M.A., and Haque, A. (2020). A novel fault classification approach for photovoltaic systems. *Energies* 13: 308. https://doi.org/10.3390/en13020308.

13 Das, M. and Agarwal, V. (2016). Design and analysis of a high-efficiency DC–DC converter with soft switching capability for renewable energy applications requiring high voltage gain. *IEEE Trans. Ind. Electron.* 63: 2936–2944. https://doi.org/10.1109/TIE.2016.2515565.

14 Shayestegan, M., Shakeri, M., Abunima, H. et al. (2018). An overview on prospects of new generation single-phase transformerless inverters for grid-connected photovoltaic (PV) systems. *Renew. Sustain. Energy Rev.* 82: 515–530. https://doi.org/10.1016/j.rser.2017.09.055.

15 Kjaer, S.B., Pedersen, J.K., Member, S., and Blaabjerg, F. (2005). A review of single-phase grid-connected inverters for photovoltaic modules. *IEEE. Trans. Ind. Appl.* 41: 1292–1306. https://doi.org/10.1109/TIA.2005.853371.

16 Khan, M.A., Haque, A., and Kurukuru, V.S.B. (2018). Control and stability analysis of H5 transformerless inverter topology. In: *2018 International Conference on Computing, Power and Communication Technology*, 310–315. IEEE https://doi.org/10.1109/GUCON.2018.8674915.

17 Durga Deenadayalan K., Jayanthi, S., Arunraja A., and Selvaraj S. (2020). IoT based remote monitoring of mass solar panels. *International Conference on Electronics and Sustainable Communication Systems (ICESC 2020)*, pp. 1009–14.

18 Priharti, W., Rosmawati, A.F.K., and Wibawa, I.P.D. (2019). IoT based photovoltaic monitoring system application. *Int. Conf. Eng., Technol. Innovat. Res. Indonesia* 1367: 012069. https://doi.org/10.1088/1742-6596/1367/1/012069.

19 Abu-siada, A. and Muyeen, S.M. (2021). Industrial IoT based condition monitoring for wind energy conversion system. *CSEE J. Power Energy Syst.* 7: 654–664. https://doi.org/10.17775/CSEEJPES.2020.00680.

20 Priyadarshi, S., Bhaduri, S., and Shiradkar, N. (2018). IoT based, inexpensive system for large scale, wireless, remote temperature monitoring of photovoltaic modules. *IEEE 7th World Conference on Photovoltaic Energy Conversion (WCPEC)*, pp. 749–752. doi: 10.1109/PVSC.2018.8547354 :749–52.

21 Sethi, P. and Sarangi, S.R. (2017). Internet of things: architectures, protocols, and applications. *J. Electr. Comput. Eng.* 2017: 1–25. https://doi.org/10.1155/2017/9324035.

22 Azizi, K., Farsadi, M., and Farhadi, K.M. (2017). Efficient approach to LVRT capability of DFIG-based wind turbines under symmetrical and asymmetrical voltage dips using dynamic voltage restorer. *Int. J. Power Electron. Drive Syst.* 8: 945. https://doi.org/10.11591/ijpeds.v8.i2.pp945-956.

23 Dhimish, M., Holmes, V., and Dales, M. (2017). Parallel fault detection algorithm for grid-connected photovoltaic plants. *Renew. Energy* 113: 94–111. https://doi.org/10.1016/j.renene.2017.05.084.

24 Subramaniam, U., Vavilapalli, S., Padmanaban, S. et al. (2020). A hybrid PV-battery system for ON-grid and OFF-grid applications—controller-in-loop simulation validation. *Energies* 13: 755. https://doi.org/10.3390/en13030755.

25 Madeti, S.R. and Singh, S.N. (2017). A comprehensive study on different types of faults and detection techniques for solar photovoltaic system. *Sol Energy* 158: 161–185. https://doi.org/10.1016/j.solener.2017.08.069.

26 Zhao, Y., de Palma, J., Mosesian, J. et al. (2013). Line fault analysis and protection challenges in solar photovoltaic arrays. *Ind. Electron. IEEE Trans.* 60: 3784–3795. https://doi.org/10.1109/TIE.2012.2205355.

27 Karmacharya, I.M. and Gokaraju, R. (2018). Fault location in ungrounded photovoltaic system using wavelets and ANN. *IEEE Trans. Power Deliv.* 33: 549–559. https://doi.org/10.1109/TPWRD.2017.2721903.

28 Ancuta, F. and Cepisca, C. (2011). Failure analysis capabilities for PV systems. *International Conference Recent Researches in Energy, Environment,* Lanzarote, pp. 109–115.

29 Mellit, A., Tina, G.M., and Kalogirou, S.A. (2018). Fault detection and diagnosis methods for photovoltaic systems: a review. *Renew. Sustain. Energy Rev.* 91: 1–17. https://doi.org/10.1016/j.rser.2018.03.062.

30 Wang, Z., McConnell, S., Balog, R.S., and Johnson, J. (2014). Arc fault signal detection – Fourier transformation vs. wavelet decomposition techniques using synthesized data. *2014 IEEE 40th Photovoltics Specialists Conference PVSC 2014,* Denver, CO, USA (8–13 June 2014), pp. 3239–44. https://doi.org/10.1109/PVSC.2014.6925625.

31 Johnson, J., Schoenwald, D., Kuszmaul, S. et al. (2011). Creating dynamic equivalent PV circuit models with impedance spectroscopy for arc fault modeling. *Conf. Rec. IEEE Photovolt Spec. Conf.* 002328–002333. https://doi.org/10.1109/PVSC.2011.6186419.

32 Appiah, A.Y., Zhang, X., Ayawli, B.B.K., and Kyeremeh, F. (2019). Review and performance evaluation of photovoltaic array fault detection and diagnosis techniques. *Int. J. Photoenergy* 2019: 1–19. https://doi.org/10.1155/2019/6953530.

33 Rao, R.R., Mani, M., and Ramamurthy, P.C. (2018). An updated review on factors and their inter-linked influences on photovoltaic system performance. *Heliyon* 4: e00815. https://doi.org/10.1016/j.heliyon.2018.e00815.

34 Sánchez-Carbajal, S. and Rodrigo, P.M. (2019). Optimum array spacing in grid-connected photovoltaic systems considering technical and economic factors. *Int. J. Photoenergy* 2019: https://doi.org/10.1155/2019/1486749.

35 Chikate, B.V. and Sadawarte, Y.A. (2015). The factors affecting the performance of solar cell. *Int. J. Comput. Appl. Sci. Technol.* 975–987.

36 Dewi, T., Risma, P., and Oktarina, Y. (2019). A review of factors affecting the efficiency and output of a pv system applied in tropical climate. *IOP Conf. Ser. Earth Environ. Sci.* 258: https://doi.org/10.1088/1755-1315/258/1/012039.

37 Çınar, Z.M., Abdussalam Nuhu, A., Zeeshan, Q. et al. (2020). Machine learning in predictive maintenance towards sustainable smart manufacturing in industry 4.0. *Sustainability* 12: 8211. https://doi.org/10.3390/su12198211.

38 Mishra, P.K. and Yadav, A. (2019). Combined DFT and fuzzy based faulty phase selection and classification in a series compensated transmission line. *Model Simul. Eng.* 2019: 1–18. https://doi.org/10.1155/2019/3467050.

39 Kurukuru, V.S.B., Haque, A., and Khan, M.A. (2019). Fault detection in single-phase inverters using wavelet transform-based feature extraction and classification. *Techniques* 649–661. https://doi.org/10.1007/978-981-13-6772-4_56.

40 Zhang, Y.M., Chamseddine, A., Rabbath, C.A. et al. (2013). Development of advanced FDD and FTC techniques with application to an unmanned quadrotor helicopter testbed. *J. Franklin Inst.* 350: 2396–2422. https://doi.org/10.1016/j.jfranklin.2013.01.009.

41 Rahman Fahim, S.K., Sarker, S., Muyeen, S.M. et al. (2020). Microgrid fault detection and classification: machine learning based approach, comparison, and reviews. *Energies* 13: 3460. https://doi.org/10.3390/en13133460.

42 Kurukuru, V.S.B., Haque, A., Khan, M.A. et al. (2021). A review on artificial intelligence applications for grid-connected solar photovoltaic systems. *Energies* 14: 4690. https://doi.org/10.3390/en14154690.

43 Gan, C., Wu, J., Yang, S. et al. (2016). Fault diagnosis scheme for open-circuit faults in switched reluctance motor drives using fast Fourier transform algorithm with bus current detection. *IET Power Electron.* 9: 20–30. https://doi.org/10.1049/iet-pel.2014.0945.

44 Laadjal, K., Sahraoui, M., and AJM, C. (2021). On-line fault diagnosis of DC-link electrolytic capacitors in boost converters using the STFT technique. *IEEE Trans. Power Electron.* 36: 6303–6312. https://doi.org/10.1109/TPEL.2020.3040499.

45 Sun, Q., Yu, X., Li, H. (2020). Open-circuit fault diagnosis based on 1D-CNN for three-phase full-bridge inverter. *Proceedings – 11th International Conference on Prognostics and System Health Management PHM-Jinan 2020*, Jinan, China (23–25 October 2020), pp. 322–7. https://doi.org/10.1109/PHM-Jinan48558.2020.00064.

46 Haque, A., Alshareef, A., Khan, A.I. et al. (2020). Data description technique-based islanding classification for single-phase grid-connected photovoltaic system. *Sensors* 20: 3320.

47 Panagopoulos, A.D., Georgiadou, E.M., and Kanellopoulos, J.D. (2007). Selection combining site diversity performance in high altitude platform networks. *IEEE Commun. Lett.* 11: 787–789. https://doi.org/10.1109/LCOMM.2007.070801.

48 Chen, Z., Han, F., Wu, L. et al. (2018). Random forest based intelligent fault diagnosis for PV arrays using array voltage and string currents. *Energ. Conver. Manage.* 178: 250–264. https://doi.org/10.1016/j.enconman.2018.10.040.

49 Chen, Z., Wu, L., Cheng, S. et al. (2017). Intelligent fault diagnosis of photovoltaic arrays based on optimized kernel extreme learning machine and I-V characteristics. *Appl. Energy* 204: 912–931. https://doi.org/10.1016/j.apenergy.2017.05.034.

50 Ando, B., Baglio, S., Pistorio, A. et al. (2015). Sentinella: smart monitoring of photovoltaic systems at panel level. *IEEE Trans. Instrum. Meas.* 64: 2188–2199. https://doi.org/10.1109/TIM.2014.2386931.

51 Gao, Z., Cecati, C., and Ding, S.X. (2015). A survey of fault diagnosis and fault-tolerant techniques-part I: fault diagnosis with model-based and signal-based approaches. *IEEE Trans. Ind. Electron.* 62: 3757–3767. https://doi.org/10.1109/TIE.2015.2417501.

52 Guo, D., Zhong, M., Ji, H. et al. (2018). A hybrid feature model and deep learning based fault diagnosis for unmanned aerial vehicle sensors. *Neurocomputing* 319: 155–163. https://doi.org/10.1016/j.neucom.2018.08.046.

53 Darhmaoui, H. and Lahjouji, D. (2013). Latitude based model for tilt angle optimization for solar collectors in the mediterranean region. *Energy Procedia* 42: 426–435. https://doi.org/10.1016/j.egypro.2013.11.043.

54 Md. Hoque, E., Rashid, F., Shahriar, S., Md. Islam, K. (2018). An automatic solar tracking system using programmable logic controller. *International Conference on Mechanical, Industrial and Energy Engineering 2018*, Khulna, Bangladesh (23–24 December 2018)

55 Chhaya, L., Sharma, P., Bhagwatikar, G., and Kumar, A. (2016). Design and implementation of remote wireless monitoring and control of smart power system using personal area network. *Indian J. Sci. Technol.* 9: https://doi.org/10.17485/ijst/2016/v9i43/104392.

56 Cetinkaya, O. and Akan, O.B. (2015). A ZigBee based reliable and efficient power metering system for energy management and controlling. In: *2015 International Conference on Computer Networks and Communication*, 515–519. IEEE https://doi.org/10.1109/ICCNC.2015.7069397.

57 Trzmiel, G., Łopatka, M., and Kurz, D. (2018). The use of the SCADA system in the monitoring and control of the performance of an autonomous hybrid power supply system using renewable energy sources. *E3S Web Conf.* 44: 00180. https://doi.org/10.1051/e3sconf/20184400180.

58 Raju, L. and Amalraj, M.A. (2020). Multi-agent systems based advanced energy management of smart micro-grid. In: *Multi Agent Systems — Strategies and Applications*. IntechOpen https://doi.org/10.5772/intechopen.90402.

59 Raju, L., Balaji, V., Keerthivasan, S., and Keerthivasan, C. (2020). Internet of things and blockchain based distributed energy management of smart micro-grids. In: *Proceeding of the International Conference on Computer Networks, Big Data and IoT (ICCBI – 2019). ICCBI 2019. Lecture Notes on Data Engineering and Communications Technologies*, vol. 49 (ed. A. Pandian, R. Palanisamy and K. Ntalianis). Cham: Springer https://doi.org/10.1007/978-3-030-43192-1_67.

60 Andreoni Lopez, M.E., Galdeano Mantinan, F.J., and Molina, M.G. (2012). Implementation of wireless remote monitoring and control of solar photovoltaic (PV) system. In: *2012 Sixth IEEE/PES Transmission and Distribution. Latin America Conference and Exposition*, 1–6. IEEE https://doi.org/10.1109/TDC-LA.2012.6319050.

61 Albishi, S., Soh, B., Ullah, A., and Algarni, F. (2017). Challenges and solutions for applications and technologies in the internet of things. *Procedia Comput. Sci.* 124: 608–614. https://doi.org/10.1016/j.procs.2017.12.196.

62 Filios, G., Katsidimas, I., Kerimakis, E. et al. (2020). An IoT based solar park health monitoring system for PID and hotspots effects. In: *2020 16th International Conference on Distributed Computing in Sensor Systems*, 396–403. IEEE https://doi.org/10.1109/DCOSS49796.2020.00069.

63 Oukennou, A., Berrar, A., Belbhar, I., and Hamri, N.E. (2019). *Low Cost IoT System for Solar Panel Power Monitoring Low Cost IoT System for Solar Panel Power Monitoring*. Colloque sur les Objets et systems Connectés, Ecole Supérieure de Technologie de Casablanca (Maroc), Casablanca, Morocco.hal-02298769. France: Institut Universitaire de Technologie d'Aix-Marseille.

64 Kumar, N.M., Atluri, K., and Palaparthi, S. (2018). Internet of things (IoT) in photovoltaic systems. In: *2018 National Power on Engineering Conference*, 1–4. IEEE https://doi.org/10.1109/NPEC.2018.8476807.

65 Shaban, M., Ben Dhaou, I., Alsharekh, M.F., and Abdel-Akher, M. (2021). Design of a partially grid-connected photovoltaic microgrid using IoT technology. *Appl. Sci.* 11: https://doi.org/10.3390/app112411651.

66 Cheddadi, Y., Cheddadi, H., Cheddadi, F. et al. (2020). Design and implementation of an intelligent low-cost IoT solution for energy monitoring of photovoltaic stations. *SN Appl. Sci.* 2: 1165. https://doi.org/10.1007/s42452-020-2997-4.

67 Luo, X.J., Oyedele, L.O., Ajayi, A.O. et al. (2019). Development of an IoT-based big data platform for day-ahead prediction of building heating and cooling demands. *Adv. Eng. Informatics* 41: 100926. https://doi.org/10.1016/j.aei.2019.100926.

68 Mellit, A. and Kalogirou, S. (2021). Artificial intelligence and internet of things to improve efficacy of diagnosis and remote sensing of solar photovoltaic systems: challenges, recommendations and future directions. *Renew.Sustain.Energy Rev.* 143: https://doi.org/10.1016/j.rser.2021.110889.

69 Adhya, S., Saha, D., Das, A., Jana, J., and Saha, H (2016). An IoT based smart solar photovoltaic remote monitoring and control unit. *IEEE 2nd International Conference on Control, Instrumentation, Energy & Communication (CIEC)*, pp. 432–436.

70 Sivagami, P., Jamunarani, D., Abirami, P. et al. (2021). Review on IoT based remote monitoring for solar photovoltaic system. *IEEE International Conference on Communication information and Computing Technology (ICCICT)*. doi: 10.1109/ICCICT50803.2021.9510163.

71 Gupta, V., Sharma, M., Pachauri, R.K., and Dinesh, K.N. (2020). Environmental effects a low-cost real-time IOT enabled data acquisition system for monitoring of PV system a low-cost real-time IOT enabled data acquisition system for monitoring of PV system Vinay. *Energy Sources, Part A Recov. Utilization Environ. Effects* https://doi.org/10.1080/15567036.2020.1844351.

72 Deriche, M., Raad, M.W., and Suliman, W. (2019). An IOT based sensing system for remote monitoring of PV panels. In: *2019 16th International Multi-Conference Systems on Signals Devices*, 393–397. IEEE https://doi.org/10.1109/SSD.2019.8893161.

73 Ramu, S.K., Christopher, G., Irudayaraj, R., and Elango, R. (2021). An IoT-based smart monitoring scheme for solar PV applications. *Electric. Electron. Dev. Circuits Mater. Technol. Challenges Solut.* 211–233.

74 Shrihariprasath, B. and Rathinasabapathy V. (2016). A smart IoT system for monitoring solar PV power conditioning unit. *World Conference on Futuristic Trends in Research and Innovation for Social Welfare (WCFTR'16)*, pp. 3–7.

75 Gad, H.E. and Gad, H.E. (2015). Development of a new temperature data acquisition system for solar energy applications. *Renew. Energy J.* 74: 337–343. http://dx.doi.org/10.1016/j.renene.2014.08.006.

76 Ciani, L., Cristaldi, L., Faifer, M. et al. (2013). Design and implementation of a on-board device for photovoltaic panels monitoring. *IEEE Int.Instrum. Meas. Technol. Conf. (I2MTC)* 2–7. 10.1109/I2MTC.2013.6555684.

77 Kekre, A. and Gawre, S.K. (2017). Solar photovoltaic remote monitoring system using IOT. *International conference on Recent Innovations in Signal Processing and Embedded Systems (RISE -2017)*. doi: 10.1109/RISE.2017.8378227.

78 Kumar NM, Atluri K, Palaparthi S. *Internet of Things (IoT) in Photovoltaic Systems* 2018. 10.1109/NPEC.2018.8476807

79 Seyyedhosseini, M. and Mohamadian, M. (2020). IOT based multi agent micro inverter for condition monitoring and controlling of PV systems. *11th Power Electronics, Drive Systems, and Technologies Conference, PEDSTC 2020.* doi: 10.1109/PEDSTC49159.2020.9088449.

80 López-vargas, A., Fuentes, M., and Vivar, M. (2019). IoT application for real-time monitoring of solar home systems based on arduino TM with 3G connectivity. *IEEE Sens. J.* 19 (2): 679–691. https://doi.org/10.1109/JSEN.2018.2876635.

81 Spanias, A.S. (2017). Solar energy management as an internet of things (IoT) application *2017 8th International Conference on Information, Intelligence, Systems & Applications (IISA)* (August 2017), pp. 1–4. doi: 10.1109/IISA.2017.8316460.

82 Paredes-Parra, J.M., García-Sánchez, A.J., Mateo-Aroca, A., and Molina-García, Á. (2019). An alternative internet-of-things solution based on LoRa for PV power plants: data monitoring and management. *Energies* 12 (5): 881. https://doi.org/10.3390/en12050881.

83 Shaw, R.N., Walde, P., and Ghosh, A. (2019). IOT based MPPT for performance improvement of solar PV arrays operating under partial shade dispersion. *2020 IEEE 9th Power India International Conference (PIICON)* (February 2020), pp. 1–4. doi: 10.1109/PIICON49524.2020.9112952.

84 Phoolwani, U.K., Sharma, T., Singh, A., and Gawre, S.K. (2020). IoT Based Solar Panel Analysis using Thermal Imaging 2020. *2020 IEEE International Students' Conference on Electrical,Electronics and Computer Science (SCEECS)* (February 2020), pp. 1–5. doi: 10.1109/SCEECS48394.2020.114

85 Aagri, D.K. and Bisht, A. (2018). Export and import of renewable energy by hybrid microgrid via IoT 2018. *2018 3rd International Conference On Internet of Things: Smart Innovation and Usages (IoT-SIU)*, Bhimtal, India (23–24 February 2018), pp. 1–4. doi: 10.1109/IoT-SIU.2018.8519873.

86 Al-kadhim, H.M. and Al-raweshidy, H.S. (2019). Energy efficient and reliable transport of data in cloud-based IoT. *IEEE Access* 7: 64641–64650. 10.1109/ACCESS.2019.2917387.

87 Kraemer, F.A., Palma, D., Braten, A.E. et al. (2020). Operationalizing solar energy predictions for sustainable. *Autonomous IoT Dev.Manag.* 7: 11803–11814.

88 Hatem, T., Ismail, Z., Elmahgary, M.G., and Ghannam, R. (2021). Optimization of organic meso-superstructured solar cells for underwater IoT2 self-powered sensors. *IEEE Trans. Electron Dev.* 68: 5319–5321. https://doi.org/10.1109/TED.2021.3101780.

Index

Fault Analysis and its Impact on Grid-connected Photovoltaic Systems Performance, First Edition.
Edited by Ahteshamul Haque and Saad Mekhilef.
© 2023 The Institute of Electrical and Electronics Engineers, Inc. Published 2023 by John Wiley & Sons, Inc.

Printed and bound by CPI Group (UK) Ltd, Croydon, CR0 4YY

16/04/2025

14658602-0001